Multifunctional Heterostructure Thin Films Handbook

Subhash Sharma
Editor

Multifunctional Heterostructure Thin Films Handbook

Fundamentals, Fabrication, and Device Applications

Springer

Editor
Subhash Sharma
SECIHTI—Centro de Nanociencias y
Nanotecnología (CNyN)
Universidad Nacional Autónoma de México
Ensenada, Baja California, Mexico

ISBN 978-3-032-18691-1 ISBN 978-3-032-18692-8 (eBook)
https://doi.org/10.1007/978-3-032-18692-8

This Springer imprint is published by the registered company Springer Nature Switzerland AG
The registered company address is: Gewerbestrasse 11, 6330 Cham, Switzerland

This book is dedicated to my parents, whose sacrifices and values shaped my journey; to my teachers and mentors, whose guidance and wisdom laid the foundation of my academic life; to Lord Shiva Ji, for spiritual strength and inspiration; and to my loving family, for their unwavering support and encouragement.

Preface

The *Multifunctional Heterostructure Thin Films Handbook* is conceived as a comprehensive and authoritative reference addressing the rapidly advancing field of heterostructure thin films and multifunctional nanomaterials. The handbook aims to provide a unified understanding of how interface engineering in thin-film architectures enables emergent physical phenomena and enhanced functionalities that are not attainable in bulk materials. Heterostructure thin films, created by the controlled stacking of dissimilar materials, form the foundation of many modern technologies. Interfacial phenomena, including band alignment, strain-mediated coupling, charge transfer, interfacial polarization, and defect engineering, play a crucial role in determining the electrical, magnetic, optical, and energy-related properties of these systems. Harnessing these effects has opened new pathways for the development of advanced electronic, spintronic, optoelectronic, and energy devices. This handbook systematically presents the fundamental principles underlying heterostructure thin films, followed by detailed discussions of advanced fabrication techniques, including pulsed laser deposition, atomic layer deposition, molecular beam epitaxy, chemical vapor deposition, and physical vapor deposition. Equal emphasis is placed on state-of-the-art characterization techniques, covering structural, compositional, electrical, magnetic, and optical methods, as well as emerging in situ, operando, and AI-assisted multimodal approaches for probing interfaces and defects in complex thin-film systems.

A central theme of this volume is the strong interconnection between interface design and multifunctional properties. The handbook explores functional phenomena such as ferroelectricity and magnetoelectric coupling, spin-dependent transport, thermoelectric and piezoelectric responses, and optoelectronic effects, linking these properties directly to device-level implementations. Case studies on multiferroic heterostructures, van der Waals systems, oxide–oxide interfaces, and hybrid organic–inorganic architectures demonstrate the power of rational interface engineering in achieving optimized performance. In addition to its fundamental and applied aspects, the handbook features focused chapters on inorganic and oxide-based perovskite solar cells, thin-film sensors, and porous anodic aluminum oxide (PAAO)-based plasmonic and metamaterial structures, highlighting scalable and application-driven innovations. The concluding sections offer forward-looking perspectives on AI-guided materials discovery, quantum and topological heterostructures, and strategies for energy-efficient and large-area device integration, providing a roadmap for future research.

This handbook is intended for graduate students, researchers, and industry professionals working in materials science, nanotechnology, solid-state physics, and electronic engineering. By combining fundamental concepts with practical fabrication strategies and device-oriented insights, this volume aims to serve as a definitive reference and to stimulate continued innovation in multifunctional heterostructure thin films.

Ensenada, Mexico Subhash Sharma

Acknowledgements

The editor sincerely acknowledges the institutional support provided by the SECIHTI–Researcher for Mexico Program (Project No. 2018–352), which has played a crucial role in supporting the research activities related to this handbook. The editor also gratefully acknowledges the support of the Center for Nanoscience and Nanotechnology (CNyN) at the Universidad Nacional Autónoma de México (UNAM) in Ensenada, for providing an excellent research environment and infrastructure. The editor expresses heartfelt gratitude to all contributing authors for their valuable chapters, scientific rigor, and timely cooperation. Special thanks are extended to the reviewers for their critical evaluations and constructive feedback, which helped maintain the high academic standards of this handbook.

The editor also appreciates the professional support and guidance provided by Springer Nature throughout the preparation and publication process. Their assistance has been instrumental in shaping this handbook into a comprehensive and high-quality reference. Finally, sincere thanks are extended to colleagues, collaborators, and students whose discussions and insights have contributed, directly or indirectly, to the development of this volume.

Contents

Editor and Contributors

About the Editor

Dr. Subhash Sharma is a materials science researcher with more than 15 years of experience, specializing in the development and optimization of advanced functional materials. His research primarily focuses on multiferroic materials, where he works on enhancing their properties for device applications by designing innovative multiferroic heterostructures. In addition, Dr. Subhash actively investigates nanomaterials for photocatalytic applications and hydrogen production, as well as the development of relaxor-type ferroelectrics for high-energy-density storage solutions. His laboratory also conducts research on superconductors and their physical properties, further broadening the scope of his scientific work. Complementing his experimental research, Dr. Subhash has strong expertise in photovoltaic device simulation using SCAPS calculations to enhance solar cell efficiency. He is also highly proficient in thin-film and heterostructure fabrication techniques, particularly RF magnetron sputtering and pulsed laser deposition (PLD). His theoretical investigations are supported by advanced density functional theory (DFT) calculations using VASP, enabling a deeper understanding of the physical properties of a wide range of materials.

Dr. Subhash's solid academic background, including a B.Sc., M.Sc., M.Tech., and Ph.D. in Physics and Materials Science (2016), has equipped him with extensive expertise in materials characterization. After completing his Ph.D. at JIIT, Noida, India, in 2016, he joined the NPDF Program as a postdoctoral researcher at the Institute of Science, Banaras Hindu University (ISC, BHU), India. He subsequently received a postdoctoral fellowship from the UNAM–DGAPA Program at CNyN–UNAM and, in December 2018, joined the Cátedras Program at the same center. Dr. Subhash has strong expertise in crystal structure analysis through Rietveld refinement and is proficient in modern microscopic and spectroscopic techniques, including SEM, AFM, TEM, Raman spectroscopy, UV–Vis spectroscopy, FTIR spectroscopy, impedance spectroscopy, and ferroelectric measurements. His scientific excellence is reflected in an extensive publication record, comprising over 90 articles published in high-impact international journals. In addition, he has supervised two undergraduate theses and six master's theses, with several additional projects currently in progress, demonstrating his strong commitment to both research and the training of the next generation of scientists.

Through his innovative research and leadership, Dr. Subhash continues to make significant contributions to the fields of multiferroic materials, nanomaterials, and energy storage, establishing himself as a leading researcher in these dynamic and interdisciplinary areas.

Contributors

Arunachalam Arulraj Departamento de Electricidad, Facultad de Ingeniería, Universidad Tecnológica Metropolitana, Santiago, Chile

Fakir Chand Department of Physics, Kurukshetra University, Kurukshetra, India

Sunil Chauhan Department of Physics & Environmental Sciences, Sharda School of Engineering & Science, Sharda University, Greater Noida, Uttar Pradesh, India; Centre for Solar Cell and Renewable Energy, Department of Physics & Environmental Sciences, Sharda School of Engineering & Science, Sharda University, Greater Noida, Uttar Pradesh, India

Gabriel Omar Mendoza Conde Coordinación de Electrónica, Instituto Nacional de Astrofísica, Óptica y Electrónica (INAOE), Puebla, México

Eduardo Flores Departamento de Física Aplicada, Centro de Investigación y de Estudios Avanzados (Cinvestav), Mérida, México

R. González-Campuzano Instituto de Investigaciones en Materiales, Universidad Nacional Autónoma de México, Ciudad de México, México

Indu Gupta Department of Physics, UIS, Chandigarh University, Mohali, India

Rahul Jain Department of Applied Sciences, CGC University, Mohali, India

Esquivel-Marin Judi Judith Unidad Académica de Ciencias Químicas, Universidad Autónoma de Zacatecas, Campus Universitario UAZ Siglo XXI Carretera Zacatecas-Guadalajara Km. 6, Zacatecas, Mexico

Pawan Kumar School of Basic and Applied Sciences, K. R. Mangalam University, Gurugram, Haryana, India

Ramesh Kumar Department of Physics, Guru Jambheshwar University of Science and Technology, Hisar, India

Sanjay Kumar Thin Film Physics Division, Department of Physics, Chemistry, and Biology (IFM), Linköping University, Linköping, Sweden

Vinod Kumar Department of Physics, The University of the West Indies St. Augustine Campus, St. Augustine, Trinidad and Tobago;
Department of Physics, Kurukshetra University, Kurukshetra, India

José Alberto Luna López Centro de Investigaciones en Dispositivos Semiconductores (CIDS-ICUAP), Benemérita Universidad Autónoma de Puebla (BUAP), Puebla, México

Matheu Mahabir Department of Physics, The University of the West Indies St. Augustine Campus, St. Augustine, Trinidad and Tobago

Piyali Maity City University of New York, New York, NY, USA

Ramalinga Viswanathan Mangalaraja Facultad de Ingeniería y Arquitectura, Universidad Arturo Prat, Iquique, Chile

D. E. Martínez-Lara Instituto de Investigaciones en Materiales, Universidad Nacional Autónoma de México, Ciudad de México, México;
Escuela Nacional Preparatoria No. 7 "Ezequiel A. Chávez", Universidad Nacional Autónoma de México, Ciudad de México, México

D. Mendoza Instituto de Investigaciones en Materiales, Universidad Nacional Autónoma de México, Ciudad de México, México

Ranjana Pande Department of Life Science, Sharda School of Bio-Science & Technology, Sharda University, Greater Noida, Uttar Pradesh, India

Kaushlendra Pandey Departamento de Electricidad, Facultad de Ingeniería, Universidad Tecnológica Metropolitana, Santiago, Chile

Dinesh Pathak Department of Physics, The University of the West Indies St. Augustine Campus, St. Augustine, Trinidad and Tobago

Priyanka Department of Physics, Kurukshetra University, Kurukshetra, India

Sandeep Kumar Pundir Department of Physics, University of Lucknow, Lucknow, Uttar Pradesh, India

Nikita Rawal Department of Life Science, Sharda School of Bio-Science & Technology, Sharda University, Greater Noida, Uttar Pradesh, India

Sangeeta Rawal Department of Physics & Environmental Sciences, Sharda School of Engineering & Science, Sharda University, Greater Noida, Uttar Pradesh, India

Ritu Department of Physics, Kurukshetra University, Kurukshetra, India

Zaira Jocelyn Hernández Simón Centro de Investigaciones en Dispositivos Semiconductores (CIDS-ICUAP), Benemérita Universidad Autónoma de Puebla (BUAP), Puebla, México

Parminder Singh Department of Physics, School of Sciences, IILM University, Greater Noida, India

Kavya K. Swamy Department of Physics, The University of the West Indies St. Augustine Campus, St. Augustine, Trinidad and Tobago

Anjali Varshney Department of Physics & Environmental Sciences, Sharda School of Engineering & Science, Sharda University, Greater Noida, Uttar Pradesh, India;
Centre for Solar Cell and Renewable Energy, Department of Physics & Environmental Sciences, Sharda School of Engineering & Science, Sharda University, Greater Noida, Uttar Pradesh, India

Upkar Kumar Verma Department of Physics, B. S. N. V. P. G. College, Lucknow, Uttar Pradesh, India

Vinay Pratap Verma Department of Physics, University of Lucknow, Lucknow, Uttar Pradesh, India

Chapter 1
Multifunctional Heterostructure Thin Films: Fundamentals, Fabrication, and Device Applications

Kavya K. Swamy, Sanjay Kumar, Matheu Mahabir, Vinod Kumar, and Dinesh Pathak

1.1 Introduction

Heterostructures are multilayer thin-film systems in which dissimilar materials are stacked to engineer properties that are not attainable in homogeneous films. By combining semiconductors with different compositions, band structures, and lattice parameters, entirely new functionalities can be realized. A notable historical example is the development of the heterojunction, in which two different semiconductors form a pn junction with superior performance compared to homojunction devices. Subsequent advances, including the transistor and the progression from thick films to thin films, have driven modern electronics and remain central to research today [1, 2].

Unlike individual thin films, layered structures exploit not only the intrinsic properties of each constituent material but also the unique interfacial effects that emerge at their junctions. These include enhanced carrier mobility, strain tolerance, mechanical flexibility, and improved electronic coupling. Despite these advantages, challenges such as lattice mismatch, fabrication cost, reproducibility, and scalability continue to limit widespread adoption.

The technological significance of heterostructures spans diverse sectors. In electronics and optoelectronics, they underpin high-performance light-emitting diodes, lasers, and high-speed transistors. In energy applications, they are integral to solar cells, batteries, and supercapacitors. In the biomedical domain, heterostructures

K. K. Swamy · M. Mahabir · V. Kumar · D. Pathak (✉)
Department of Physics, The University of the West Indies St. Augustine Campus, St. Augustine, Trinidad and Tobago
e-mail: dineshpathak80@gmail.com

S. Kumar
Thin Film Physics Division, Department of Physics, Chemistry, and Biology (IFM), Linköping University, Linköping SE-581 83, Sweden

S. Sharma (ed.), *Multifunctional Heterostructure Thin Films Handbook*,
https://doi.org/10.1007/978-3-032-18692-8_1

contribute to X-ray detectors, biosensors, protective coatings, and implantable devices. Magnetic heterostructures, formed from ferroelectric and ferromagnetic materials, enable strong coupling effects and new mechanisms for electric-field control of magnetic states [3].

Heterostructures can take many forms: semiconductor–semiconductor junctions, metal–insulator–semiconductor structures that enhance capacitance, or insulator–insulator stacks that serve as optical coatings and dielectric layers. In photovoltaic systems, multilayer heterostructures facilitate maximum absorption across the solar spectrum [1, 4–6]. This chapter will examine the theoretical foundations of heterostructures, methods for characterizing their interfaces and material properties, mechanisms of energy conversion, fabrication techniques along with their advantages and limitations, and both established and emerging thin-film combinations for next-generation device platforms.

1.2 Fundamentals of Heterostructure Thin Films

1.2.1 Definition and Classification

A heterostructure is formed when two or more dissimilar materials are joined together to create an interface with distinct electronic, optical, or structural properties. The constituent layers may differ in lattice parameters, atomic composition, crystal structure, or carrier type, and their combination gives rise to new functionalities not available in the individual materials [7]. A classic example is the heterojunction, in which two different semiconductors, such as GaAs and AlGaAs, form a pn junction with enhanced carrier confinement and transport compared to a homojunction, see Fig. 1.1. More complex multilayer devices, such as the heterojunction bipolar transistor (HBT) with its emitter–base–collector stack, further demonstrate the power of heterostructure engineering [8, 9]. Heterostructures can be broadly classified according to the types of materials combined:

(a) Semiconductor–semiconductor heterostructures, used in lasers, LEDs, and solar cells [10].
(b) Metal–semiconductor heterostructures, important for Schottky diodes and high-speed transistors [11].
(c) Metal–insulator–semiconductor (MIS) heterostructures, used in capacitors and field-effect devices [12].

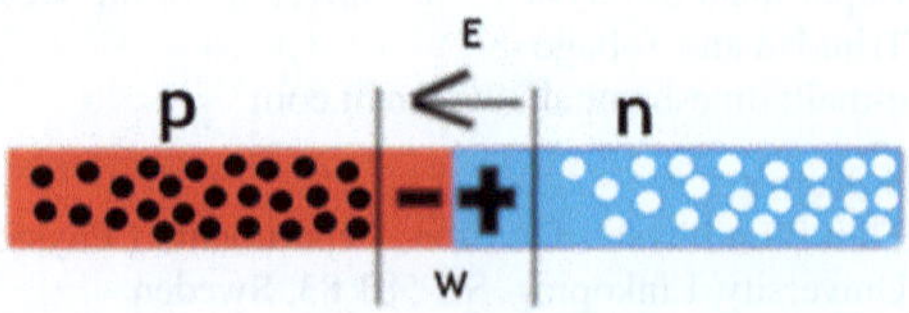

Fig. 1.1 Schematic of a pn heterojunction illustrating carrier distribution across the depletion region (width *w*) and the built-in electric field (*E*)

(d) Insulator–insulator heterostructures, common in optical coatings and dielectric mirrors [13].

This wide range of possible combinations provides a versatile platform for tailoring properties to meet the needs of advanced microelectronic, optoelectronic, energy, and biomedical applications.

1.2.2 Interface Physics and Band Alignment

Before a junction is produced, both the p-type and n-type semiconductors are electrically neutral, each containing balanced positive and negative charges. When the two are brought into contact, electrons diffuse from the n-type region, where they are abundant, into the p-type region, while holes migrate in the opposite direction. This exchange of carriers leaves behind immobile charged ions, creating a space-charge or depletion region. Within this region, an internal electric field and potential barrier develop, preventing further carrier motion once equilibrium is reached. The magnitude of this built-in potential depends on the doping concentrations of the two materials and on temperature. In silicon, for example, the barrier is typically about 0.6–0.7 V at room temperature [14, 15]. When a forward bias is introduced, the applied potential diminishes the energy barrier, leading to a reduction in the width of the depletion zone. This allows electrons and holes to cross the junction more easily, resulting in current flow sustained by continuous replenishment of carriers from the external circuit. Under reverse bias, the external field strengthens the barrier, widening the depletion layer and suppressing charge transport. From an energy-band perspective, the Fermi level of the n-type semiconductor initially lies closer to the conduction band, while that of the p-type is closer to the valence band. Upon contact, the two levels align to form a common Fermi level across the junction. The accompanying band bending across the depletion width reflects the built-in potential and defines the energy barrier that carriers must overcome, see Fig. 1.2.

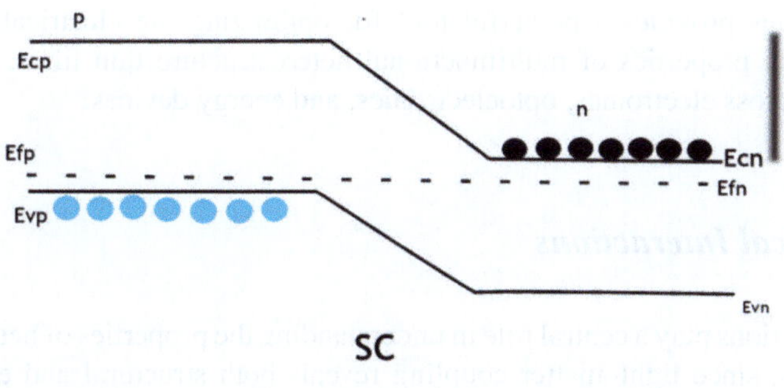

Fig. 1.2 Energy band diagram of a pn junction, showing band bending, depletion region formation, and built-in potential

1.2.3 Strain Engineering

Strain engineering is a broadly used approach for tailoring the properties of heterostructure thin films. When a film is deposited on a substrate with a different lattice constant, the atoms in the film adjust their positions to match the periodic arrangement of the substrate. This adjustment introduces strain at the interface. If the lattice constants are nearly identical, as in the case of GaAs and AlAs, the strain is minimal and the layers form a coherent, well-matched interface. However, larger lattice mismatches force the film to distort, leading to significant strain and, in some cases, the formation of dislocations to relieve the stress [16, 17].

The presence of strain has a direct impact on the electronic structure of the material. By altering interatomic distances, strain modifies the potential landscape experienced by charge carriers. This in turn changes the bandgap energy, carrier mobility, and optical transition properties. For example, tensile strain stretches the film lattice and often reduces the bandgap, while compressive strain squeezes the lattice and can increase the bandgap or lift degeneracies in the band structure. These effects are deliberately exploited in devices such as strained-silicon transistors, high-performance lasers, and advanced photovoltaic cells.

On the mechanical side, small lattice distortions are generally elastic, meaning that the material can return to its original state if the strain is removed. The extent to which a crystal can sustain elastic deformation depends on its intrinsic stiffness constants, which define how it responds to stress applied along different directions. These constants are critical for predicting the limits of strain before relaxation mechanisms, such as dislocation formation, begin to occur.

Strain can also be classified as tensile or compressive depending on how the film lattice compares with that of the substrate. If the film is forced to expand to match a smaller substrate lattice, tensile strain is introduced. Conversely, if the film is compressed to match a larger substrate lattice, compressive strain results. The degree of mismatch determines whether the film can remain coherently strained (pseudomorphic) or whether it will relax through the generation of defects [18–20].

By carefully choosing material combinations and controlling growth conditions, researchers can use strain as a design parameter to enhance performance. Strain engineering thus provides a powerful tool for optimizing the electrical, optical, and mechanical properties of multifunctional heterostructure thin films, enabling applications across electronics, optoelectronics, and energy devices.

1.2.4 Optical Interactions

Optical interactions play a central role in understanding the properties of heterostructure thin films, since light–matter coupling reveals both structural and electronic characteristics. One of the most widely used techniques for this purpose is ellipsometry, which analyzes alterations in the polarization of light upon reflection from a film surface. By measuring the amplitude ratio and the phase difference between the two orthogonal components of polarized light, commonly referred to as *p*- (parallel) and *s*- (perpendicular) polarization, ellipsometry provides detailed information about film thickness, refractive index, surface roughness, and optical constants. Measurements are typically carried out across the ultraviolet, visible, and infrared spectral ranges, making the technique versatile for both semiconductors and dielectrics [21, 22].

The quality of optical measurements is strongly influenced by the surface topography of the film. A smooth surface leads to predominantly specular reflection, while increased surface roughness introduces scattering that reduces the accuracy of the detected signal. For this reason, precise control over thin-film growth and surface morphology is essential when using ellipsometry or related optical characterization methods.

The behavior of light in thin films is often described in terms of the complex refractive index, the dielectric function, and the absorption coefficient. Together, these parameters reveal fundamental properties of the material. Measurements in the UV–visible range are commonly used to extract the optical bandgap, alloy composition, crystalline phase, and grain size. In contrast, measurements in the infrared region provide insights into free-carrier concentration, carrier mobility, and overall electrical conductivity. Thus, optical interactions offer a direct link between a film's structural characteristics and its electronic behavior.

The refractive index itself is one of the most important optical parameters. It describes how light slows down as it passes through a medium compared to its velocity in vacuum. In absorbing materials, such as conductors or semiconductors with significant optical losses, the refractive index becomes a complex quantity. The real part corresponds to the phase velocity of light, while the imaginary part indicates the degree of absorption. A nonzero imaginary component implies that photons are absorbed as they propagate through the material, causing the transmitted intensity to decrease exponentially with depth. This absorption behavior is commonly expressed through the Beer–Lambert law, which relates the incident and transmitted light intensities to the thickness of the film and its absorption coefficient. By combining these optical measurements with structural and electrical analyses, researchers can obtain a comprehensive understanding of heterostructure thin films. Optical characterization not only helps determine fundamental parameters such as bandgap and carrier concentration, but also provides essential feedback for optimizing growth conditions, improving surface quality, and designing films for advanced optoelectronic and energy applications.

1.2.5 Dielectric Constant

Dielectric materials are electrical insulators in which charges cannot move freely, but the atoms or molecules can undergo polarization in response to an applied electric field. When such a field is present, bound charges shift slightly from their equilibrium positions, producing local dipoles. The collective effect of these dipoles generates regions of positive and negative polarity within the material, reducing the effective internal electric field. This phenomenon is referred to as dielectric polarization [23].

Depending on the mechanism, polarization can arise from the displacement of electron clouds relative to nuclei (electronic polarization), the relative shift of positive and negative ions in a lattice (ionic polarization), or the reorientation of permanent molecular dipoles (orientational polarization). These contributions are often visualized. Given sufficient time in the field, the dipoles align with the external electric field, stabilizing the induced polarization.

A key parameter describing this behavior is the dielectric constant, or relative permittivity, which quantifies the extent of polarization relative to that in vacuum. In simple terms, it measures how much a dielectric material reduces the effective electric field compared to free space. For example, in a capacitor, inserting a dielectric between the plates increases capacitance because the internal field is weakened by polarization. The degree of polarization in a material can be understood as the sum of all dipole moments created under the influence of the applied field. The dipole moment of an individual entity is given by the product of its charge and the separation distance between positive and negative charges. When these moments add collectively across the material, the macroscopic dielectric response emerges.

Dielectric properties are fundamental for heterostructure thin films, as they influence charge storage, energy dissipation, and interfacial electric fields. By selecting and engineering dielectric layers, researchers can design structures with optimized capacitance, insulation strength, and frequency response, enabling their integration into high-performance electronic, optoelectronic, and energy devices.

1.3 Material Systems and Combinations

Heterostructure thin films can be realized through a wide variety of material combinations, each offering unique advantages for electronic, optical, and energy applications. The following subsections highlight representative systems and their functionalities.

1.3.1 Oxide–Oxide Heterostructures

Metal oxides are widely studied as thin-film semiconductors due to their transparency, scalability for large-area production, and tunable electronic properties. When combined into oxide–oxide heterostructures, these materials often exhibit superior performance compared to single-layer oxides. For example, heterojunction transistors based on indium zinc tin oxide (InZnSnO) and indium zinc oxide (InZnO) have been demonstrated as transparent phototransistors. Similarly, InSnO/GaInZnO thin-film transistors have shown extremely high mobilities, attributed to the complementary roles of the two layers: the bottom layer provides high conductivity, enabling enhanced mobility, while the upper layer offers lower conductivity, which helps control the threshold voltage. Comparable behavior has been observed in ITO/TZO heterostructures, where the balance between high carrier concentration in ITO and lower conductivity in TZO yields favorable electronic characteristics. Overall, oxide–oxide heterostructures typically display improved optical transparency and electronic tunability relative to single-layer films [24, 25].

1.3.2 Semiconductor–Dielectric Interfaces

Dielectrics play a critical role in heterostructures designed for charge storage and transistor applications. While dielectrics are insulating, they can be penetrated by electric fields, enabling polarization and screening effects. When a semiconductor is interfaced with a dielectric, the Coulomb interaction between charge carriers is reduced, leading to enhanced mobility and improved charge transport across the interface. High-κ dielectric oxides are particularly valuable, as they allow stronger gate coupling and reduced operating voltages in thin-film transistors. In this configuration, the dielectric layer not only modulates the channel but also acts as an insulating barrier between the gate and the semiconductor, minimizing leakage currents and enhancing device reliability [26, 27].

1.3.3 Metal Oxide/2D Material Heterostructures

Metal oxide semiconductors are central to display technologies and transparent electronics due to their high mobility, optical transparency, and suitability for large-scale processing [28]. In multilayer heterostructures, the front (channel) layer is typically more conductive, ensuring high carrier density and mobility, while the back-channel layer is less conductive, helping to control threshold behavior and suppress leakage. Introducing dielectric interlayers, such as between InSnZnO and InGaZnO, has been shown to further boost mobility. Notably, In_2O_3/ZnO heterojunction transistors exhibit mobilities up to two orders of magnitude higher than their single-layer counterparts. This enhancement arises from band alignment effects and the formation of high-density electron gases at the interface. In superlattice structures composed of In_2O_3, Ga_2O_3, and ZnO, conduction shifts from trap-limited behavior in single oxides to percolation-dominated conduction, resembling transport in metals [29, 30].

1.3.4 Emerging Combinatorial Approaches

Beyond oxides and conventional semiconductors, a new class of heterostructures involves two-dimensional (2D) materials such as graphene, transition-metal dichalcogenides (e.g., MoS_2, WS_2), hexagonal boron nitride (h-BN), black phosphorus, and organic perovskites. These materials are bonded by van der Waals forces, which eliminates the strict lattice-matching requirements typical of conventional epitaxial growth. This makes it possible to stack diverse materials without significant strain or defect formation, greatly expanding the design space for heterostructures [31].

Graphene is a particularly notable example: as a 2D conductor with exceptionally low resistivity and high carrier mobility, it serves as an ideal electrode or channel material. When sandwiched between wide bandgap materials, graphene and related 2D materials form quantum wells that confine carriers and minimize scattering from bulk defects. Combinatorial strategies include 0D-2D hybrids (quantum dots integrated with 2D sheets), 1D-2D hybrids (nanowires coupled with graphene or dichalcogenides), and 2D–3D hybrids (layered 2D materials integrated with bulk perovskites). For example, graphene/MoS_2 heterostructures demonstrate unique charge-transfer properties, enabling applications in high-performance photodetectors and flexible electronics [32]. These emerging approaches highlight the future potential of multifunctional heterostructures, where novel combinations of dimensionalities and materials are used to achieve performance well beyond that of traditional thin-film systems.

1.4 Fabrication Techniques

The performance of heterostructure thin films depends strongly on the fabrication method. The choice of deposition technique influences composition, thickness, crystallinity, surface quality, and interface sharpness, all of which are critical for tailoring device performance. This section first introduces the fundamentals of thin-film growth, followed by descriptions of major physical and chemical deposition techniques [33].

1.4.1 Growth Fundamentals

Thin-film growth begins with the processes of nucleation and crystallization. A deposited material may form either amorphous or crystalline structures depending on deposition conditions such as temperature, pressure, and precursor supply. Crystallization requires supersaturation, where the concentration of atoms or molecules exceeds equilibrium solubility. Once this state is achieved, nuclei form on the substrate surface and serve as seeds for further crystal growth [32].

Supersaturation can be controlled by adjusting the substrate temperature, vapor pressure of the precursors, or solution concentration in liquid-phase methods. For example, heating a solvent can temporarily increase solubility, while cooling induces nucleation. In vapor-phase growth, the balance between vapor pressure, substrate temperature, and deposition rate determines whether the film is crystalline, polycrystalline, or amorphous. Thin films typically grow according to three classical mechanisms:

(a) Volmer–Weber (island growth): atoms preferentially bond to each other, forming clusters or islands on the surface [34].
(b) Frank-van der Merwe (layer-by-layer growth): atoms adhere more strongly to the substrate than to each other, forming smooth monolayers [35].
(c) Stranski–Krastanov (layer-plus-island growth): initial layers grow smoothly, but accumulated strain eventually leads to island formation [36].

Understanding these mechanisms is crucial, as growth mode determines surface roughness, defect density, and strain management in heterostructures.

1.4.2 Physical Vapor Deposition (PVD)

In PVD, solid materials are vaporized under vacuum and condensed onto a substrate. Evaporation can occur by heating to induce sublimation, minimizing contamination compared to melting. Metals such as Al, Cu, Co, Cr, and Ni are often deposited this way, though each has a specific evaporation temperature [37].

For compounds or alloys, exact stoichiometry is difficult to maintain, since the vapor composition may differ from the solid source. Nevertheless, PVD is widely used in industry to produce protective coatings that enhance hardness, abrasion resistance, chemical stability, and toughness [38]. Films fabricated by PVD generally exhibit uniform coverage with minimal pinholes.

A widely employed variant is magnetron sputtering, in which an inert gas (commonly argon) is ionized into plasma. The ions are accelerated toward a solid target, knocking atoms from its surface. These atoms then deposit on the substrate. Magnetron sputtering offers high deposition rates, uniform films, and does not require substrate heating, making it suitable for temperature-sensitive substrates. However, precise composition control is more challenging than in chemical deposition methods [39].

1.4.3 Chemical Vapor Deposition (CVD)

CVD is a bottom-up process in which volatile precursors react or decompose on a heated substrate to form a thin film. Precursor gases are introduced into a reactor, carried by an inert gas such as nitrogen or argon. Inside the reactor, chemical reactions at the substrate surface yield the desired material, while by-products and unreacted gases are exhausted [40].

CVD is particularly important for synthesizing graphene and 2D materials, owing to its scalability and ability to produce large-area films. Key process parameters include temperature, pressure, carrier gas composition, and precursor flow rates. Fine-tuning these allows control over stoichiometry, crystallinity, and thickness. A major variant is low-pressure CVD (LPCVD), where reduced chamber pressure enhances precursor diffusion to the substrate, resulting in higher-quality films and faster deposition rates.

1.4.4 Pulsed Laser Deposition (PLD)

In PLD, high-energy laser pulses ablate material from a solid target in a vacuum environment. The absorbed laser energy generates a plasma plume of ions, atoms, and clusters that expands toward the substrate, where it condenses to form a thin film [41].

The characteristics of the plume depend on laser parameters such as fluence, pulse duration, and spot size. Short-pulse (femtosecond) lasers tend to produce denser plumes and generate nanoparticles, providing fine control of film morphology. One advantage of PLD is that the composition of the deposited film often closely matches that of the target, making it highly suitable for complex oxides and multilayer heterostructures. However, scaling PLD for large-area uniform coatings remains challenging [42].

1.4.5 Atomic Layer Deposition (ALD)

ALD is a vapor-phase technique that enables angstrom-level thickness control. Unlike CVD, ALD proceeds through self-limiting surface reactions, with alternating precursor pulses introduced in cycles. In the first half-cycle, precursor molecules chemisorb onto the substrate; excess molecules are purged by a carrier gas. In the second half-cycle, a second precursor reacts with the adsorbed layer, forming a monolayer of the desired material. Repetition of these cycles builds the film layer by layer [43].

This sequential process ensures exceptional uniformity, conformal coverage on 3D surfaces, and precise stoichiometry control. ALD can operate over a wide temperature range, from room temperature to elevated conditions, offering versatility beyond CVD [44]. Growth can proceed through different modes:

(a) Island growth (Volmer–Weber), forming clusters [35].
(b) Layer-by-layer growth (Frank–van der Merwe), forming continuous monolayers [35].
(c) Layer-plus-island growth (Stranski–Krastanov), where initial smooth layers transition into island formation due to strain [35].

These growth modes strongly influence film density, surface roughness, and mechanical stability.

1.4.6 Molecular Beam Epitaxy (MBE)

MBE is one of the most precise deposition methods, capable of producing ultra-thin layers with atomic-scale control. The process is carried out in an ultra-high vacuum (UHV) chamber, where beams of atoms or molecules are directed at a heated substrate. The deposited layer grows epitaxially [45], adopting the crystal orientation of the substrate [46].

Reflection high-energy electron diffraction (RHEED) is commonly used to monitor MBE growth in real time. A shallow-angle electron beam reflects from the film surface, and oscillations in the detected intensity provide feedback on growth rate, layer completion, and surface smoothness [47]. Variants of MBE include solid-source MBE, where elemental materials are evaporated by thermal or electron-beam heating, and gas-source MBE, which uses molecular precursors. While MBE provides unmatched control of purity, interface sharpness, and doping, it is less scalable and more costly than PVD or CVD.

1.4.7 Challenges in Layer Control and Uniformity

The quality and performance of thin films [48] are strongly dependent on the deposition method used [49]. While some techniques, such as PVD, offer excellent film density and smoothness, they often provide limited stoichiometric control. In contrast, solution-based methods are cost-effective and scalable, but achieving reproducibility and uniformity remains difficult. Thus, the choice of deposition method must balance precision, cost, scalability, and application requirements.

Vacuum-based techniques (such as sputtering and evaporation) generally yield smoother and more uniform films. However, deposition is line-of-sight: atoms or ions travel directly from the source to the substrate. If the substrate has complex geometries or shadowed regions, those areas may receive less material, resulting in thickness variations. While vacuum processes minimize pinholes and defects, perfect uniformity is rarely achievable across large areas.

Solution-based methods such as spin coating, dip coating [50], and droplet deposition introduce different challenges. In spin coating, film thickness depends on rotation speed, but other factors such as solvent viscosity, humidity, and vapor pressure also play critical roles. Highly viscous solvents may prevent complete substrate coverage, leading to pinholes or rough surfaces. Dip coating, where a substrate is immersed in a precursor solution and then withdrawn, depends on surface tension, immersion time, and withdrawal speed. Non-uniform immersion can produce films of variable thickness or grain size [51].

In droplet-based deposition methods, film formation depends on how droplets spread, adhere, or evaporate upon impact. Spreading ensures good coverage, while splashing or rebounding leads to surface irregularities. The substrate temperature also plays a crucial role: higher temperatures promote rapid solvent evaporation and supersaturation, which can improve crystallization but may also increase viscosity and hinder uniform spreading.

These challenges highlight that while each deposition method offers distinct advantages, none can produce defect-free films under all conditions. Careful optimization of process parameters, such as precursor composition, substrate temperature, pressure, and ambient environment, is essential for achieving high-quality, uniform thin films [52] suitable for heterostructure applications [47].

1.5 Characterization Methods

Characterization of heterostructure thin films is essential to understanding their structural, compositional, electrical, and optical properties. A wide range of techniques is available, each offering complementary insights into film quality, interfaces, and performance.

1.5.1 Structural Characterization

X-Ray Diffraction (XRD)

XRD is one of the most widely used methods for studying crystal structure. When X-rays interact with the periodic lattice planes of a crystal, they undergo constructive or destructive interference depending on the interplanar spacing. According to Bragg's law, constructive interference produces diffraction peaks, allowing calculation of lattice parameters, crystallite size, and strain. Crystalline materials yield sharp peaks, while amorphous films produce broad diffuse patterns. The presence of forbidden reflections, determined by the structure factor, helps identify lattice symmetries (e.g., FCC vs BCC) [47].

Transmission Electron Microscopy (TEM) and Scanning TEM (STEM)

TEM uses high-energy electrons transmitted through a thin specimen to reveal crystal structure, grain boundaries, and defects at atomic resolution. Recent advances enable sub-angstrom resolution, making TEM ideal for thin films. STEM enhances contrast by using bright-field and dark-field imaging modes. When combined with energy-dispersive X-ray spectroscopy (EDS), TEM can map elemental composition at interfaces. Electron energy loss spectroscopy (EELS) further provides information on bonding, valence states, and local electronic structure [53, 54].

Atomic Force Microscopy (AFM)

AFM provides topographic information with nanometer resolution by scanning a sharp probe across the film surface. Unlike electron microscopy, AFM does not require conductive samples. It operates in several modes: contact mode (continuous tip contact), tapping mode (intermittent contact at resonance frequency), and non-contact mode (probe interaction via van der Waals forces). AFM can measure surface roughness, adhesion forces, elasticity, and local mechanical properties, making it highly useful for surface-sensitive thin films [55, 56].

1.5.2 Compositional Analysis

Electron Microscopy with Spectroscopy

High-resolution TEM and cross-sectional TEM (XTEM) provide structural and chemical information at the atomic scale. When combined with EDS, they identify elemental composition and impurity distribution. EELS complements this by probing electronic states and chemical bonding [57, 58].

Time-Of-Flight Secondary Ion Mass Spectrometry (ToF–SIMS)

ToF–SIMS bombards the surface with primary ions and analyzes the emitted secondary ions by mass-to-charge ratio. It offers extremely high sensitivity (parts per million) for surface and near-surface analysis (1–3 monolayers) and enables chemical mapping of defects and interfaces with nanometer resolution [59].

Photoemission Electron Microscopy (PEEM)

PEEM uses UV or soft X-rays to excite photoelectrons from a surface. The technique provides surface-sensitive imaging (2–5 nm depth) and is particularly valuable for studying magnetic thin films, as it reveals electronic states associated with 3d and 4f orbitals [60, 61].

Spin-Polarized Low-Energy Electron Microscopy (SPLEEM)

SPLEEM employs low-energy spin-polarized electrons, which are backscattered from the surface. The interaction between the electron spin and local magnetic moments generates contrast, allowing visualization of magnetic domains, domain walls, and spin textures in ferromagnetic and multilayer heterostructures [62, 63].

1.5.3 Electrical Characterization

Hall Effect

The Hall effect is widely used to determine carrier type, mobility, and concentration. When a magnetic field is applied perpendicular to a current-carrying thin film, charge carriers are deflected, creating a measurable Hall voltage. The sign of the Hall coefficient indicates whether the majority carriers are electrons or holes [64, 65].

Four-Point Probe Method

This technique measures sheet resistivity by passing current through two outer probes and measuring the voltage drop across two inner probes. It eliminates the effect of contact resistance and is widely used for resistivity measurements in thin films [66, 67].

Van Der Pauw Technique

A variant of four-probe measurements, the van der Pauw method determines sheet resistance and carrier concentration for arbitrarily shaped films. When combined with a magnetic field, it also yields Hall coefficients [68–70].

Capacitance–Voltage (C–V) and Impedance Spectroscopy

These techniques probe dielectric properties, charge storage, and interface traps. C–V measurements reveal doping profiles, while impedance spectroscopy provides frequency-dependent information about conduction and polarization mechanisms [71–73].

1.5.4 Optical Characterization

Optical measurements are essential for analyzing absorption, transmission, and reflection properties of thin films. Absorption links photon energy to vibrational and electronic transitions, with opaque materials showing higher absorption coefficients. Transmission depends on thickness, morphology, and doping levels, while reflection provides insights into refractive index and surface quality [74].

For inhomogeneous thin films with defects, advanced models such as the Wentzel–Kramers–Brillouin–Jeffreys (WKBJ) approximation are applied. If refractive index gradients are large, the film may be modeled as a multilayer system composed of uniform sublayers, each contributing to the overall optical response [75].

1.5.5 Interface and Defect Analysis

Defects such as dislocations, vacancies, and interfacial traps strongly influence device performance by scattering carriers and creating recombination centers. Several techniques are specialized for analyzing defects and buried interfaces:

(a) Coherent Diffractive Imaging (CDI): reconstructs 3D images of dislocations and irregular atomic arrangements based on high-resolution diffraction patterns [76, 77].
(b) X-ray Absorption Near Edge Structure (XANES) and Extended X-ray Absorption Fine Structure (EXAFS): provide element-specific information about local bonding environments and interatomic distances [78].
(c) X-ray Photoelectron Spectroscopy (XPS): primarily a surface-sensitive technique, but also probes near-surface layers. It reveals oxidation states, chemical bonding, and electronic configurations, and is widely used to study buried interfaces in nanostructured materials [79, 80].

Collectively, these methods provide comprehensive insights into structure, composition, electrical behavior, and interfacial quality, enabling optimization of heterostructure thin films for multifunctional applications.

1.6 Multifunctional Properties and Mechanisms

Heterostructure thin films are attractive for multifunctional applications because they can simultaneously exhibit coupled ferroic, electronic, optical, and energy-harvesting phenomena. By carefully controlling lattice structure, strain, and interfaces, it is possible to achieve properties not present in bulk materials. The following subsections highlight key multifunctional effects and their underlying mechanisms [81].

1.6.1 Ferroelectricity and Multiferroicity

Ferroelectricity refers to the presence of a spontaneous electric polarization that can be reversed by an external electric field. This phenomenon arises from the absence of inversion symmetry in the crystal lattice, often caused by asymmetric ion displacements that stabilize a permanent dipole moment. A well-known example is barium titanate ($BaTiO_3$), where off-center displacement of titanium ions within oxygen octahedra produces polarization. Ferroelectricity is strongly temperature dependent, vanishing above the Curie temperature [82].

Multiferroic materials combine two or more ferroic orders, most commonly ferroelectricity, ferromagnetism, and ferroelasticity, in a single phase. This coexistence is rare because electronic configurations that favor magnetism (partially filled d-orbitals) typically conflict with those that enable ferroelectricity (empty d-orbitals). However, compounds such as $BiFeO_3$ and $YMnO_3$, as well as artificially engineered composites, demonstrate multiferroic behavior [83].

Of particular interest is magnetoelectric coupling, which enables the control of magnetic states by electric fields or vice versa. This cross-coupling provides a platform for low-power, multifunctional devices such as non-volatile memories, logic gates, and spintronic systems. In thin-film heterostructures, strain, charge modulation, and interfacial effects further enhance coupling strength. Mechanisms such as spin–orbit interactions, charge ordering, and geometric frustration can stabilize multiferroicity even in systems that are not ferroic in bulk. Advances in thin-film deposition now allow precise tuning of thickness, strain, and interfaces, enabling multiferroic properties to be engineered on demand [84, 85].

1.6.2 Magnetoelectric Coupling

Magnetoelectric (ME) coupling describes the interaction between electric and magnetic order parameters, where an applied electric field modifies magnetic behavior or vice versa. ME effects are categorized as intrinsic (single-phase materials such as $BiFeO_3$) or extrinsic (engineered heterostructures). Intrinsic systems exhibit limited coupling strength, as ferroelectricity and magnetism require conflicting orbital configurations [86].

Extrinsic systems overcome this limitation by combining different materials. A common approach integrates a piezoelectric or ferroelectric layer (e.g., PZT) with a magnetostrictive material (e.g., $CoFe_2O_4$ or Terfenol-D). Here, strain acts as the mediator: electric fields induce mechanical deformation in the piezoelectric layer, which transfers to the magnetostrictive layer, altering its magnetization, see Fig. 1.3. This strain-mediated mechanism enables voltage control of magnetism with much lower energy consumption than conventional magnetic-field switching.

Charge-mediated coupling has also emerged as a promising route. In such systems, carrier concentrations at magnetic interfaces are tuned by electric fields, leading to dynamic modulation of coercivity, magnetoresistance, and spin polarization. Achieving strong and reproducible ME effects requires atomically sharp interfaces and precise control of film quality, achievable through deposition methods such as pulsed laser deposition and atomic layer deposition. Incorporating ME heterostructures into spintronics and memory devices offers pathways toward ultralow-power, multifunctional electronics [87].

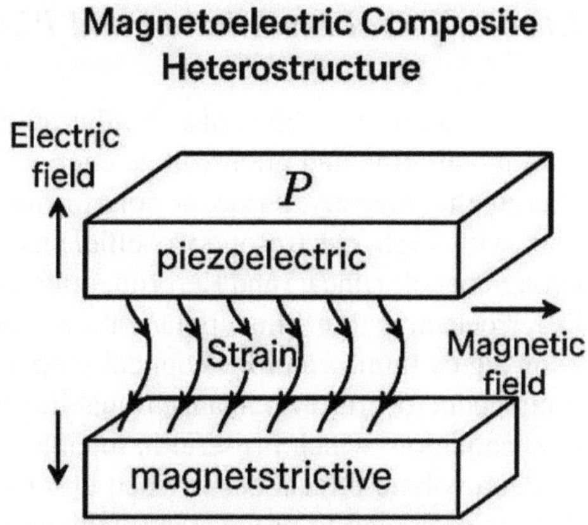

Fig. 1.3 Schematic illustration of strain-mediated magnetoelectric coupling in a ferroelectric–magnetostrictive heterostructure

1.6.3 Thermoelectric and Piezoelectric Effects

Thermoelectric and piezoelectric properties enable the direct conversion of thermal and mechanical energy into electrical energy, making them central to energy-harvesting and self-powered technologies [88]. Thermoelectric materials generate voltage from temperature gradients (Seebeck effect) and conversely produce heating or cooling under current flow (Peltier effect). Their efficiency is expressed by the dimensionless figure of merit (ZT), which depends on a high Seebeck coefficient, high electrical conductivity, and low thermal conductivity. Achieving high ZT values is challenging because these parameters are often interdependent. Nanostructuring, heterostructuring, and quantum confinement techniques have significantly improved performance by scattering phonons while preserving charge carrier transport. Superlattices of materials such as Bi_2Te_3 and PbTe exemplify this approach, reducing thermal conductivity without compromising electronic properties [89].

Piezoelectric materials generate an electric charge when mechanically deformed, a property that arises from non-centrosymmetric crystal structures. Widely used piezoelectrics include ZnO, AlN, $BaTiO_3$, and polymer-based PVDF. Their performance depends on piezoelectric coefficients, dielectric properties, and mechanical compliance. Thin films, nanowires, and flexible composites enhance their responsiveness, enabling applications in sensors, actuators, ultrasonic transducers, and wearable energy harvesters. Coupling piezoelectrics with flexible substrates further extends their use in electronic skin and biomedical implants [90]. Hybrid heterostructures that integrate thermoelectric and piezoelectric layers can simultaneously harvest thermal and mechanical energy. Such multifunctional composites are highly attractive for powering low-energy autonomous devices.

1.6.4 Optical Modulation and Photovoltaic Effects

Heterostructure thin films also exhibit multifunctional optical responses, including light modulation and photovoltaic energy conversion. Optical modulation involves altering the intensity, phase, or polarization of light through external stimuli. Materials with high electro-optic coefficients, such as $LiNbO_3$ and $BaTiO_3$, exhibit strong Pockels (linear) and Kerr (quadratic) effects, enabling high-speed modulation. In ferroelectric thin films, polarization switching changes local refractive indices, allowing dynamic control of optical properties. Integrating ferroelectric layers with semiconductors further enhances tunability by modifying interfacial fields and carrier recombination, which is useful in tunable photodetectors and adaptive optics [91].

Photovoltaic effects occur when absorbed photons generate electron–hole pairs that are separated to produce current, see Fig. 1.4. In addition to conventional p–n junction solar cells, some ferroelectric materials exhibit the bulk photovoltaic effect (BPVE), where asymmetric crystal structures generate shift currents, sometimes producing photovoltages exceeding the bandgap [92].

Heterostructures improve photovoltaic efficiency by facilitating charge separation at built-in interfacial fields and reducing carrier recombination. Hybrid perovskite materials such as $CH_3NH_3PbI_3$ have achieved power conversion efficiencies above 25%, and when combined with 2D materials (e.g., MoS_2, graphene), further enhancements in carrier extraction, stability, and spectral response are realized. These advances point to compact, high-efficiency, multifunctional solar energy technologies [93, 94].

Collectively, the multifunctional properties described above highlight the potential of heterostructure thin films to combine ferroic, electronic, and optical phenomena in a single platform. By leveraging interfacial coupling mechanisms and advanced fabrication methods, researchers can design next-generation devices that are energy-efficient, miniaturized, and adaptable for diverse applications.

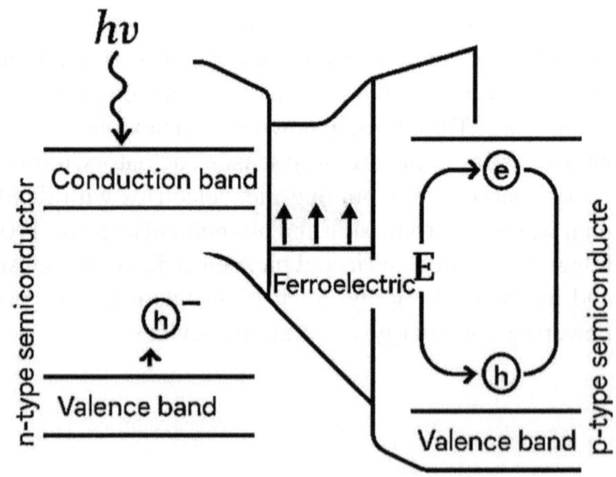

Fig. 1.4 Band diagram illustrating charge separation in a ferroelectric-assisted heterojunction solar cell

1.7 Device Applications

1.7.1 Sensors and Actuators

Sensors and actuators provide the interface between electronic systems and the physical world. In multifunctional devices, heterostructure engineering enables these components to integrate sensing/actuation with energy harvesting, signal conditioning, and adaptive responses [95, 96].

Sensors

External stimuli, pressure, temperature, magnetic fields, and chemical species are transduced into electrical signals. Coupled phenomena in heterostructures enhance sensitivity and selectivity. For example, piezoelectric sensors generate voltage under mechanical stress for pressure mapping and vibration monitoring. Magnetoelectric sensors convert magnetic inputs into electrical outputs via strain-mediated coupling between piezoelectric and magnetostrictive layers, enabling high sensitivity with low power and without bulky coils [97, 98].

Actuators

Actuators transform electrical inputs into mechanical motion. Ferroelectric and piezoelectric films produce displacement or stress under applied voltage, allowing precise control at micro- and nanoscales. Electrostrictive and magnetostrictive materials offer complementary mechanisms, with deformation driven by electric or magnetic fields [99, 100].

Co-Integration

In heterostructures, sensors and actuators can be co-fabricated to create compact, self-sufficient systems, crucial for biomedical implants, MEMS/NEMS, and soft robotics, see Fig. 1.5. Advances in thin films and flexible substrates have enabled printed and wearable implementations (e.g., PVDF- or 2D-material-based sensors for real-time health tracking). Arrays of flexible actuators serve as artificial muscles and adaptive surfaces. Integrating multifunctional sensing and actuation on a single platform reduces system complexity and improves portability [100, 101].

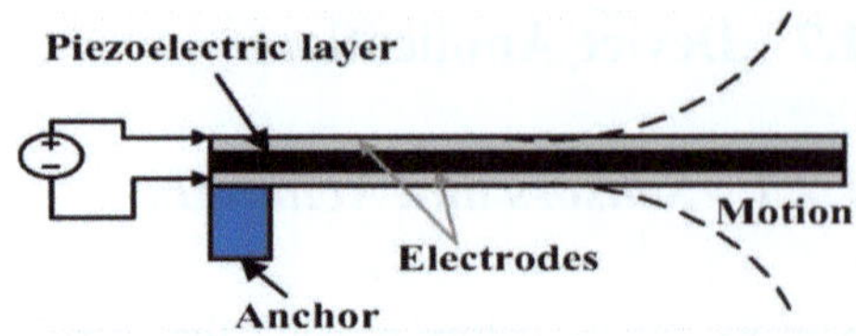

Fig. 1.5 Schematic of a piezoelectric actuator integrated with electrodes and anchor support

1.7.2 Spintronic

Spintronics exploits the device's spin degree of freedom of electrons alongside charge to process and store information, offering non-volatility, fast operation, and low power. Key building blocks include spin valves, magnetic tunnel junctions (MTJs), and spin-transfer torque (STT) devices. In an MTJ, two ferromagnets separated by an insulator exhibit tunneling magnetoresistance (TMR), where resistance depends on relative magnetization, forming the basis of MRAM [102, 103]. Heterostructures elevate performance in several ways:

(a) Multiferroic tunnel junctions (MFTJs) use a ferroelectric barrier between magnetic electrodes, enabling four distinct resistance states through coupled ferroelectric/magnetic switching, see Fig. 1.6 [104, 105].
(b) Spin–orbit coupling (SOC) in heavy metals or topological insulators adjacent to ferromagnets generates spin currents (e.g., via the spin Hall effect), enabling voltage-driven magnetization switching without external magnetic fields [106–108].
(c) 2D materials (graphene, TMDs) offer long spin diffusion lengths and tunable interfaces; combined with magnetic insulators, they support ultra-thin spin transistors and spin logic [109, 110].

Applications extend beyond memory to spin logic, neuromorphic elements, and quantum information, leveraging coherent spin states.

1.7.3 Non-volatile Memory (RRAM, FERAM)

Resistive RAM (RRAM) stores information by toggling the resistance of an oxide layer (e.g., HfO_2, TiO_2) through the formation/rupture of nanoscale conductive paths. Benefits include fast switching, scalability, and CMOS compatibility. Heterostructures improve filament control, reduce leakage, and enhance endurance via interface engineering and nanoscale laminates [111].

Ferroelectric RAM (FeRAM) relies on bistable polarization states in a thin ferroelectric (e.g., PZT or HfO_2-based films). Writing switches polarization with an electric field; reading senses the associated displacement current, see Fig. 1.7. FeRAM delivers low energy operation, high endurance, and nanosecond-scale switching. Ferroelectric FETs (FeFETs) embed a ferroelectric gate for direct logic-memory

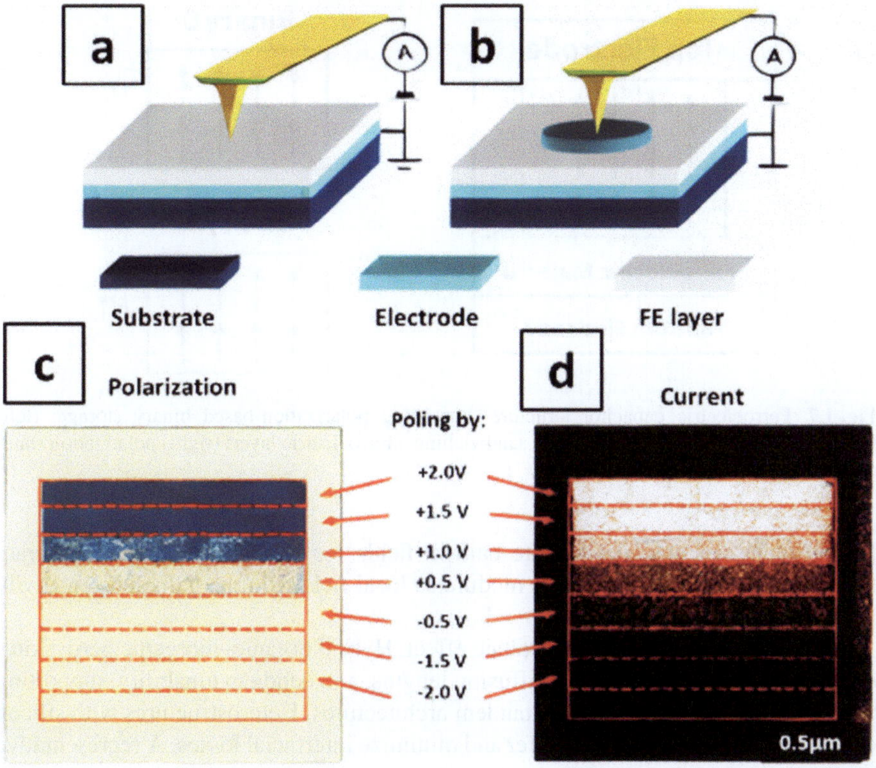

Fig. 1.6 Ferroelectric-assisted spintronic heterostructures: **a**–**b** schematic of local probing of ferroelectric polarization and current response in a layered device; **c** polarization mapping under applied voltages; **d** corresponding current distribution, demonstrating multiferroic tunnel junction (MFTJ) behavior

integration and in-memory computing [112, 113]. Both RRAM and FeRAM benefit from buffer/seed layers (for crystallinity and stability), carefully engineered electrodes, and scaling strategies (< 10 nm) addressing retention, variability, and thermal budgets [111–113].

1.7.4 Photodetectors and Solar Cells

Photodetectors convert light across UV-IR into electrical signals. Perovskites, graphene, TMDs, and quantum dots provide high mobilities, tunable band gaps, and strong light–matter coupling. Heterostructures improve carrier separation and reduce recombination, enabling higher responsivity and faster response. Vertical graphene/

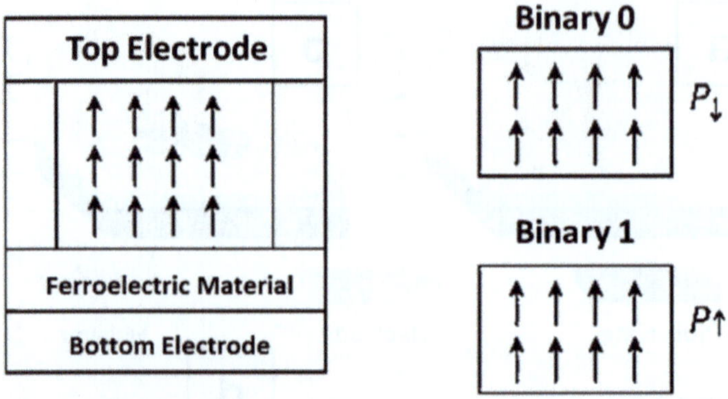

Fig. 1.7 Ferroelectric capacitor structure illustrating polarization-based binary storage: (left) schematic of top and bottom electrode sandwiching a ferroelectric layer; (right) polarization states corresponding to binary '0' and binary '1'

MoS_2 stacks, for instance, create built-in fields for efficient photocarrier extraction. Ferroelectric gating further modulates local fields to tune gain and sensitivity [114, 115].

Solar cells harness the photovoltaic effect. Hybrid organic–inorganic perovskites exhibit strong absorption, long diffusion lengths, and bandgap tunability, supporting $> 25\%$ efficiencies and enabling tandem architectures. Heterostructures with silicon or organics enhance charge transfer and minimize interfacial losses. A representative mechanism of carrier separation in ferroelectric-assisted heterojunction solar cells is shown in Fig. 1.8. Flexible thin-film photovoltaics on lightweight substrates support wearables and portable power, with ongoing work on scalable processing, moisture resistance, and durability [116, 117].

Fig. 1.8 Band diagram of a ferroelectric-assisted heterojunction solar cell showing charge separation under illumination, where the ferroelectric polarization enhances carrier separation and suppresses recombination

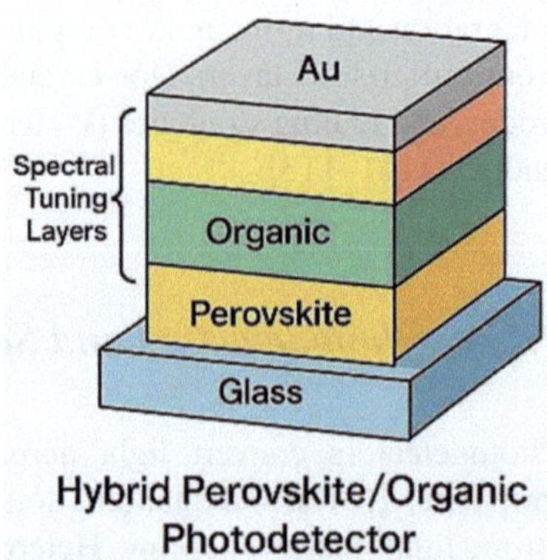

1.7.5 Flexible and Wearable Electronics

Flexible/wearable systems couple mechanical compliance with robust electronic performance. Heterostructures of conductive polymers, metal oxides, CNTs, graphene, and TMDs on PI or PDMS substrates preserve function under bending and stretching. Wearable sensors (piezoelectric/ferroelectric) harvest biomechanical energy and monitor physiological signals. Flexible oxide/2D-material TFTs maintain mobility under strain for displays and communications. Challenges include fatigue resistance, interconnect reliability, and layer integration at low thickness and weight. Printed electronics (inkjet, aerosol jet) and hybrid roll-to-roll + ALD workflows are accelerating manufacturability toward fully integrated, body-conformable systems [118, 119].

1.8 Recent Advances and Case Studies

1.8.1 Interface-Engineered Devices

Interfaces govern charge transfer, strain coupling, and polarization continuity. Atomic-scale control via MBE, PLD, and ALD yields sharp, coherent interfaces with emergent behavior. In $LaAlO_3/SrTiO_3$, polar discontinuity induces a high-mobility 2DEG with gate-tunable conductivity and low-temperature superconductivity, see Fig. 1.9. Tailoring interfacial thickness, chemistry, and dipoles optimizes carrier injection, recombination suppression, and magnetoelectric coupling. Buffer layers, passivation, and targeted doping enhance stability and reliability [120, 121].

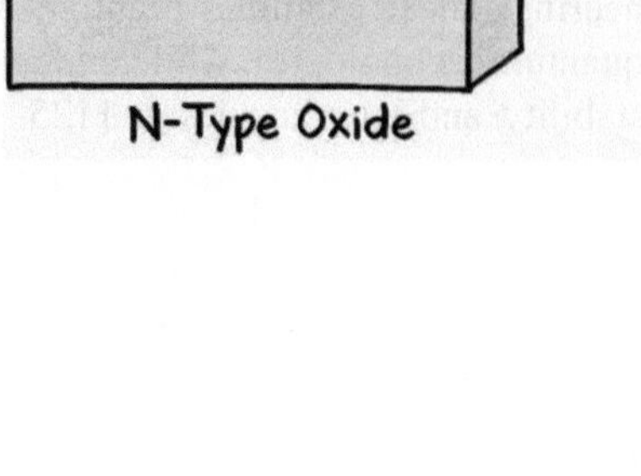

Fig. 1.9 Interface diagram showing two-dimensional electron gas (2DEG) formation at an oxide heterojunction, driven by polar discontinuity

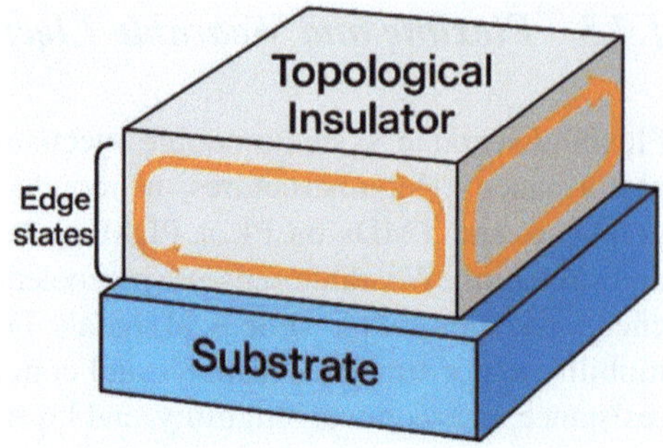

Fig. 1.10 Edge states in a topological-insulator-based heterostructure, illustrating symmetry-protected conduction channels that are robust against disorder

1.8.2 Quantum and Topological Effects in Heterostructures

Quantum confinement in wells, wires, and dots reshapes electronic density of states, enabling lasers, LEDs, resonant tunneling, and quantum-interference devices. Topological materials contribute symmetry-protected surface/edge states robust against disorder, see Fig. 1.10 [48], attractive for low-dissipation electronics and spintronics.

Proximity coupling to superconductors or ferromagnets in heterostructures can host exotic quasiparticles (e.g., Majorana modes) relevant to quantum computation. In 2D stacks (TMDs, graphene, topological insulators), spin/valley degrees of freedom and interlayer twist engineering unlock spintronic and valleytronic functionalities [122, 123].

1.8.3 Hybrid Organic–Inorganic Structures

Hybrid systems combine inorganic semiconductors (mobility, stability, strong absorption) with organics (mechanical flexibility, solution processability), see Fig. 1.11 [124]. Hybrid perovskites pair organic cations with inorganic halide frameworks, delivering high absorption, long diffusion lengths, and compositional bandgap tuning. Beyond photovoltaics, hybrids enable concurrent emission/detection for integrated optoelectronics. Key challenges, interface stability, moisture sensitivity, and long-term reliability, are addressed via passivation, encapsulation, and additive engineering. These examples highlight versatility across memory, sensing, energy, and quantum technologies, while underscoring common themes: scalability, operational stability, and interface design [125–127].

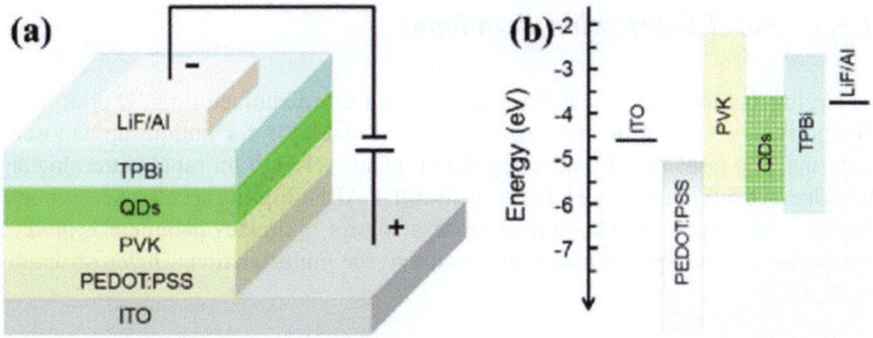

Fig. 1.11 **a** Schematic device structure of a hybrid organic–inorganic heterostructure incorporating quantum dots (QDs) with organic transport layers. **b** Corresponding energy band alignment of the multilayer device

1.9 Challenges and Future Perspectives

1.9.1 Scalability and Integration

Transitioning from laboratory prototypes to products requires balancing precision with throughput. Techniques like ALD, PLD, and MBE offer atomic control but face cost and throughput constraints and may not align seamlessly with CMOS or large/flexible substrates. Scalable approaches, roll-to-roll coating, printing (inkjet, aerosol jet), solution processes, improve cost and area coverage but demand advanced interface control to rival epitaxial quality. Co-integration with silicon requires thermal/chemical compatibility, robust interconnects, and uniformity across large areas [98, 128].

1.9.2 Interface Stability and Reliability

Interfaces concentrate defects, interdiffusion, and stress that degrade performance over time. High fields and temperatures promote chemical reactions (e.g., oxygen-vacancy migration in oxides), altering polarization or magnetization and increasing leakage. Mechanical mismatch in flexible devices can cause cracking or delamination. Environmental factors, moisture, oxidation, contaminants, further destabilize interfaces, especially in hybrid materials. Mitigation strategies include buffer layers (strain/diffusion control), surface passivation, targeted anneals, and barrier encapsulation. Accelerated aging (bending, humidity, thermal cycling) is essential for reliability qualification [129, 130].

1.9.3 Novel Fabrication Paradigms

Emerging methods target scalable precision: additive manufacturing/3D printing for geometry + composition control, directed self-assembly for bottom-up nanoscale ordering, and laser-based processing (LIFT, nano-welding) for rapid patterning and low-damage interconnects. Hybrid roll-to-roll + ALD/printing accelerates high-area flexible electronics. AI/ML-guided process control optimizes parameters, predicts performance, and speeds materials discovery for multifunctional heterostructures [131, 132].

1.9.4 Roadmap for Next-Generation Devices

Priorities include room-temperature multiferroics with strong magnetoelectric coupling, topological/quantum heterostructures for low-loss and quantum functions, and flexible neuromorphic systems integrating ferroic and optical responses for edge AI. AI-assisted design and diagnostics will enhance yield and enable self-healing operation. Cross-disciplinary collaboration, standardized benchmarks, and manufacturable interface engineering will accelerate translation to commercial products [133–136].

1.10 Conclusion

1.10.1 Summary of Key Insights

This chapter surveyed multifunctional heterostructure thin films from fundamentals to applications. We outlined interface physics, band alignment, and strain engineering; reviewed material systems (oxide–oxide, semiconductor–dielectric, metal-oxide, and emerging 2D/van der Waals stacks); and detailed fabrication (PVD, CVD, PLD, ALD, MBE) with growth fundamentals and uniformity challenges. We organized characterization across structural, compositional, electrical, optical, and interfacial methods to link microstructure with properties. We then examined multifunctional phenomena, ferroelectricity, magnetoelectric coupling, thermoelectricity, piezoelectricity, optical/photovoltaic effects, and mapped these to devices including sensors/actuators, spintronics, non-volatile memory, photodetectors/solar cells, and flexible electronics. Case studies in interface engineering, quantum/topological effects, and hybrid systems illustrate rapid progress, alongside standing challenges in scalability, stability, and integration.

1.10.2 Outlook and Emerging Opportunities

Advances in nanoscale synthesis, computation, and in situ/operando characterization are expanding the heterostructure design space. Trends such as 2D integration, quantum/topological stacks, and bio-inspired systems promise new device paradigms. Materials informatics + AI will accelerate discovery and process optimization. Flexible/wearable platforms will leverage multifunctionality for personalized health, HMI, and soft robotics. Success hinges on CMOS-compatible scaling, robust interfaces/passivation, and standardized reliability metrics. With coordinated, interdisciplinary effort, multifunctional heterostructures will underpin the next wave of energy-efficient, adaptive electronics across sensing, information processing, and energy conversion.

References

1. V. Kumar, D. Pathak, S. Kumar, D. P. Sharma, J.M. Nunzi, *An overview of photovoltaic technology*. Elsevier Inc., 2026. https://doi.org/10.1016/B978-0-443-29250-7.00010-4.
2. S. Kumar, A. Sharma, D. Pathak, J. M. Nunzi, *Bulk heterojunction solar cells*. Elsevier Inc., 2026. https://doi.org/10.1016/B978-0-443-29250-7.00012-8.
3. Z. I. Alferov, "The double heterostructure: Concept and its applications in physics, electronics and technology," *Int. J. Mod. Phys. B*, vol. 16, no. 5, pp. 647–675, 2002, https://doi.org/10.1142/S0217979202010233.
4. R. Zhang *et al.*, "The More, the Better–Recent Advances in Construction of 2D Multi-Heterostructures," *Adv. Funct. Mater.*, vol. 31, no. 26, pp. 1–18, 2021, https://doi.org/10.1002/adfm.202102049.
5. T. L. Chu and S. S. Chu, "Thin film II-VI photovoltaics," *Solid State Electron.*, vol. 38, no. 3, pp. 533–549, 1995, https://doi.org/10.1016/0038-1101(94)00203-R.
6. A. Hinderhofer and F. Schreiber, "Organic-organic heterostructures: Concepts and applications," *ChemPhysChem*, vol. 13, no. 3, pp. 628–643, 2012, https://doi.org/10.1002/cphc.201100737.
7. A. Pant *et al.*, "Fundamentals of lateral and vertical heterojunctions of atomically thin materials," *Nanoscale*, vol. 8, no. 7, pp. 3870–3887, 2016, https://doi.org/10.1039/c5nr08982d.
8. A. E. Material, "GaAlAs/GaAs Heterojunction Bipolar Transistors: Issues and Prospects for Application," vol. 36, no. 10, pp. 2032–2042, 1989.
9. S. S. Iyer, G. L. Patton, J. M. Stork, B. S. Meyerson, and D. L. Harame, "Heterojunction Bipolar Transistors Using Si-Ge Alloys," *IEEE Trans. Electron Devices*, vol. 36, no. 10, pp. 2043–2064, 1989, https://doi.org/10.1109/16.40887.
10. Z. I. Alferov, "The history and future of semiconductor heterostructures from the point of view of a Russian scientist," *Phys. Scr. T*, vol. 68, no. January, pp. 32–45, 1996, https://doi.org/10.1088/0031-8949/1996/t68/005.
11. S. K. Dutta, S. K. Mehetor, and N. Pradhan, "Metal semiconductor heterostructures for photocatalytic conversion of light energy," *J. Phys. Chem. Lett.*, vol. 6, no. 6, pp. 936–944, 2015, https://doi.org/10.1021/acs.jpclett.5b00113.
12. S. Wei *et al.*, "Metal-insulator-semiconductor photoelectrodes for enhanced photoelectrochemical water splitting," *Chem. Soc. Rev.*, vol. 53, no. 13, pp. 6860–6916, 2024, https://doi.org/10.1039/d3cs00820g.
13. J. Liu and T. Hesjedal, "Magnetic Topological Insulator Heterostructures: A Review," *Adv. Mater.*, vol. 35, no. 27, 2023, https://doi.org/10.1002/adma.202102427.

14. J. Zhang, W. Xie, J. Zhao, and S. Zhang, "Band alignment of two-dimensional lateral heterostructures," *2D Mater.*, vol. 4, no. 1, 2017, https://doi.org/10.1088/2053-1583/aa50cc.
15. A. Giampietri, G. Drera, and L. Sangaletti, "Band Alignment at Heteroepitaxial Perovskite Oxide Interfaces. Experiments, Methods, and Perspectives," *Adv. Mater. Interfaces*, vol. 4, no. 11, pp. 1–13, 2017, https://doi.org/10.1002/admi.201700144.
16. P. A. Vermeulen, J. Mulder, J. Momand, and B. J. Kooi, "Strain engineering of van der Waals heterostructures," *Nanoscale*, vol. 10, no. 3, pp. 1474–1480, 2018, https://doi.org/10.1039/c7nr07607j.
17. B. Liu *et al.*, "Strain-Engineered van der Waals Interfaces of Mixed-Dimensional Heterostructure Arrays," *ACS Nano*, vol. 13, no. 8, pp. 9057–9066, 2019, https://doi.org/10.1021/acsnano.9b03239.
18. Y. Zhang, Y. Zhu, W. Wang, Z. Ning, S. Feng, and K. Höeg, "Compressive and tensile stress–strain-strength behavior of asphalt concrete at different temperatures and strain rates," *Constr. Build. Mater.*, vol. 311, no. November, pp. 10–17, 2021, https://doi.org/10.1016/j.conbuildmat.2021.125362.
19. C. Y. Lin and J. H. Kang, "Mechanical properties of compact bone defined by the stress-strain curve measured using uniaxial tensile test: A concise review and practical guide," *Materials (Basel).*, vol. 14, no. 15, 2021, https://doi.org/10.3390/ma14154224.
20. G. E. Dieter, "Mechanical Behavior Under Tensile and Compressive Loads[1]," *Mech. Test. Eval.*, vol. 8, pp. 99–108, 2018, https://doi.org/10.31399/asm.hb.v08.a0003261.
21. N. Mori and T. Ando, "Electronoptical-phonon interaction in single and double heterostructures," *Phys. Rev. B*, vol. 40, no. 9, pp. 6175–6188, 1989, https://doi.org/10.1103/PhysRevB.40.6175.
22. L. Britnell *et al.*, "Strong Light-Matter Interactions Thin Films," *Science (80-.).*, vol. 340, no. June, pp. 1311–1314, 2013.
23. O. Auciello, "Science and technology of thin films and interfacial layers in ferroelectric and high-dielectric constant heterostructures and application to devices," *J. Appl. Phys.*, vol. 100, no. 5, 2006, https://doi.org/10.1063/1.2337005.
24. A. Bhattacharya and S. J. May, "Magnetic oxide heterostructures," *Annu. Rev. Mater. Res.*, vol. 44, pp. 65–90, 2014, https://doi.org/10.1146/annurev-matsci-070813-113447.
25. J. A. Sulpizio, S. Ilani, P. Irvin, and J. Levy, "Nanoscale phenomena in oxide heterostructures," *Annu. Rev. Mater. Res.*, vol. 44, pp. 117–149, 2014, https://doi.org/10.1146/annurev-matsci-070813-113437.
26. C. Jia *et al.*, "Interface-Engineered Plasmonics in Metal/Semiconductor Heterostructures," *Adv. Energy Mater.*, vol. 6, no. 17, pp. 1–11, 2016, https://doi.org/10.1002/aenm.201600431.
27. H. H. Wieder, "Dielectric/semiconductor interfaces," *J. Vac. Sci. Technol. B Microelectron. Nanom. Struct. Process. Meas. Phenom.*, vol. 11, no. 4, pp. 1331–1335, 1993, https://doi.org/10.1116/1.586937.
28. S. Kumar, S. Saralch, U. Jabeen, and D. Pathak, *Metal oxides for energy applications*. Elsevier Inc., 2020. https://doi.org/10.1016/b978-0-12-813357-6.00017-6.
29. A. R. Fareza, F. A. A. Nugroho, F. F. Abdi, and V. Fauzia, "Nanoscale metal oxides-2D materials heterostructures for photoelectrochemical water splitting—a review," *J. Mater. Chem. A*, vol. 10, no. 16, pp. 8656–8686, 2022, https://doi.org/10.1039/d1ta10203f.
30. J. Azadmanjiri, V. K. Srivastava, P. Kumar, J. Wang, and A. Yu, "Graphene-supported 2D transition metal oxide heterostructures," *J. Mater. Chem. A*, vol. 6, no. 28, pp. 13509–13537, 2018, https://doi.org/10.1039/c8ta03404d.
31. B. Kundu *et al.*, "Synthesis of lateral heterostructure of 2D materials for optoelectronic devices: challenges and opportunities," *Emergent Mater.*, vol. 4, no. 4, pp. 923–949, 2021, https://doi.org/10.1007/s42247-021-00219-0.
32. Z. Zhang, P. Lin, Q. Liao, Z. Kang, H. Si, and Y. Zhang, "Graphene-Based Mixed-Dimensional van der Waals Heterostructures for Advanced Optoelectronics," *Adv. Mater.*, vol. 31, no. 37, pp. 1–15, 2019, https://doi.org/10.1002/adma.201806411.
33. V. Kumar, A. Sharma, S. Kumar, D. Pathak, and D. P. Sharma, *Chemical methods for specialized nanostructure*. INC, 2024. https://doi.org/10.1016/B978-0-443-13819-5.00005-7.

34. L. Guo, G. Oskam, A. Radisic, P. M. Hoffmann, and P. C. Searson, "Island growth in electrodeposition," *J. Phys. D. Appl. Phys.*, vol. 44, no. 44, 2011, https://doi.org/10.1088/0022-3727/44/44/443001.
35. K. A. Lozovoy, A. G. Korotaev, A. P. Kokhanenko, V. V. Dirko, and A. V. Voitsekhovskii, "Kinetics of epitaxial formation of nanostructures by Frank–van der Merwe, Volmer–Weber and Stranski–Krastanow growth modes," *Surf. Coatings Technol.*, vol. 384, no. December 2019, p. 125289, 2020, https://doi.org/10.1016/j.surfcoat.2019.125289.
36. A. Yanase and Y. Segawa, "Nucleation from ionic molecules in Stranski-Krastanov growth of CuCl on MgO(001)," *Appl. Surf. Sci.*, vol. 130–132, pp. 566–571, 1998, https://doi.org/10.1016/S0169-4332(98)00116-0.
37. S. M. Rossnagel, "Thin film deposition with physical vapor deposition and related technologies," *J. Vac. Sci. Technol. A Vacuum, Surfaces, Film.*, vol. 21, no. 5, pp. S74–S87, 2003, https://doi.org/10.1116/1.1600450.
38. D. M. Mattox, "Physical vapor deposition (PVD) processes," *Met. Finish.*, vol. 100, Supplement 1, pp. 394–408, 2002. https://doi.org/10.1016/S0026-0576(02)82043-8.
39. D. M. Mattox, "Physical vapor deposition (PVD) processes," *Met. Finish.*, vol. 98, no. 1, pp. 410–423, 2000, https://doi.org/10.1016/S0026-0576(00)80350-5.
40. Y. M. Manawi, Ihsanullah, A. Samara, T. Al-Ansari, and M. A. Atieh, "A review of carbon nanomaterials' synthesis via the chemical vapor deposition (CVD) method," *Materials (Basel).*, vol. 11, no. 5, 2018, https://doi.org/10.3390/ma11050822.
41. G. K. Hubler, "Pulsed Laser Deposition," *MRS Bull.*, vol. 17, no. 2, pp. 26–29, 1992, https://doi.org/10.1557/s0883769400040586.
42. J. A. Greer and M. D. Tabat, "Large-area pulsed laser deposition: Techniques and applications," *J. Vac. Sci. Technol. A Vacuum, Surfaces, Film.*, vol. 13, no. 3, pp. 1175–1181, 1995, https://doi.org/10.1116/1.579857.
43. M. Leskelä and M. Ritala, "Atomic layer deposition (ALD): From precursors to thin film structures," *Thin Solid Films*, vol. 409, no. 1, pp. 138–146, 2002, https://doi.org/10.1016/S0040-6090(02)00117-7.
44. R. W. Johnson, A. Hultqvist, and S. F. Bent, "A brief review of atomic layer deposition: From fundamentals to applications," *Mater. Today*, vol. 17, no. 5, pp. 236–246, 2014, https://doi.org/10.1016/j.mattod.2014.04.026.
45. J. R. Arthur, "Molecular beam epitaxy," *Surf. Sci.*, vol. 500, no. 1–3, pp. 189–217, 2002, https://doi.org/10.1016/S0039-6028(01)01525-4.
46. A. Y. Cho, "Advances in molecular beam epitaxy (MBE)," *J. Cryst. Growth*, vol. 111, no. 1–4, pp. 1–13, 1991, https://doi.org/10.1016/0022-0248(91)90938-2.
47. A. Y. Cho, "Recent Developments in Molecular Beam Epitaxy (Mbe).," *J Vac Sci Technol*, vol. 16, no. 2, pp. 275–284, 1979, https://doi.org/10.1116/1.569926.
48. Ankita, S. Kumar, S. Saralch, D. Pathak, "Nanopowder and Thin Films of ZnO by Sol Gel Approach," *J. Nano- Electron. Phys.*, vol. 11, no. 4, pp. 1–5, 2019, https://doi.org/10.21272/jnep.11(4).04027.
49. D. Pathak, S. Kumar, S. Andotra, J. Thomas, N. Kaur, P. Kumar, V. Kumar, "New tailored organic semiconductors thin films for optoelectronic applications," *Eur. Phys. J. Appl. Phys.*, vol. 95, no. 1, pp. 10201, 2021, https://doi.org/10.1051/epjap/2021210090.
50. S. Kumar, S. Saralch, and D. Pathak, "Low-cost deposition of cupric oxide thin films for optoelectronic applications," *EPJ Appl. Phys.*, vol. 84, no. 2, pp. 1–6, 2018, https://doi.org/10.1051/epjap/2018180194.
51. Meenakshi, S. Kumar, S. Saralch, N. Dhiman, M. Kumar, and D. Pathak, "Dip coated ZnO films for transparent window applications," *J. Nano- Electron. Phys.*, vol. 10, no. 5, pp. 1–5, 2018, https://doi.org/10.21272/jnep.10(5).05038.
52. K. Beepat, S. Kumar, A. Sharma, D. P. Sharma, D. Pathak, J. M. Nunzi, "Perovskite materials for photovoltaics: a review", *Eur. Phys. J. Appl. Phys.*, vol. 98, no. 43, pp. 37, 2023, https://doi.org/10.1051/epjap/2021210090.
53. C. Sun, E. Müller, M. Meffert, and D. Gerthsen, "On the Progress of Scanning Transmission Electron Microscopy (STEM) Imaging in a Scanning Electron Microscope," *Microsc. Microanal.*, vol. 24, no. 2, pp. 99–106, 2018, https://doi.org/10.1017/S1431927618000181.

54. C. Sun, E. Müller, M. Meffert, and D. Gerthsen, "Analysis of crystal defects by scanning transmission electron microscopy (STEM) in a modern scanning electron microscope," *Adv. Struct. Chem. Imaging*, vol. 5, no. 1, pp. 1–9, 2019, https://doi.org/10.1186/s40679-019-0065-1.
55. F. J. Giessibl, "Advances in atomic force microscopy," *Rev. Mod. Phys.*, vol. 75, no. 3, pp. 949–983, 2003, https://doi.org/10.1103/RevModPhys.75.949.
56. D. Johnson, D. L. Oatley-Radcliffe, and N. Hilal, *Atomic Force Microscopy (AFM).* Elsevier B.V., 2017. https://doi.org/10.1016/B978-0-444-63776-5.00007-3.
57. S. J. Pennycook and C. Colliex, "Spectroscopic imaging in electron microscopy," *MRS Bull.*, vol. 37, no. 1, pp. 13–18, 2012, https://doi.org/10.1557/mrs.2011.332.
58. Q. Luo, "Electron microscopy and spectroscopy in the analysis of friction and wear mechanisms," *Lubricants*, vol. 6, no. 3, pp. 20–25, 2018, https://doi.org/10.3390/lubricants6030058.
59. R. N. S. Sodhi, S. I. Ontario, and A. Chemistry, "i—section education," pp. 483–487, 2004.
60. L. Douillard and F. Charra, "Photoemission electron microscopy, a tool for plasmonics," *J. Electron Spectros. Relat. Phenomena*, vol. 189, no. SUPPL., pp. 24–29, 2013, https://doi.org/10.1016/j.elspec.2013.03.013.
61. A. Scholl, "Applications of photoemission electron microscopy (PEEM) in magnetism research," *Curr. Opin. Solid State Mater. Sci.*, vol. 7, no. 1, pp. 59–66, 2003, https://doi.org/10.1016/S1359-0286(03)00003-2.
62. E. Bauer, T. Duden, and R. Zdyb, "Spin-polarized low energy electron microscopy of ferromagnetic thin films," *J. Phys. D. Appl. Phys.*, vol. 35, no. 19, pp. 2327–2331, 2002, https://doi.org/10.1088/0022-3727/35/19/301.
63. N. Rougemaille and A. K. Schmid, "Magnetic imaging with spin-polarized low-energy electron microscopy," *EPJ Appl. Phys.*, vol. 50, no. 2, pp. 201011–2010118, 2010, https://doi.org/10.1051/epjap/2010048.
64. K. von Klitzing, "Quantized Hall Effect.," *Phys. B Phys. Condens. Matter C At. Mol. Plasma Physics, Opt.*, vol. 126 B-C, no. 1–3, pp. 242–249, 1984.
65. S. B. Mitta *et al.*, "Electrical characterization of 2D materials-based field-effect transistors," *2D Mater.*, vol. 8, no. 1, 2021, https://doi.org/10.1088/2053-1583/abc187.
66. A. P. Schuetze, W. Lewis, C. Brown, and W. J. Geerts, "A laboratory on the four-point probe technique," *Am. J. Phys.*, vol. 72, no. 2, pp. 149–153, 2004, https://doi.org/10.1119/1.1629085.
67. R. S. Waremra and P. Betaubun, "Analysis of Electrical Properties Using the four point Probe Method," *E3S Web Conf.*, vol. 73, pp. 1–4, 2018, https://doi.org/10.1051/e3sconf/20187313019.
68. D. W. Koon, A. A. Bahl, and E. O. Duncan, "Measurement of contact placement errors in the van der Pauw technique," *Rev. Sci. Instrum.*, vol. 60, no. 2, pp. 275–276, 1989, https://doi.org/10.1063/1.1140423.
69. C. Kasl and M. J. R. Hoch, "Effects of sample thickness on the van der Pauw technique for resistivity measurements," *Rev. Sci. Instrum.*, vol. 76, no. 3, 2005, https://doi.org/10.1063/1.1866232.
70. D. W. Koon, "Effect of contact size and placement, and of resistive inhomogeneities on van der Pauw measurements," *Rev. Sci. Instrum.*, vol. 60, no. 2, pp. 271–274, 1989, https://doi.org/10.1063/1.1140422.
71. E. Von Hauff, "Impedance Spectroscopy for Emerging Photovoltaics," *J. Phys. Chem. C*, vol. 123, no. 18, pp. 11329–11346, 2019, https://doi.org/10.1021/acs.jpcc.9b00892.
72. K. M. Guenther, H. Witte, A. Krost, S. Kontermann, and W. Schade, "Extracting accurate capacitance voltage curves from impedance spectroscopy," *Appl. Phys. Lett.*, vol. 100, no. 4, 2012, https://doi.org/10.1063/1.3679380.
73. O. Gharbi, M. T. T. Tran, B. Tribollet, M. Turmine, and V. Vivier, "Revisiting cyclic voltammetry and electrochemical impedance spectroscopy analysis for capacitance measurements," *Electrochim. Acta*, vol. 343, p. 136109, 2020, https://doi.org/10.1016/j.electacta.2020.136109.

74. C. B. De Araújo, A. S. L. Gomes, and G. Boudebs, "Techniques for nonlinear optical characterization of materials: A review," *Reports Prog. Phys.*, vol. 79, no. 3, 2016, https://doi.org/10.1088/0034-4885/79/3/036401.
75. M. D. Verweij, "Transient acoustic wave modeling: Higher-order Wentzel–Kramers–Brillouin–Jeffreys asymptotics and symbolic manipulation," *J. Acoust. Soc. Am.*, vol. 92, no. 4, pp. 2223–2238, 1992, https://doi.org/10.1121/1.405217.
76. S. Marchesini *et al.*, "X-ray image reconstruction from a diffraction pattern alone," *Phys. Rev. B - Condens. Matter Mater. Phys.*, vol. 68, no. 14, pp. 2344–2353, 2003, https://doi.org/10.1103/PhysRevB.68.140101.
77. J. Miao, "Computational microscopy with coherent diffractive imaging and ptychography," *Nature*, vol. 637, no. 8045, pp. 281–295, 2025, https://doi.org/10.1038/s41586-024-08278-z.
78. Y.-X. Li, N.-S. Chiu, W.-H. Lee, S. H. Bauer, and B. H. Davis, "X-ray Absorption Near Edge Structure (XANES) and Extended X-ray Absorption Fine Structure (EXAFS) Spectra of Pt—Sn—Alumina Catalysts," pp. 328–341, 1989, https://doi.org/10.1021/bk-1989-0411.ch030.
79. M. Aziz and A. F. Ismail, *X-Ray Photoelectron Spectroscopy (XPS)*. Elsevier B.V., 2017. https://doi.org/10.1016/B978-0-444-63776-5.00005-X.
80. C. J. Powell and A. Jablonski, "Progress in quantitative surface analysis by X-ray photoelectron spectroscopy: Current status and perspectives," *J. Electron Spectros. Relat. Phenomena*, vol. 178–179, no. C, pp. 331–346, 2010, https://doi.org/10.1016/j.elspec.2009.05.004.
81. M. Opel *et al.*, "Novel multifunctional materials based on oxide thin films and artificial heteroepitaxial multilayers," *Phys. Status Solidi Appl. Mater. Sci.*, vol. 208, no. 2, pp. 232–251, 2011, https://doi.org/10.1002/pssa.201026403.
82. A. Barbier, *Single and heterostructure multiferroic thin films*. Elsevier Inc., 2018. https://doi.org/10.1016/B978-0-12-811180-2.00023-2.
83. J. M. Hu, L. Q. Chen, and C. W. Nan, "Multiferroic Heterostructures Integrating Ferroelectric and Magnetic Materials," *Adv. Mater.*, vol. 28, no. 1, pp. 15–39, 2016, https://doi.org/10.1002/adma.201502824.
84. M. Markov, L. Alaerts, H. Pereira, C. Miranda, G. Petretto, and W. Chen, "Ferroelectricity and multiferroicity in anti—Ruddlesden—Popper structures," pp. 1–6, 2021, https://doi.org/10.1073/pnas.2026020118/-/DCSupplemental.y.
85. M. Trassin, E. Gradauskaite, P. Meisenheimer, M. Müller, and J. Heron, "Multiferroic heterostructures for spintronics," *Phys. Sci. Rev.*, vol. 6, no. 2, 2021, https://doi.org/10.1515/psr-2019-0072.
86. G. Lawes and G. Srinivasan, "Introduction to magnetoelectric coupling and multiferroic films," *J. Phys. D. Appl. Phys.*, vol. 44, no. 24, 2011, https://doi.org/10.1088/0022-3727/44/24/243001.
87. R. Gupta and R. K. Kotnala, "A review on current status and mechanisms of room-temperature magnetoelectric coupling in multiferroics for device applications," *J. Mater. Sci.*, vol. 57, no. 27, pp. 12710–12737, 2022, https://doi.org/10.1007/s10853-022-07377-4.
88. L. Guo and Q. Lu, "Potentials of piezoelectric and thermoelectric technologies for harvesting energy from pavements," *Renew. Sustain. Energy Rev.*, vol. 72, no. December 2015, pp. 761–773, 2017, https://doi.org/10.1016/j.rser.2017.01.090.
89. Z. H. Jin and J. S. Yang, "Energy Conversion Efficiency of a Piezo-Thermoelectric Material," *J. Electron. Mater.*, vol. 47, no. 8, pp. 4533–4538, 2018, https://doi.org/10.1007/s11664-018-6383-6.
90. Z. H. Jin and J. S. Yang, "Analysis of a sandwiched piezoelectric semiconducting thermoelectric structure," *Mech. Res. Commun.*, vol. 98, pp. 31–36, 2019, https://doi.org/10.1016/j.mechrescom.2019.05.004.
91. S. Rajput, V. Kaushik, S. Jain, P. Tiwari, A. K. Srivastava, and M. Kumar, "Optical Modulation in Hybrid Waveguide Based on Si-ITO Heterojunction," *J. Light. Technol.*, vol. 38, no. 6, pp. 1365–1371, 2020, https://doi.org/10.1109/JLT.2019.2953690.
92. C. Wang *et al.*, "Structure and ferroelectric photovoltaic effect modulation in the epitaxial $BiFeO_3/La_{0.5}Sr_{0.5}CO_3$ heterostructures," *Micro and Nanostructures*, vol. 195, no. May, p. 207956, 2024, https://doi.org/10.1016/j.micrna.2024.207956.

93. Z. Zhang *et al.*, "Enhanced Electro-Optical and Heat Regulation of Intelligent Dimming Films Using the Photovoltaic Effect of p–n Heterostructures," *Adv. Funct. Mater.*, vol. 34, no. 45, pp. 1–12, 2024, https://doi.org/10.1002/adfm.202406858.
94. S. Kumar, V. Anand, U. Jabeen, and D. Pathak, *Solar power energy derived from nanotools and devices*. INC, 2021. https://doi.org/10.1016/B978-0-12-821709-2.00008-6.
95. P. V. Pham *et al.*, "2D Heterostructures for Ubiquitous Electronics and Optoelectronics: Principles, Opportunities, and Challenges," *Chem. Rev.*, vol. 122, no. 6, pp. 6514–6613, 2022, https://doi.org/10.1021/acs.chemrev.1c00735.
96. Z. Huo, Y. Wei, Y. Wang, Z. L. Wang, and Q. Sun, "Integrated Self-Powered Sensors Based on 2D Material Devices," *Adv. Funct. Mater.*, vol. 32, no. 41, pp. 1–30, 2022, https://doi.org/10.1002/adfm.202206900.
97. A. Bag and N. E. Lee, "Gas sensing with heterostructures based on two-dimensional nanostructured materials: A review," *J. Mater. Chem. C*, vol. 7, no. 43, pp. 13367–13383, 2019, https://doi.org/10.1039/c9tc04132j.
98. S. Zhang, J. Liu, M. M. Kirchner, H. Wang, Y. Ren, and W. Lei, "Two-dimensional heterostructures and their device applications: progress, challenges and opportunities—review," *J. Phys. D. Appl. Phys.*, vol. 54, no. 43, 2021, https://doi.org/10.1088/1361-6463/ac16a4.
99. Y. Hu *et al.*, "Light-Driven Self-Oscillating Actuators with Phototactic Locomotion Based on Black Phosphorus Heterostructure," *Angew. Chemie*, vol. 133, no. 37, pp. 20674–20680, 2021, https://doi.org/10.1002/ange.202108058.
100. V. Yakovlevich Prinz, V. Alexandrovich Seleznev, A. Victorovich Prinz, and A. Vladimirovich Kopylov, "3D heterostructures and systems for novel MEMS/NEMS," *Sci. Technol. Adv. Mater.*, vol. 10, no. 3, 2009, https://doi.org/10.1088/1468-6996/10/3/034502.
101. S. Ghosh *et al.*, "Nonreciprocal acoustoelectric interaction of surface waves and fluorine plasma-treated AlGan/GaN 2DEG," *TRANSDUCERS 2017—19th Int. Conf. Solid-State Sensors, Actuators Microsystems*, pp. 1939–1942, 2017, https://doi.org/10.1109/TRANSDUCERS.2017.7994448.
102. J. F. Sierra, J. Fabian, R. K. Kawakami, S. Roche, and S. O. Valenzuela, "Van der Waals heterostructures for spintronics and opto-spintronics," *Nat. Nanotechnol.*, vol. 16, no. 8, pp. 856–868, 2021, https://doi.org/10.1038/s41565-021-00936-x.
103. Q. Hao *et al.*, "2D Magnetic Heterostructures and Emergent Spintronic Devices," *Adv. Electron. Mater.*, vol. 8, no. 11, pp. 1–16, 2022, https://doi.org/10.1002/aelm.202200164.
104. E. Y. Tsymbal, A. Gruverman, V. Garcia, M. Bibes, and A. Barthélémy, "Ferroelectric and multiferroic tunnel junctions," *MRS Bull.*, vol. 37, no. 2, pp. 138–143, 2012, https://doi.org/10.1557/mrs.2011.358.
105. X. Yu, X. Zhang, and J. Wang, "Fully Electrically Controlled van der Waals Multiferroic Tunnel Junctions," *ACS Nano*, vol. 17, no. 24, pp. 25348–25356, 2023, https://doi.org/10.1021/acsnano.3c08747.
106. X. Wang, C. O. Pauyac, and A. Manchon, "Spin-orbit-coupled transport and spin torque in a ferromagnetic heterostructure," *Phys. Rev. B—Condens. Matter Mater. Phys.*, vol. 89, no. 5, pp. 1–11, 2014, https://doi.org/10.1103/PhysRevB.89.054405.
107. J. Fu *et al.*, "Spin-orbit coupling in wurtzite heterostructures," *Phys. Rev. B*, vol. 101, no. 13, pp. 1–27, 2020, https://doi.org/10.1103/PhysRevB.101.134416.
108. E. Rossi and C. Triola, "Van Der Waals Heterostructures with Spin-Orbit Coupling," *Ann. Phys.*, vol. 532, no. 2, pp. 1–17, 2020, https://doi.org/10.1002/andp.201900344.
109. K. S. Novoselov, A. Mishchenko, A. Carvalho, and A. H. Castro Neto, "2D materials and van der Waals heterostructures," *Science (80-.).*, vol. 353, no. 6298, 2016, https://doi.org/10.1126/science.aac9439.
110. J. H. Garcia, M. Vila, A. W. Cummings, and S. Roche, "Spin transport in graphene/transition metal dichalcogenide heterostructures," *Chem. Soc. Rev.*, vol. 47, no. 9, pp. 3359–3379, 2018, https://doi.org/10.1039/c7cs00864c.
111. D. Li *et al.*, "Ferroelectric diode-like resistive switching behavior in $Bi_{0.95}Er_{0.05}FeO_3$/$CuFe_2O_4$ heterostructures for non-volatile memories," *Appl. Mater. Today*, vol. 36, no. January, 2024, https://doi.org/10.1016/j.apmt.2024.102074.

112. G. Feng *et al.*, "A ferroelectric fin diode for robust non-volatile memory," *Nat. Commun.*, vol. 15, no. 1, pp. 1–11, 2024, https://doi.org/10.1038/s41467-024-44759-5.
113. H. Wong, W. Li, J. Zhang, W. Bao, L. Wu, and J. Liu, "Driving for More Moore on Computing Devices with Advanced Non-Volatile Memory Technology," pp. 1–45, 2025.
114. H. Geng *et al.*, "Graphene van der Waals heterostructures for high-performance photodetectors," *J. Mater. Chem. C*, vol. 7, no. 36, pp. 11056–11067, 2019, https://doi.org/10.1039/c9tc03213d.
115. R. Liu *et al.*, "Band Alignment Engineering in Two-Dimensional Transition Metal Dichalcogenide-Based Heterostructures for Photodetectors," *Small Struct.*, vol. 2, no. 3, pp. 1–27, 2021, https://doi.org/10.1002/sstr.202000136.
116. E. Assmann, P. Blaha, R. Laskowski, K. Held, S. Okamoto, and G. Sangiovanni, "Oxide heterostructures for efficient solar cells," *Phys. Rev. Lett.*, vol. 110, no. 7, pp. 1–5, 2013, https://doi.org/10.1103/PhysRevLett.110.078701.
117. M. J. Molaei, M. Younas, and M. Rezakazemi, "Van der Waals heterostructures in ultrathin 2D solar cells: State-of-the-art review," *Mater. Sci. Eng. B*, vol. 285, no. July, p. 115936, 2022, https://doi.org/10.1016/j.mseb.2022.115936.
118. H. Kim, S. Seong, and X. Gong, "Heterostructure Engineering of Solution-Processable Semiconductors for Wearable Optoelectronics," *ACS Appl. Electron. Mater.*, vol. 5, no. 10, pp. 5278–5290, 2023, https://doi.org/10.1021/acsaelm.2c01791.
119. S. Aftab, H. H. Hegazy, and F. Kabir, "Emerging Trends in 2D Flexible Electronics," *Adv. Mater. Technol.*, vol. 8, no. 11, pp. 1–16, 2023, https://doi.org/10.1002/admt.202201897.
120. Z. Liao *et al.*, "Controlled lateral anisotropy in correlated manganite heterostructures by interface-engineered oxygen octahedral coupling," *Nat. Mater.*, vol. 15, no. 4, pp. 425–431, 2016, https://doi.org/10.1038/nmat4579.
121. Y. Han *et al.*, "Interface engineered WS2/ZnS heterostructures for sensitive and reversible NO_2 room temperature sensing," *Sensors Actuators, B Chem.*, vol. 296, no. April, p. 126666, 2019, https://doi.org/10.1016/j.snb.2019.126666.
122. G. Zhang *et al.*, "Quantum phase transitions and topological proximity effects in graphene nanoribbon heterostructures," *Nanoscale*, vol. 6, no. 6, pp. 3259–3267, 2014, https://doi.org/10.1039/c3nr05284b.
123. J. D. Sau, R. M. Lutchyn, S. Tewari, and S. Das Sarma, "Generic new platform for topological quantum computation using semiconductor heterostructures," *Phys. Rev. Lett.*, vol. 104, no. 4, pp. 1–4, 2010, https://doi.org/10.1103/PhysRevLett.104.040502.
124. D. Pathak *et al.*, "New tailored organic semiconductors thin films for optoelectronic applications," *EPJ Appl. Phys.*, vol. 95, no. 1, pp. 1–10, 2021, https://doi.org/10.1051/epjap/2021210090.
125. J. Sun *et al.*, "2D–Organic Hybrid Heterostructures for Optoelectronic Applications," *Adv. Mater.*, vol. 31, no. 34, pp. 1–30, 2019, https://doi.org/10.1002/adma.201803831.
126. M. Gobbi, E. Orgiu, and P. Samorì, "When 2D Materials Meet Molecules: Opportunities and Challenges of Hybrid Organic/Inorganic van der Waals Heterostructures," *Adv. Mater.*, vol. 30, no. 18, pp. 1–20, 2018, https://doi.org/10.1002/adma.201706103.
127. C. Li, K. P. Loh, and K. Leng, "Organic-inorganic hybrid perovskites and their heterostructures," *Matter*, vol. 5, no. 12, pp. 4153–4169, 2022, https://doi.org/10.1016/j.matt.2022.11.002.
128. Z. U. D. Babar, A. Raza, A. Cassinese, and V. Iannotti, "Two Dimensional Heterostructures for Optoelectronics: Current Status and Future Perspective," *Molecules*, vol. 28, no. 5, pp. 1–31, 2023, https://doi.org/10.3390/molecules28052275.
129. S. Gbadamasi *et al.*, "Interface chemistry of two-dimensional heterostructures-fundamentals to applications," *Chem. Soc. Rev.*, vol. 50, no. 7, pp. 4684–4729, 2021, https://doi.org/10.1039/d0cs01070g.
130. X. Zhou, Q. Shen, Y. Wang, Y. Dai, Y. Chen, and K. Wu, "Surface and interfacial sciences for future technologies," *Natl. Sci. Rev.*, vol. 11, no. 9, 2024, https://doi.org/10.1093/nsr/nwae272.

131. S. K. Chakraborty, B. Kundu, B. Nayak, S. P. Dash, and P. K. Sahoo, "Challenges and opportunities in 2D heterostructures for electronic and optoelectronic devices," *iScience*, vol. 25, no. 3, p. 103942, 2022, https://doi.org/10.1016/j.isci.2022.103942.
132. Y. Liu, S. Zhang, J. He, Z. M. Wang, and Z. Liu, "Recent Progress in the Fabrication, Properties, and Devices of Heterostructures Based on 2D Materials," *Nano-Micro Lett.*, vol. 11, no. 1, pp. 1–24, 2019, https://doi.org/10.1007/s40820-019-0245-5.
133. Z. Wang and J. K. Huang, "Latest Advancements in Next-Generation Semiconductors: Materials and Devices for Wide Bandgap and 2D Semiconductors," *Micromachines*, vol. 14, no. 11, pp. 14–17, 2023, https://doi.org/10.3390/mi14111992.
134. N. Zhang, D. Ouyang, Y. Li, and T. Zhai, "The roadmap of two-dimensional materials toward next-generation image sensor," *Natl. Sci. Rev.*, vol. 11, no. 12, pp. 10–12, 2024, https://doi.org/10.1093/nsr/nwae431.
135. A. Liu *et al.*, *The Roadmap of 2D Materials and Devices Toward Chips*, vol. 16, no. 1. Springer Nature Singapore, 2024. https://doi.org/10.1007/s40820-023-01273-5.
136. J. E. Jeyanthi, T. S. A. Samuel, A. S. Geege, and P. Vimala, "A Detailed Roadmap from Single Gate to Heterojunction TFET for Next Generation Devices," *Silicon*, vol. 14, no. 7, pp. 3185–3197, 2022, https://doi.org/10.1007/s12633-021-01148-7.

Dinesh Pathak is a Faculty Member in Physics at The University of the West Indies, St. Augustine Campus, Trinidad, where he is affiliated with the Advanced Functional Materials Laboratory. His academic and research activities focus on the development and characterization of advanced functional materials, with an emphasis on their physical properties and potential applications in next-generation devices.

Chapter 2
Advanced Characterization Techniques for Thin Films

Indu Gupta, Kaushlendra Pandey, Rahul Jain, Ramalinga Viswanathan Mangalaraja, and Arunachalam Arulraj

2.1 Introduction

Thin films are the cornerstone in modern-day materials science and device engineering and the basis for a range of applications such as microelectronics, optoelectronics, photovoltaics, catalysis, and sensing technologies [1–3]. Their unique properties—often deviating from bulk material values—arise due to factors such as reduced dimensionality, interface strain, and compositional variations [4]. To optimize their functionalities, it is essential to understand the structural, morphological, chemical, and electronic properties at atomic and nanoscale length scales. In the recent decade, advanced characterization techniques have advanced significantly, enabling the scientist to probe thin films in situ with resolution in the range down to nanometers or even atomic sizes [5, 6]. Such probing techniques provide valuable information concerning crystallinity, defects, dopant profile, surface morphology, and charge transport kinetics. In this chapter, an account of the key advanced techniques, categorized structurally, along with their fundamental principles, capabilities, and practical applications, is discussed. Herein, the chapter systematically covers the main structural, compositional, electronic, and in-situ/operando characterization techniques suitable for thin films and highlights their operational principles, data processing, and some typical applications in advanced material systems.

I. Gupta (✉)
Department of Physics, UIS, Chandigarh University, Mohali, India
e-mail: indugupta17887@gmail.com

K. Pandey · A. Arulraj
Departamento de Electricidad, Facultad de Ingeniería, Universidad Tecnológica Metropolitana, Santiago, Chile

R. Jain
Department of Applied Sciences, CGC University, Mohali, India

R. V. Mangalaraja
Facultad de Ingeniería y Arquitectura, Universidad Arturo Prat, Iquique, Chile

S. Sharma (ed.), *Multifunctional Heterostructure Thin Films Handbook*,
https://doi.org/10.1007/978-3-032-18692-8_2

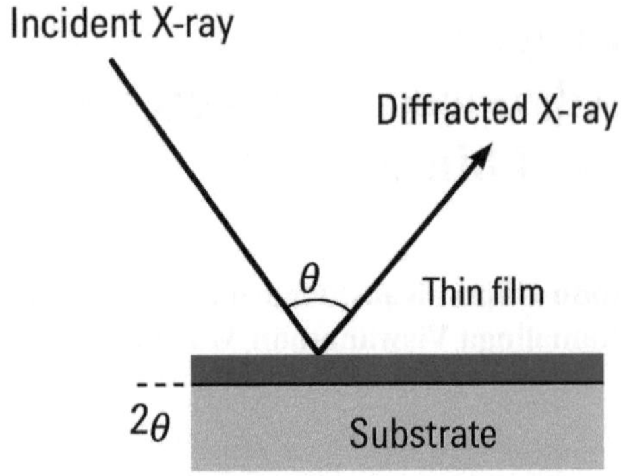

Fig. 2.1 Schematic of grazing incidence XRD geometry showing incident and diffracted X-rays interacting with a thin film on a substrate

2.2 Structural Characterization

2.2.1 X-ray Diffraction (XRD)

X-ray diffraction (XRD) is still the absolute technique in determining the crystalline structure and phase uniformity of the thin films [7]. The diffraction interaction between X-rays and a periodic lattice puts constructive interference at defined angles, in accordance with Bragg's law:

$$n\lambda = 2d \sin \theta$$

XRD enables the crystalline phases, preferred orientation, and identification of strain [8]. For thin films, owing to low intensities of diffracted rays, the grazing incidence XRD (GIXRD) enhances the surface sensitivity. Reciprocal space mapping (RSM) provides strain and mosaicity in epitaxial layers with high resolution, and the measurement of the rocking curve estimates crystalline perfection quantitatively as mentioned in Fig. 2.1 [9].

Applications: Determination of perovskite oxide phases, TiN coating texture measurement, and characterization of strain in semiconductor heterostructures [10].

2.2.2 Transmission Electron Microscopy (TEM)

TEM provides atomic-scale structural and compositional information [11]. High-resolution TEM (HRTEM) images show lattice fringes, allowing direct observation of defects, dislocations, and interfaces. Selected area electron diffraction (SAED) patterns can aid in the identification of crystalline phases and orientation relationships [12]. Scanning TEM (STEM) with energy-dispersive X-ray spectroscopy (EDS) and electron energy-loss spectroscopy (EELS) allows high-resolution elemental mapping [13]. Techniques are pivotal in correlating microstructure with property, such as in

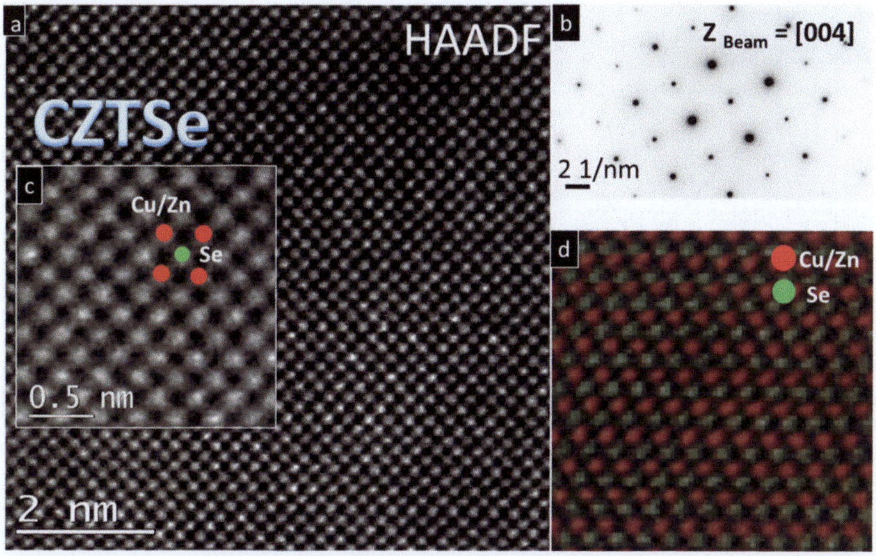

Fig. 2.2 **a** Atomic-resolution STEM image of CZTSe acquired along the [100] zone axis of CIGS, **b** with crystallographic orientation verified via selected area electron diffraction (SAED). **c** High-magnification view of the image highlights, **d** alternating Cu/Zn and S atomic columns, as identified through STEM-EELS column-by-column chemical analysis [14]. Copyright 2016. Reproduced with permission from Wiley Advanced

the grain boundary characterization of CIGS or CZTS thin films as shown in Fig. 2.2 [14].

2.2.3 Atomic Force Microscopy (AFM)

AFM can perform nanoscale topographical and quantitative roughness measurements [15]. Depending on the operating mode-contact, tapping, or non-contact—AFM can probe morphological surfaces, mechanical properties, and adhesion. 3D AFM surface morphology of a CdS thin film grown by chemical bath technique showing uniform grain distribution as shown in Fig. 2.3 [16].

Advanced variations are:

- Conductive AFM (C-AFM): maps local current–voltage response [17].
- Kelvin probe force microscopy (KPFM): To measure the grain boundary characteristics.
- Photo-conductive AFM (pc-AFM): correlates photocurrent with local structure [18].

Applications: Investigating grain-boundary conductivity in perovskite films and charge distribution at the nanoscale in oxide semiconductors.

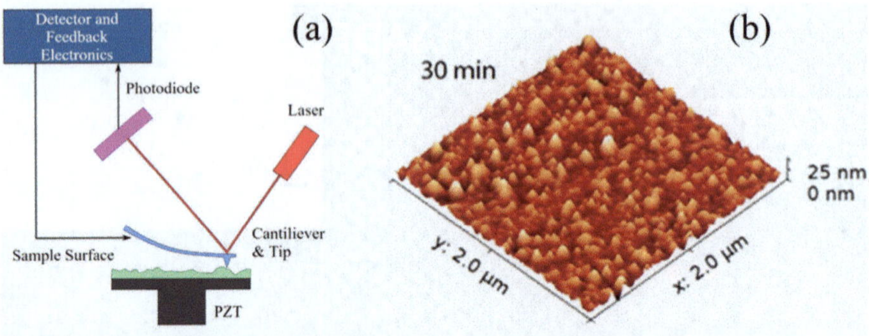

Fig. 2.3 **a** Schematic of an atomic force microscope (AFM) showing its basic working principle. **b** 3D AFM surface morphology of a CdS thin film showing uniform grain distribution [16]. Copyright 2016. Reproduced with permission from Springer nature

2.3 Compositional and Chemical Characterization

2.3.1 X-ray Photoelectron Spectroscopy (XPS)

X-ray photoelectron spectroscopy (XPS) is an extremely accurate analytical technique used to determine the elemental composition, chemical states, and electronic surroundings of atoms positioned in the top few nanometers of a material's surface. The procedure consists of irradiating a material with monochromatic X-rays to produce the removal of core electrons from the constituent atoms via the photoelectric effect. Their emitted electrons' resultant kinetic energies are then gauged and used to compute their corresponding binding energies, which give rise to a distinctive fingerprint per element as given in schematics of Fig. 2.4 [19]. By examining shifts in binding energies, XPS is able to divulge oxidation states, bonding arrangements, and chemical settings. Depth profiling may be achieved by combining XPS with controllable argon ion sputtering that allows subsurface composition analysis and buried interfaces to be determined [20]. Angle-resolved XPS (ARXPS) extracts considerable depth sensitivity with no requirement to sputter, and it is especially suitable in examining layered and nanostructured films [21]. XPS is extensively applied to valence state assessment in metal oxides (e.g., Fe^{3+}/Fe^{2+} in Fe_2O_3), determination of surface contaminants, and dopant incorporation and distribution analysis. Its quantitative and surface selectivity make it indispensible in thin film research, in the process of processing semiconductors, and in catalyst investigation studies.

2.3.2 Secondary Ion Mass Spectrometry (SIMS)

Secondary ion mass spectrometry (SIMS) is a very sensitive analytical tool employed to identify and quantify elemental and isotopic compositions in surfaces and thin

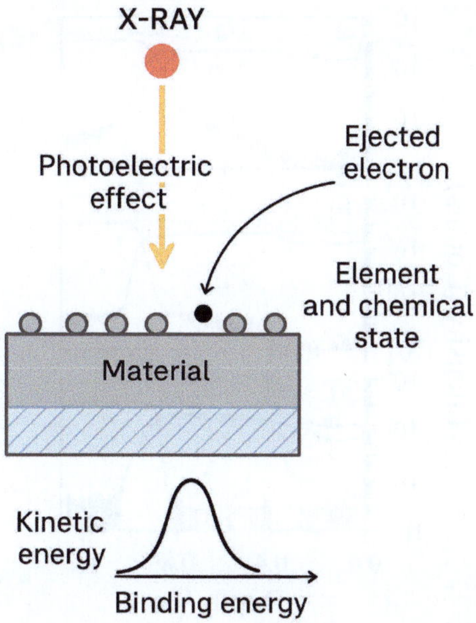

Fig. 2.4 Schematic representation of the X-ray photoelectron spectroscopy (XPS) process. Monochromatic X-rays irradiate the sample surface, causing the ejection of core electrons via the photoelectric effect

films. In SIMS, a highly concentrated primary ion beam (e.g., Cs^+ or O_2^+) impinges on the sample's surface, sputtering atoms and molecular units off from the very topmost surfaces [22]. A small portion of these sputtered units get ionized—called secondary ions—and are resolved with a mass spectrometer to identify their mass-to-charge ratios. This allows detection of the elements and isotopes at ppm to ppb concentrations [23]. Through continuous sputtering of the surface, SIMS yields depth profiles that illustrate how the composition changes with depth through the thickness of the film, and hence it's very important in investigating doping grades, interface diffusion, and segregation of impurities. As illustrated in Fig. 2.5, SIMS profiling was performed on Mo films deposited onto both commercially available and in-house composition-tuned glass substrates to investigate their diffusion characteristics. Time-of-flight SIMS (ToF–SIMS) expands these capabilities by adding high mass resolution and three-dimensional chemical imaging at submicron lateral resolution [24]. This enables visualization of lateral and vertical distribution of dopants or contaminants within submicron and nanostructured films. SIMS is therefore an indispensable tool in characterizing thin semiconductor films, in diffusion analysis in oxide heterostructures, and in identifying trace impurities that affect the performance and reliability of the device [25].

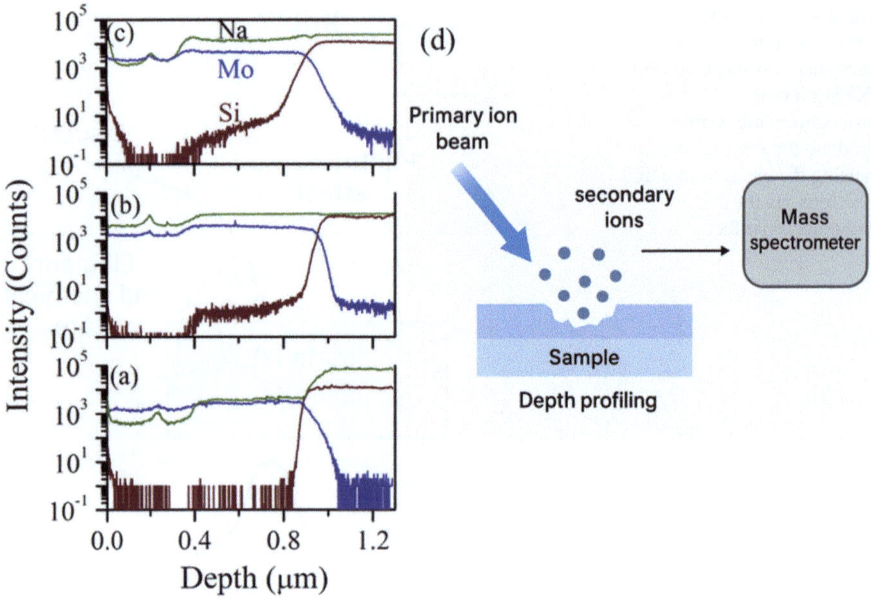

Fig. 2.5 TOF–SIMS depth profiles of Na, Mo, and Si for **a** as-deposited Mo sample on commercial SLG slides, **b** sample on SLG slides annealed at 500 °C for 60 min in flowing Ar, and **c** sample on in-house glass substrates annealed under identical conditions. All samples were deposited at 50 W using 30 min HPD followed by 30 min LPD [24], and **d** schematic showing a primary ion beam sputtering a sample surface, with secondary ions analyzed by a mass spectrometer for depth profiling. Copyright 2019. Reproduced with permission from Elsevier

2.3.3 Rutherford Backscattering Spectrometry (RBS)

Rutherford backscattering spectrometry (RBS) is an extremely efficient, non-destructive analytical technique used to determine the composition, thickness, and areal density of thin films with high accuracy. During RBS, a high-energy ion beam (mostly He^+) is directed at the substrate's surface, and the subsequent energy level of the backscattered ions is measured as illustrated in Fig. 2.6 [26]. From the resultant energy spectrum, information on mass and depth distribution of the constituent elements—thicker elements produce higher energy scatterer ions, while lighter ones produce lower energy signals—is extracted. This allows for the quantitative determination of elemental make-up and layer thickness [27]. RBS is also notable for its high depth resolution and finds common application with nuclear reaction analysis (NRA) to increase sensitivity to light elements like oxygen and nitrogen. Individually and collectively, these techniques play a prominent role in confirming film stoichiometry, assessing diffusion or intermixing at material interfaces, and confirming careful composition control in such coatings as Al_2O_3, TiN, and complex oxide thin films [28].

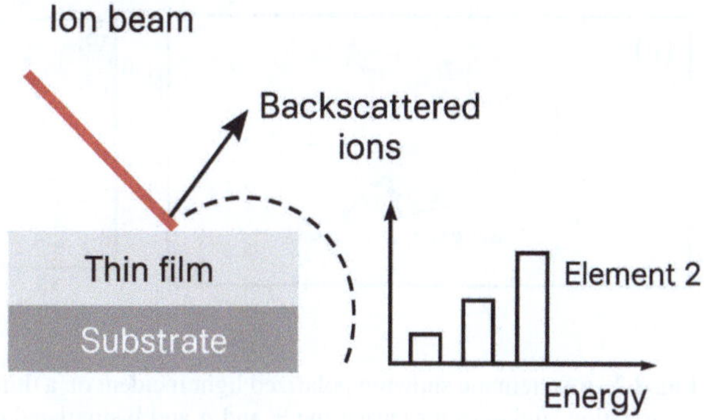

Fig. 2.6 A schematic showing ion beam incident on a thin film, with backscattered ions detected at an angle and an energy spectrum illustrating elemental peaks

2.4 Optical and Electronic Characterization

2.4.1 UV–Visible–NIR Spectroscopy

Optical spectroscopy is a very important approach to the investigation of light–matter interaction, and it provides fundamental knowledge of thin films and semiconductors in terms of electronic and optical properties. From the transmittance (T) and reflectance (R) spectra in the span of a wavelength spectrum, it becomes possible to determine parameters such as absorption coefficient (α), refractive index, and thickness of the film [29]. Significantly, the absorption spectrum reveals the optical bandgap (Eg) of the material, which represents the minimum amount of energy needed to excite electrons from the valence band to the conduction band. Tauc plot analysis is commonly used to approximate the optical bandgap by plotting $(\alpha h\nu)^n$ versus photon energy (hν), where the value of the exponent n represents the nature of the transition between the electrons and the holes (½ corresponds to direct and 2 corresponds to indirect transition) [30]. Extrapolation in the straight line part yields the value of E_g. This kind of analysis serves to understand photon absorption efficiency, tailor-made optical coatings, and optimize photovoltaic absorber materials to be used in solar energy conversion. In this sense, optical spectroscopy is a quick, non-destructive, and deeply enlightening tool to characterize thin-film optoelectric materials [31].

2.4.2 Spectroscopic Ellipsometry (SE)

Ellipsometry is a non-destructive optical technique applied to the characterization of thin films based on the measurement of changes in the polarization state of light upon reflecting off the surface of a specimen. This measurement determines two fundamental parameters—Ψ (amplitude ratio) and Δ (phase difference)—that are self-consistently related to the complex reflectance ratio between p- and s-polarized

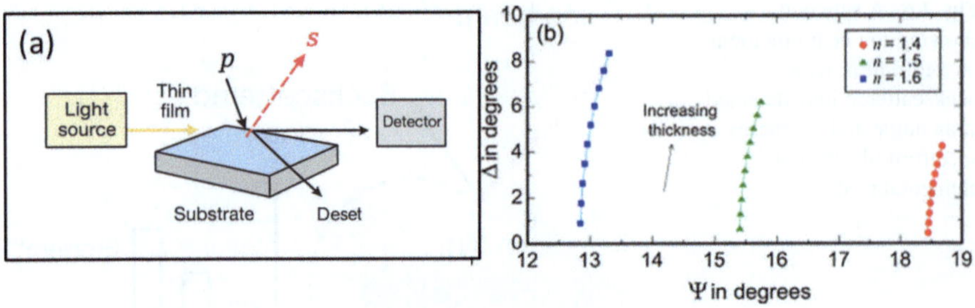

Fig. 2.7 **a** Schematic showing polarized light incident on a thin film, reflected beam with changed polarization, and detector measuring Ψ and Δ and **b** simulated reflectance curves for free-standing films at an incident angle of 70° and a wavelength of 633 nm, with film thicknesses up to 10 nm and refractive indices of 1.4, 1.5, and 1.6 [33]. Copyright 2017. Reproduced with permission from Elsevier

light [32]. Through correlation with optical models of these registered values, it becomes possible to determine important material parameters, such as refractive index (n), extinction coefficient (k), and thickness of the film. Spectroscopic ellipsometry (SE) encompasses this approach over a very wide spectral band and hence makes it possible to study dispersion and absorption characteristics [33]. SE demonstrates high sensitivity and finds it possible to detect sub-nanometer grading in multilayered and graded films. Furthermore, in-situ ellipsometry enables real-time in situ observation of growing processes in thin films and changes in optical characteristics and finds applications in optical modeling, refractive index modulating in doped oxides, and quality control during deposition processes as shown in Fig. 2.7a–b [33, 34].

2.4.3 *Electrical Characterization: Hall and Four-Point Probe*

Transport measurements via electricity are central to the determination of the electronic properties of thin films that are semiconducting and oxide in nature [35]. Important parameters such as carrier concentration, mobility, and resistivity are determined via methods such as the four-probe or Van der Pauw setup, which minimizes geometrical errors and contact resistance satisfactorily. These evaluations provide important knowledge on how far dopants have changed the electrical properties of materials. Temperature-dependent experiments that follow on provide insight into conduction processes by establishing whether charge transportation occurs through thermally activated band conduction, small polaron hopping, or variable-range hopping [36]. From the examination of temperature dependences of resistivity and mobility, it is possible to deduce activation energies and determine carrier scattering processes. Such evaluations are especially important for estimating doping efficiency and carrier kinetics in materials, including ZnO, SnO_2, and perovskite oxide thin films, where the increase in electrical transport is significant in applications such as thermoelectric devices, sensors, and transparent conductors [37].

2.5 Advanced and Synchrotron-Based Techniques

2.5.1 X-ray Absorption Spectroscopy (XAS)

X-ray absorption spectroscopy (XAS) is an element-specific probe that yields rich structural and electronic local properties of materials [38]. Upon absorption of X-rays of tunable energy, core electrons get excited to produce characteristic absorption features. The absorption spectrum consists of two regions: the X-ray absorption near edge structure (XANES) and the extended X-ray absorption fine structure (EXAFS) [39]. XANES encodes oxidation states, unoccupied states, and local asymmetry around the edge atom, while the EXAFS encodes interatomic distances, coordination number, and disorder in a quantitative manner. Synchrotron sources allow XAS to probe with high brightness, tunable energy, and high resolution to characterize complex materials accurately. In rare-earth-doped oxide materials, XAS is widely applied to probe dopant valence states, dopant and host atom hybridization, and dopant local coordination environment [40]. It also assists in the determination of cation distribution and structural distortions in mixed-metal oxides and nanoscale thin films to understand their functions and thermoelectric properties.

2.5.2 Scanning Transmission X-ray Microscopy (STXM)

Scanning transmission X-ray microscopy (STXM) is a strong synchrotron-based technique that blends high-resolution X-ray imaging with X-ray absorption spectroscopy (XAS). This combination offers element-specific and chemically sensitive maps of materials at the nanoscale [41]. In STXM, researchers scan a focused monochromatic X-ray beam—typically with a spot size of 20–30 nm—across a thin specimen. They record the transmitted intensity based on position and photon energy. By adjusting the incident energy around specific absorption edges, they can gather detailed chemical, electronic, and structural information with high spatial precision. STXM helps visualize oxidation states, bonding environments, and chemical differences across interfaces and grain boundaries. This makes it especially useful for studying functional oxide thin films [42]. It allows mapping of phase segregation, cation ordering, and electronic phase separation in complex systems like manganites, cobaltites, and nickelates [43]. Furthermore, STXM can be conducted under controlled temperature, gas, or electrical conditions. This feature supports in-situ studies of redox reactions and electrochemical processes. This capability connects

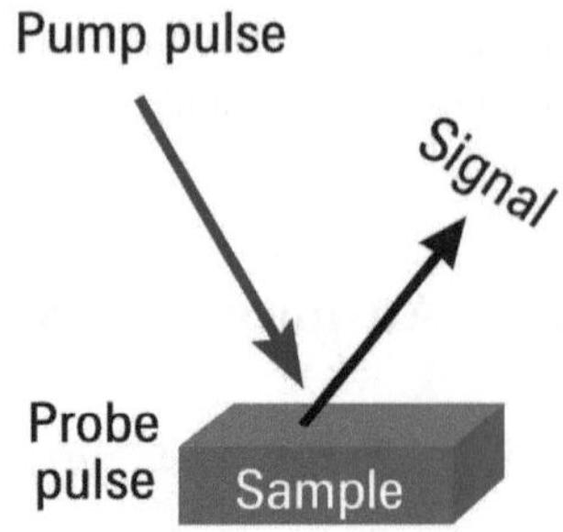

Fig. 2.8 Schematic diagram of an ultrafast pump-probe setup used to study charge carrier dynamics in thin films

nanoscale imaging and spectroscopy, making STXM an essential tool for understanding structure–property relationships in next-generation energy and electronic materials.

2.5.3 *Ultrafast Spectroscopy*

Ultrafast laser spectroscopy, especially the pump–probe technique, is a powerful tool for exploring photoinduced carrier dynamics in thin films with femtosecond to picosecond time resolution [44]. In this method, an initial pump pulse excites electrons in the material, creating non-equilibrium charge carriers as presented in Fig. 2.8. A delayed probe pulse then monitors the transient optical response as these carriers relax, recombine, or transfer energy. By varying the delay time between the pump and probe pulses, researchers can create a time-resolved picture of electronic and structural processes following photoexcitation. This technique gives detailed insights into carrier recombination rates, trapping effects, exciton dissociation, and energy transfer pathways, which are crucial for understanding and improving optoelectronic performance [45]. It has been widely used on perovskite and oxide semiconductor thin films, allowing precise measurements of charge carrier lifetimes, diffusion lengths, and recombination processes [46]. This information helps identify loss pathways, optimize defect passivation, and guide material design for high-efficiency photovoltaic, photocatalytic, and photoelectrochemical devices. Ultimately, ultrafast spectroscopy connects fundamental photophysics with practical device engineering, offering a dynamic view of processes that control light–matter interactions in functional thin-film materials.

2.6 In-Situ and Operando Characterization

In-situ and operando characterization techniques provide real-time insights into the structural, chemical, and morphological changes of thin films under realistic operating conditions. This helps connect static, ex-situ analysis with actual material performance [47]. These methods allow for direct observation of dynamic processes like crystallization, phase transitions, oxidation and reduction reactions, and degradation pathways. For example, in-situ X-ray diffraction (XRD) is commonly used to monitor phase changes and stress relaxation during thermal annealing or film deposition. Operando X-ray photoelectron spectroscopy (XPS) tracks changes in chemical states due to applied electrical bias or reactive environments. This makes it essential for studying electrocatalysts and oxide semiconductors. Likewise, environmental transmission electron microscopy (ETEM) provides real-time imaging of morphological changes during heating, gas exposure, or ion interaction [48]. These techniques are vital for understanding reaction intermediates, interface dynamics, and failure mechanisms that affect device reliability and efficiency [49]. The ability to link structure and function under realistic conditions speeds up materials optimization and helps develop predictive models for thin-film behavior in applications such as photovoltaics, catalysis, and energy storage.

2.7 Correlative and Multi-modal Characterization

Correlative characterization approaches combine techniques like AFM and Raman, TEM and EELS, and XPS and SIMS. This integration creates detailed datasets that include structural, chemical, and electronic information [50]. This multimodal strategy helps researchers find better connections between microstructure and functional properties. It offers a complete understanding of thin-film behavior. The increasing use of machine learning and data-driven analysis allows for automated pattern recognition, feature extraction, and predictive modeling. This speeds up interpretation and reduces human bias [51]. These advanced frameworks are especially useful for managing the large and complex datasets from high-resolution and time-resolved analyses. Future directions focus on 4D and multimodal tomography, which combine spatial, temporal, and chemical mapping. This technology visualizes dynamic processes like diffusion, phase transformation, and defect evolution in real time [52]. Altogether, these advancements mark a significant change toward AI-supported, holistic materials characterization.

2.8 Challenges and Future Outlook

Despite significant progress in characterizing thin films, several key challenges still exist. Achieving non-destructive depth profiling with nanometer-scale resolution remains a major hurdle, particularly for complex multilayered or buried interfaces where compositional gradients are important. Additionally, beam-induced damage during electron or ion-based imaging creates serious issues for soft, organic, or hybrid thin films. This damage often alters intrinsic material properties and microstructures. Another major challenge is integrating real-time, in situ characterization during film growth. This integration would allow for dynamic monitoring of nucleation, crystallization, and defect evolution under actual processing conditions. The large amount of data generated by modern instruments also requires artificial intelligence (AI) and big data analytics for effective interpretation and correlation of multidimensional datasets [53]. Looking forward, emerging technologies like four-dimensional scanning transmission electron microscopy (4D-STEM), ptychography, and hyperspectral imaging are set to change how we analyze materials at the nanoscale [54]. These methods promise atomic-scale spatial and temporal precision. They will enable direct mapping of strain, phase transitions, and electronic structure in functional thin films. The combination of machine learning and advanced microscopy is expected to speed up materials discovery and optimization, leading to the next generation of thin-film technologies for energy, electronics, and quantum applications.

2.9 Conclusion

Characterization techniques are crucial in thin-film materials research. They provide valuable insights into how structure, composition, and performance work together. Techniques like X-ray diffraction (XRD) and atomic force microscopy (AFM) offer essential information about crystallinity, morphology, and surface roughness. At the same time, synchrotron-based methods, such as X-ray absorption spectroscopy (XAS) and scanning transmission X-ray microscopy (STXM), allow for element-specific analysis at the nanoscale. These methods uncover details about chemical states and local structures. Ultrafast spectroscopies further enhance this understanding by observing dynamic events, such as carrier relaxation and charge transfer, on femtosecond timescales. The increasing use of in-situ and operando techniques enables researchers to track structural and electronic changes during growth or device operation. This provides a clear connection between synthesis conditions and performance. Furthermore, combining multimodal analysis with machine learning and artificial intelligence is changing how researchers interpret data. It helps identify hidden trends and connections between structure and properties. This data-driven approach is advancing the design and optimization of next-generation thin films for sustainable energy, flexible electronics, and new quantum technologies. Together, these developments signify a shift toward smarter, predictive, and efficient materials characterization in modern thin-film science.

Funding The authors, KP and AA, duly acknowledge the financial support received from the ANID-FONDECYT Postdoctoral project no. 3250236, Government of Chile, and Universidad Tecnológica Metropolitana, at Santiago, Chile.

References

1. Ohring, M. Materials Science of Thin Films, Academic Press, 2002.

2. Smith, D. L., Thin-Film Deposition: Principles and Practice, McGraw-Hill, 1995.
3. Chopra, K. L., Thin Film Phenomena, McGraw-Hill, 1969.
4. Mannhart, J. et al., Science, 2010, 327, 1607–1611.
5. Birkholz, M., Thin Film Analysis by X-ray Scattering, Wiley-VCH, 2006.
6. Cullity, B. D., Elements of X-ray Diffraction, Pearson, 2014.
7. Jenkins, R., Introduction to X-ray Powder Diffractometry, Wiley, 1996.
8. Fewster, P. F., X-ray Scattering from Semiconductors, Imperial College Press, 2003.
9. Holy, V. et al., High Resolution X-ray Scattering from Thin Films and Multilayers, Springer, 1999.
10. Hodes, G., Thin Film Growth, Springer, 2019.
11. Williams, D. B. & Carter, C. B., Transmission Electron Microscopy, Springer, 2009.
12. Reimer, L., Transmission Electron Microscopy: Physics of Image Formation, Springer, 1997.
13. Pennycook, S. J., Nature Materials, 2008, 7, 935–945.
14. Aguiar, J. A., et al., Advanced Science, 2016 3(6), 1500320.
15. Eaton, P. & West, P., Atomic Force Microscopy, Oxford University Press, 2010.
16. Gupta, I., & Mohanty, B. C. Scientific Reports, 2016, 6(1), 33136.
17. Collins, L. et al., Nature Communications, 2018, 9, 4988.
18. Qian, D. et al., Nano Energy, 2020, 75, 104962.
19. Briggs, D. & Grant, J. T., Surface Analysis by Auger and XPS, IM, 2003.
20. Watts, J. F. & Wolstenholme, J., An Introduction to Surface Analysis by XPS, Wiley, 2003.
21. Seah, M. P., Surface and Interface Analysis, 2012, 44, 1353–1359.
22. Vickerman, J. C. & Gilmore, I. S., Surface Analysis: The Principal Techniques, Wiley, 2009.
23. Benninghoven, A., Secondary Ion Mass Spectrometry, Wiley, 1987.
24. Bansal, N., et al., Vacuum, 2019, 161, 347-352.
25. Barbour, J. C., Applied Surface Science, 2001, 179, 1–8.
26. Chu, W. K. et al., Backscattering Spectrometry, Academic Press, 1978.
27. Mayer, M., Nuclear Instruments and Methods B, 1999, 194, 177–186.
28. Chu, P. K. & Li, L., Materials Chemistry and Physics, 2006, 96, 253–277.
29. Swanepoel, R., J. Phys. E., 1983, 16, 1214–1222.
30. Tauc, J., Mater. Res. Bull., 1968, 3, 37–46.
31. Hsu, C. et al., Solar Energy Materials & Solar Cells, 2015, 140, 421–429.
32. Tompkins, H. G. & Irene, E. A., Handbook of Ellipsometry, Springer, 2005.
33. Hilfiker, J. N., et al., Appl. Surf. Sci., 2017, 421, 508–512.
34. Herzinger, C. M. et al., Appl. Optics, 1998, 37, 5287–5300.
35. van der Pauw, L. J., Philips Res. Reports, 1958, 13, 1–9.
36. Seto, J. Y. W., J. Appl. Phys., 1975, 46, 5247–5254.
37. Fortunato, E. et al., Thin Solid Films, 2004, 487, 205–211.
38. Koningsberger, D. C. & Prins, R., X-ray Absorption: Principles, Applications, Wiley, 1988.
39. Bunker, G., Introduction to XAFS, Cambridge University Press, 2010.
40. Farges, F., Coordination Chemistry Reviews, 2005, 249, 1937–1960.
41. Ade, H. & Stoll, H., Nature Materials, 2009, 8, 281–290.
42. Hitchcock, A. P., J. Electron Spectrosc., 2012, 185, 1–19.
43. Kilcoyne, A. L. D. et al., J. Synchrotron Radiat., 2003, 10, 125–136.
44. Shah, J., Ultrafast Spectroscopy of Semiconductors, Springer, 1999.
45. Dutton, G. et al., Phys. Rev. B, 2010, 81, 155212.
46. Grätzel, M., Nature, 2001, 414, 338–344.
47. Tao, F., Chemical Reviews, 2012, 112, 163–220.
48. Zhang, L. et al., Micron, 2018, 110, 18–41.
49. Liu, X. et al., Adv. Funct. Mater., 2018, 28, 1707500.
50. Kalinin, S. V. & Spurgeon, S. R., ACS Nano, 2019, 13, 9841–9848.
51. Kusne, A. G. et al., Nature Communications, 2020, 11, 5966.
52. Zhang, H. et al., Science Advances, 2021, 7, eabe3727.
53. Wang, D. et al., Matter, 2022, 5, 286–309.
54. Ophus, C., Microscopy and Microanalysis, 2019, 25, 563–582.

Indu Gupta is an Assistant Professor in the Department of Physics, University Institute of Sciences (UIS), Chandigarh University, Mohali, India. She holds a Ph.D. in thin-film solar. Her research focuses on the growth, characterization, and application of advanced semiconductor thin films, particularly kesterite material, for photovoltaic devices. Dr. Gupta has published numerous peer-reviewed research papers, books, and book chapters, making significant contributions to the fields of renewable energy and thin-film photovoltaics.

Chapter 3
Thin Film Fabrication Techniques

Esquivel-Marin Judi Judith

3.1 Introduction

Thin films have recently acquired a fundamental role, as they are strategically important in the advancement of cutting-edge technologies. Their application extends to a wide variety of fields—for instance, in energy, for coatings in solar cells and batteries; in electronics, for integrated circuits and displays; in optics, for antireflective coatings and filters; and in biomedicine, for antibacterial coatings and biosensors. These coatings consist of extremely thin layers of material, with thicknesses ranging from just a few nanometers to several micrometers, deposited on different types of substrates. These substrates can vary from rigid inorganic materials such as glass, silicon, or ceramics (e.g., Al_2O_3 and ZrO_2) to metallic ones, including copper, aluminum, and their alloys. The purpose of this process is not only to protect the surface but also to modify or enhance essential properties, such as electrical conductivity, optical behavior, or mechanical resistance, depending on the needs of each application [1, 2]. Thin film deposition techniques are generally classified into physical, chemical, and hybrid methods, each with distinct principles and applications, Fig. 3.1.

Therefore, the development of diverse thin film fabrication techniques has been promoted, as the demand for innovative technologies continues to grow in order to address increasingly specific needs worldwide. Thus, within the fabrication methods, we find chemical, physical, and hybrid techniques, each with their own principles, advantages, and limitation [3]. In this sense, the choice of the deposition method will depend on multiple factors, such as the type of material to be deposited, substrate conditions, purity requirements, process efficiency, and the associated energy costs.

E.-M. J. Judith (✉)
Unidad Académica de Ciencias Químicas, Universidad Autónoma de Zacatecas, Campus Universitario UAZ Siglo XXI Carretera Zacatecas-Guadalajara Km. 6, Zacatecas, Mexico
e-mail: marin.judi@uaz.edu.mx

S. Sharma (ed.), *Multifunctional Heterostructure Thin Films Handbook*,
https://doi.org/10.1007/978-3-032-18692-8_3

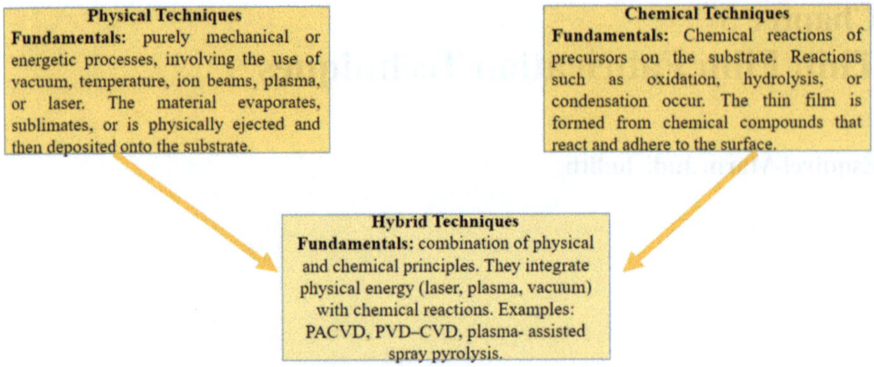

Fig. 3.1 General classification of thin film fabrication techniques

3.2 Conceptualization and Relevance of Thin Films

3.2.1 What Are Thin Films?

Thin films are coatings applied to a solid substrate with the aim of modifying or improving its properties. These thicknesses range from a few nanometers to several micrometers. These thin layers can contain metals, oxides, polymers, semiconductors, or composite materials. Their functions can range from protection against environmental factors (such as corrosion, abrasion, humidity, etc.) to electrical conduction in electronic devices or manipulation of optical response, for example, in antireflective coatings [4]. One of the distinctive features of thin films is that the functional properties of the system on which they are deposited are controlled, thanks to the precise use of their thickness, microstructure, morphology, and composition. Thin film synthesis techniques are specific to each technique and its objectives. Thus, they control the deposition conditions (e.g., temperature, pressure, time, and energy), allowing the transport and condensation of atoms, molecules, or particles on the substrate surface [5, 6]. These methods can be classified as physical, chemical, or hybrid, depending on the energy source used and the manner in which mass transfer occurs to the substrate, which are described in this chapter.

The use of thin films lies in their use in a wide range of emerging technologies, where a high area-to-volume ratio, efficient use of materials, and integration with microfabrication processes are required (e.g., in solar cells, biomedical sensors, antireflective coatings, microelectronics, etc.). In this context, thin films not only represent a functional solution but also a strategy for optimizing resources and reducing environmental impacts, especially when designed to perform specific functions with a minimal amount of material [7].

3.2.2 Importance of Their Manufacturing

The synthesis of thin films is an essential tool when it comes to designing advanced systems, where structural precision, energy efficiency, and material optimization are key. Thin films allow us to modify the optical, electrical, mechanical, and chemical properties of surfaces without changing the internal structure of the material, which in itself is a major advantage [8].

Thin films have become quite important for modern technologies. They can be used in flexible electronic devices, sensors, and also as coatings for energy systems and solar applications [9]. Depending on how their composition and microstructure are adjusted, their performance and durability can really improve. In general, making thin films is not only a technical process but also a practical way to design materials that have commercial value. Some types of thin films show antimicrobial or water-repellent behavior and even change with temperature. Because of that, they are useful in areas such as biomedicine or aerospace, where materials often face demanding working conditions [10, 11].

3.2.3 General Applications

Thin films have become increasingly common in recent years because they are used in many kinds of modern technologies. They can provide special properties to a surface while keeping the main structure of the material unchanged, which makes them very useful in different applications. Their applications range from optical coatings and piezoelectric sensors to active components in solar cells (such as the absorber or window layer), microbatteries, photonic systems, medical devices, and barrier coatings [12].

In the energy sector, thin films are used to improve the efficiency of photovoltaic conversion by controlling their optical and electronic properties. In the biomedical industry, for example, they allow the development of bioactive or antimicrobial surfaces that promote integration with living tissues or prevent bacterial colonization [13, 14]. In the automotive and aerospace industries, the implementation of thin films allows for component weight reduction thanks to very thin protective coatings that exhibit high mechanical and thermal resistance, a significant advantage [15]. This broad spectrum of applications demonstrates the extensive functional flexibility of thin films, as well as their crucial role in the innovation of advanced materials, in a context where energy efficiency and multifunctionality have become fundamental requirements.

3.3 General Classification of Manufacturing Techniques

3.3.1 Physical Techniques

Thin film deposition using physical techniques encompasses several methods, including thermal evaporation, electron beam evaporation, and physical vapor deposition, to name a few. The methodology for film synthesis using chemical methods is basically as follows: the material is drawn from a solid or liquid source to the substrate using physical mechanisms, without the intervention of chemical reactions. These processes require controlled vacuum conditions, high temperatures, or high-energy beams. The physical vapor deposition method is a set of physical methods used to deposit thin films on various substrates. It is based on the following principles: a material is vaporized within a vacuum chamber or in an inert atmosphere, transporting the atoms, ions, or molecules to the substrate surface, where they are condensed to form the film [16, 17]. The coating is generated atom by atom during the process, ensuring good adhesion and morphological control. The thickness of the film formed can range from a few nanometers to micrometers. The properties depend on the evaporated material and the deposition conditions (vacuum, gases present, plasma energy, and substrate temperature). The most representative physical techniques are described below [18].

Thermal Evaporation

Thermal evaporation is one of the oldest and also one of the most used physical methods to produce thin films. In this technique, the material that will form the coating is heated, usually inside a small crucible, until it starts to evaporate and then condenses on the substrate under vacuum conditions, forming a thin layer. This process is quite simple and, compared with other methods, it is also inexpensive. However, sometimes the coverage is not completely uniform, especially when the surface has irregular shapes. Other approaches, like electron beam evaporation, are more sophisticated and allow better control of the film's thickness and purity, but they also require more complex equipment and are much more expensive [19, 20].

The thermal evaporation works by heating a solid inside a vacuum chamber until it reaches its evaporation or sublimation point. The atoms released travel more or less in a straight line through the vacuum and then settle on the cooler substrate surface, creating a thin solid layer. This phenomenon can be explained by the Hertz–Knudsen equation, since the number of atoms that evaporate depends mainly on the vapor pressure and the temperature of the material. Keeping a good vacuum is very important because it avoids reactions with oxygen or other gases and reduces collisions between particles, which helps to get a cleaner and more even deposition [21, 22].

The process usually needs a few basic parts to work properly. There has to be a heating source, which can be electric wires, filaments, a small tungsten crucible,

Fig. 3.2 Schematic of the thermal evaporation process

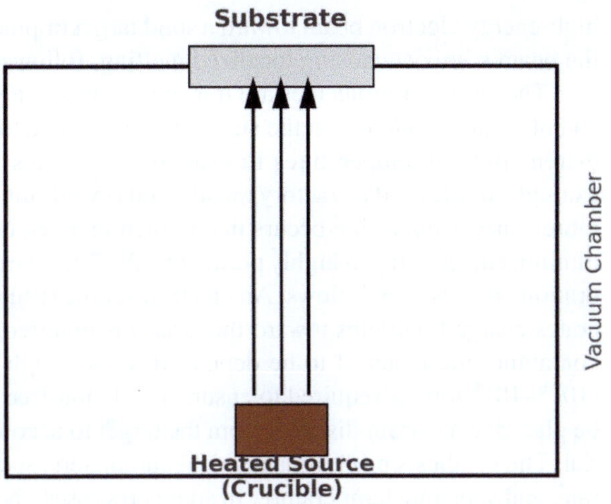

or even an electron beam. A vacuum chamber is also required; in it, the pressure is lowered to around 10^{-5} or 10^{-6} torr. Then comes the substrate, placed in front of the source so it can receive the vaporized material. The main steps are quite straightforward: first the material is loaded into the crucible, and then the chamber is pumped down until a good vacuum is reached. After that, the material is slowly heated until it starts to evaporate, and finally, the vapor condenses on the surface of the substrate, forming the thin film [23–25].

A schematic representation of the thermal evaporation process is shown in Fig. 3.2, where the heated source (crucible) releases atoms that travel under vacuum and condense on the substrate to form the thin film. Figure 3.2 shows schematic representation of the thermal evaporation technique, where the heated source (crucible) releases atoms that travel under vacuum and condense on the substrate to form a thin film.

Thermal evaporation is used in microelectronics for the production of metallic contacts in semiconductors, in the optics industry for anti-reflective coatings and high-reflectance mirrors, in photovoltaic devices for the deposition of metallic layers in solar cells, and in the sensor sector for the manufacture of thin conductive or dielectric layers. However, it has some limitations, such as limited control of stoichiometry in multicomponent compounds. Furthermore, it is difficult to coat surfaces with very complex geometries (because it is a directional process). One of these, and very important, is that it requires a high vacuum, which results in high operating costs compared to chemical methods [26, 27].

E-Beam Evaporation

Electron beam evaporation is a physical vacuum deposition (PVD) technique used to obtain high-purity thin films. Its principle is as follows: it consists of directing a high-energy electron beam toward a solid target or precursor material. This results in the beam's impact causing localized melting, followed by evaporation of the material. The atoms or molecules are transported in a straight line within a high-vacuum chamber and condense on the substrate, giving rise to the film. This technique uses extremely high temperatures to evaporate materials with high melting points, for example, oxides and refractory metals, and is well suited for these materials. Furthermore, since evaporation occurs in a limited manner, contamination of the coating is minimized, resulting in highly pure films [28, 29]. The procedure for an E-beam evaporation process is as follows. An electron source (usually an electron gun that accelerates charged particles toward the target) is required, along with a cooled crucible containing the material to be deposited. And a high or ultra-high-vacuum system (10^{-6}–10^{-8} torr) is required to ensure a collision-free trajectory. The substrate must be placed at a certain distance from the target to accommodate the evaporated material. The synthesis process has critical parameters, such as beam power, evaporation rate, and substrate temperature. In some processes, the atmosphere of reactive gases influences the morphology, crystal orientation, and, very importantly, the optical and electrical properties of the films obtained [30, 31].

The Cathodic Sputtering Technique

The cathodic sputtering technique is a popular synthesis methodology. Its versatility for the deposition of conductive metals, semiconductors, and ceramics makes it widely used in different technological areas. Its basis is that energetic ions strike a target material and eject atoms, which are deposited on the substrate. The deposited film is uniform, with strong adhesion, although the technique is expensive. Variants of this technique are as follows: DC sputtering, with direct current, is suitable for conductive materials. RF sputtering, with radio frequency, allows working with insulating materials such as nitrides and oxides. Magnetron sputtering uses high-power pulses that generate very dense plasmas, achieving high-quality films (with better adhesion and higher density). In reactive sputtering, a reactive gas (O_2 or N_2) is introduced, allowing the formation of compounds such as oxides, nitrides, or carbides directly on the substrate [32–34]. A schematic representation of the cathodic sputtering process is shown in Fig. 3.3, where Ar^+ plasma ions bombard the target material, ejecting atoms that condense on the substrate to form the thin film.

The sputtering technique is highly versatile and widely used in various industries. In microelectronics, it is used to produce metallic contacts, diffusion barriers, and dielectrics. In optics, it is used to produce anti-reflective coatings and high-reflectance mirrors. In the energy field, this method is often used to make active layers for solar cells and thermoelectric materials, as well as for developing sensor coatings. In biomedical applications, it serves for biocompatible layers that cover implants and other medical parts. In nanotechnology, it has become very common for growing nanocolumns and thin films with specific and controlled structures [35].

The main advantages are easy to notice: strong adhesion, good control of thickness, the possibility of applying it to large surfaces, and high reproducibility. On the

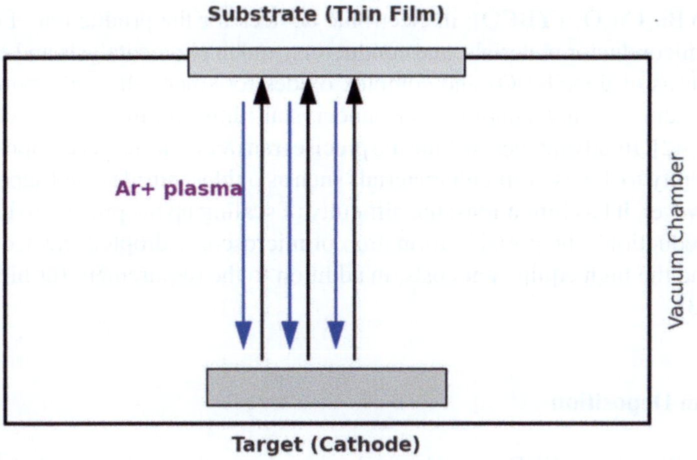

Fig. 3.3 Schematic representation of the cathodic sputtering technique

other hand, there are also some limitations. The process is a bit slower than thermal evaporation, which sometimes allows minor contamination from off-gases. It can also be difficult to control the exact proportion of several elements at the same time, and the equipment needed is rather expensive [36].

Laser Ablation

Laser ablation is quite a special technique because it uses short and intense laser pulses that hit a solid target. The energy from the laser makes the material evaporate very quickly and generates a bright plasma that contains atoms, molecules, and ions. This plasma moves toward the substrate, and when the particles reach the surface, they cool down and form a thin layer, everything happens in just a few nanoseconds. One interesting point is that the film keeps almost the same composition as the original material, which is very important when working with complex oxides, ceramics, or superconductors, where the right stoichiometry must be preserved [37, 38]. The procedure requires a laser source, typically excimer lasers capable of delivering high-energy pulses; a vacuum chamber that can operate under high-vacuum conditions or with controlled atmospheres of oxygen, nitrogen, or other reactive gases; a solid target, which corresponds to the material to be deposited (e.g., oxides, nitrides, metals); and a substrate, which is placed directly in front of the plasma. The characteristics of the procedure are: laser pulse energy (energy density in J/cm^2), pulse duration and frequency, target-substrate distance, deposition atmosphere, and substrate temperature, which influence film crystallization [39, 40].

This technique has proven versatile in obtaining advanced thin films. It is widely used in optics for the synthesis of anti-reflective coatings and transparent oxide films. In the field of superconductors for the deposition of high-critical-temperature oxides, such as $YBa_2Cu_3O_7$ (YBCO); in electronic devices for the production of dielectric films, semiconductor materials, and conductors; and in photocatalysis and energy for the synthesis of doped TiO_2 and complex oxides for solar cells and environmental catalysis; and in nanotechnology for nanocrystal films and multilayer heterostructures [41, 42]. Its advantages include the precise transfer of the target composition and the flexibility to deposit difficult materials such as oxides, nitrides, and superconductors. However, it has limitations: the difficulty of scaling up the process to large areas (mass production), the possible formation of microscopic droplets ejected from the target, and the high equipment costs, in addition to the requirement for high-energy lasers [43].

Ion Beam Deposition

Ion beam deposition (IBD) is a physical vapor deposition technique that combines the evaporation or sputtering of materials with the simultaneous irradiation of the substrate surface by a beam of energetic ions. Its rationale is that the atoms or molecules that form the film are typically deposited on the substrate, while the ion beam bombards the growing surface, transferring kinetic energy to the surface atoms and thus modifying the morphology, structure, and properties of the film. Unlike thermal evaporation (E-beam), where the process focuses on evaporating the material, in IBD, there is a direct interaction of the ions with the forming film, which increases atomic mobility on the surface and promotes a denser and more homogeneous organization [44]. The process requires: (i) a primary deposition source (e.g., sputtering, evaporation, or conventional PVD); (ii) a source that generates ions (usually Ar^+) that impinge on the substrate at controlled energies; (iii) a vacuum chamber that controls pressure, atmosphere, and temperature. Critical parameters include: ion energy, since low ions generate surface mobility without damaging the lattice, while high energies can induce collision cascades and structural changes; the angle of incidence, which determines the direction of momentum transfer; and ion flux and deposition rate regulate homogeneity and final thickness [45]. This technique is applied in many different industrial areas. For example, in the synthesis of metallic glasses, it helps improve hardness, stability, and wear resistance, all without needing to raise the substrate temperature. In microelectronics, it serves to make dense coatings that block diffusion in contacts and barriers. In optics and photonics, it helps adjust the refractive index and create anti-reflective layers or precision mirrors. In biomedical uses, it allows the formation of biocompatible coatings that also have high mechanical strength. And in advanced devices, it contributes to modifying properties such as magnetic anisotropy, wear behavior, and dielectric response [46].

Some of its main advantages are quite clear: it allows better control of the microstructure and film properties, even when working at low temperatures, and it also makes it possible to handle amorphous materials like metallic glasses. The coatings obtained usually have high hardness, good stability, and strong adhesion. In any case, there are also drawbacks. The equipment tends to be complex and expensive, which makes the process harder to implement; for example, it is an also important to emntion that, if the irradiation is not well controlled, residual stresses can appear, and the scalability of the method is still limited [47].

Arch Discharge

The way this method works is actually quite simple to describe: An electric arc is created between two electrodes—often graphite or another conductive metal—and this arc gives off a plasma so hot that it can easily pass 3000 °C. At those temperatures, some of the electrode material evaporates. The atoms and ions that come out spread into the surrounding space, which might be a vacuum, an inert gas, or sometimes even a liquid like deionized water. As they cool down, they stick together again and end up forming nanoparticles, nanotubes, or really thin coatings on nearby surfaces. The process is eminently physical, as it does not depend on chemical reactions but rather on the transfer of energy from the electric arc to the material. Its greatest advantage is that the technique has the ability to produce nanometric structures under relatively simple conditions [48].

A typical arc discharge system includes electrodes (anode and cathode) made of the material to be deposited (or graphite with metal inserts). A power source supplies high currents (10–20 A) and voltages between 50–100 V. The working medium can be vacuum, argon, helium, or deionized water, depending on the application. The process occurs in several stages. First, an electric arc is established between the electrodes. Then, the intense heat generated evaporates the anode material. Following this, the vapor ionizes, forming a glowing plasma. Finally, the ionized and neutral species condense outside the arc zone, nucleating and growing in the form of nanoparticles, nanotubes, or thin films [49].

Its applications include the following: in the synthesis of carbon nanomaterials, for example, to obtain carbon nanowires and nanotubes (SWCNTs and MWCNTs) from fullerenes; in the synthesis of hybrid nanocomposites: ZnO/MWCNTs, TiO_2/CNTs, and other combinations with enhanced optical and catalytic properties; in the environmental field, to obtain nanomaterials for the adsorption and removal of metal contaminants in water; and in the optics and electronics industries, to manufacture conductive films and coatings based on nanocarbons and oxides. In the energy industry, it is used to create electrodes for supercapacitors, batteries, and storage devices. In biomedicine, it is used to prepare biocompatible nanomaterials for controlled drug release and biosensors [50]. The main advantage of the technique is its simplicity and versatility. Furthermore, it does not require complex vacuum atmospheres (it can even be performed in liquids). High-purity materials can be obtained. Its limitations include the difficulty in controlling the size and morphology of the nanoparticles, high electrode consumption during the process, and low scalability for industrial production.

Molecular Beam Epitaxy (MBE)

Molecular beam epitaxy is a deposition technique within physical processes that enables the growth of thin films with atomic-level control. It is based on the controlled evaporation of elemental sources (metals or compounds) within an ultra-high-vacuum system. The evaporated atoms or molecules form molecular beams that are directed toward a heated substrate, where they condense into epitaxial layers (arranged according to the substrate's crystal lattice). What distinguishes MBE from other deposition techniques is its ability to independently control the flows of each source and perform layer-by-layer growth, monitored in real time, resulting in considerably smooth surfaces with excellent stoichiometric precisión [51]. In the 1960s and 1970s, the need to manufacture ultra-pure semiconductors for microelectronics applications led to the emergence of this technique. Since then, it has been consolidated as an essential method for the fabrication of semiconductor heterostructures, giving rise to devices such as semiconductor lasers and high-mobility transistors [52]. The procedure consists of the following: an ultra-high-vacuum chamber (10^{-9}–10^{-11} Torr), which prevents contamination and allows collision-free trajectories for molecular beams; evaporation sources (effusion cells, electron beams, or organometallic); in addition to a heated substrate on which the film grows; and in situ monitoring systems (e.g., monitored in real time by high-energy electron diffraction reflection or residual mass spectroscopy). Growth occurs atom by atom. It follows an epitaxial regime that replicates the substrate's crystal lattice, allowing the production of materials at the atomic scale [53, 54].

The technique is widely used in areas such as microelectronics for the growth of semiconductor heterostructures, high-mobility transistors, LEDs, and lasers. It is also used in optics and photonics for the synthesis of filters and modulators. It is also used to fabricate epitaxial films for electron spin-based devices. It is also used in the synthesis of quantum materials, for example, to obtain superconductors, perovskites, and heterostructures with two-dimensional electron gases, and to obtain hybrid nanomaterials, in the integration of oxides with semiconductors, self-supporting membranes, and multifunctional coatings [55]. However, oxide growth still presents several challenges, such as the low vapor pressure of some metals, which requires very high temperatures to evaporate them. Differential affinity for oxygen can cause partial or unstable oxidation. And the oxidation of internal components (crucible, supports, filaments), which compromises film purity [56].

3.3.2 *Chemical Techniques*

Chemical techniques for thin film deposition are based on the formation of layers through chemical reactions of precursors on a substrate, allowing for uniform coatings with structural control at the molecular level. These techniques are particularly valued because the synthesized films are highly pure, with controlled thicknesses,

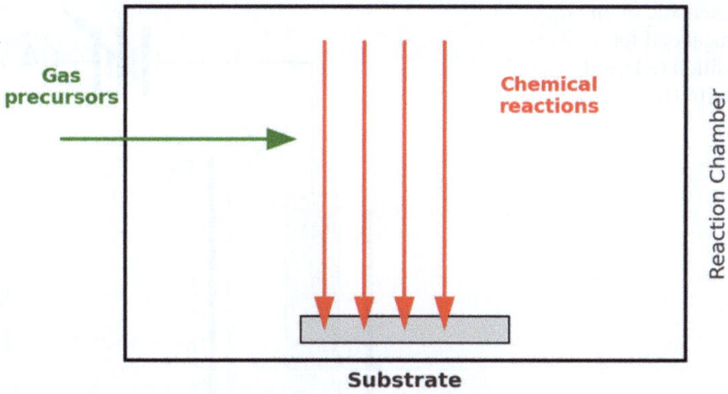

Fig. 3.4 Schematic of the chemical vapor deposition (CVD) process

and can even be deposited on surfaces with complex geometries, while still maintaining high purity and homogeneity [57]. The chemical techniques are described below:

Chemical Vapor Deposition (CVD)

One of the most widely used methodologies is chemical vapor deposition. In this method, the chemical precursors are introduced into a reaction chamber, where a series of transformations take place and the desired film begins to form. It can be applied to obtain coatings made of different materials—metals, oxides, or nitrides—that generally show good adhesion to the surface and can be tuned at the microscopic scale [58]. The chemical vapor deposition, or CVD, works through chemical reactions that occur directly on the surface of a solid substrate, using gases or simple compounds that contain the required elements.

Depending on the working conditions, several versions of the process exist: atmospheric pressure CVD (APCVD), low-pressure CVD (LPCVD), ultra-high-vacuum CVD (UHVCVD), plasma-assisted or plasma-enhanced CVD (PECVD, PACVD), and even laser-activated CVD (LECVD) [59]. Figure 3.4 shows schematic illustration of the CVD process, showing how gaseous precursors flow into the chamber, react inside, and gradually form a thin film on the substrate.

Sol–Gel

The sol–gel method basically starts with organic or inorganic precursors that, after a few reactions, turn into a kind of colloidal sol. These reactions are mainly hydrolysis and condensation, which slowly change the liquid mixture until it begins to form the film. After that, the material is usually spread or deposited on a surface and heated, and this step makes the coating solidify into a thin layer. It is a very practical

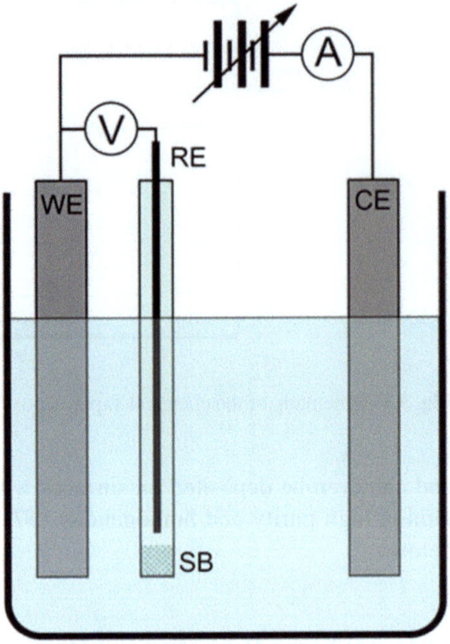

Fig. 3.5 Schematic of an electrochemical cell for electrodeposition (adapted from MDPI, [69]]

and low-cost process, and that is why it is used both in labs and in many industrial setups. In simple terms, this route is one of the easiest ways to prepare ceramic or glass-type materials at very small scales. What makes it interesting is that it allows control over the structure of the final product without expensive equipment. The transformation goes from a sol to a three-dimensional gel through two main reactions: first hydrolysis, then condensation. Typical precursors are silicon alkoxides such as tetramethoxysilane (TMOS) or tetraethoxysilane (TEOS). During hydrolysis, the-OR groups are replaced by-OH, and later, as condensation proceeds, siloxane bonds (Si-O-Si) form and join together to create a solid silica network [60].

Chemically, what happens is that hydrolysis converts the alkoxide groups into silanols (Si-OH), which are linked by condensation reactions to form siloxane (Si-O-Si) bonds. This process can be modulated by the pH of the solution. In an acidic medium, more linear structures will predominate, while in basic conditions the formation of highly branched networks is favored, which influences the gelation speed and the final texture of the material [61].

Atomic Layer Deposition

Another technique for film synthesis, atomic layer deposition (ALD), has emerged as a high-precision technique for the formation of ultra-thin films through the sequential and self-limiting adsorption of chemical precursors. This allows angstrom-level control of thickness, a quality that has made it very popular in the synthesis of semiconductors and advanced functional coatings. The technique allows for the controlled deposition of ultra-thin films with atomic-level precision, as it operates through self-limiting chemical reactions that occur successively, ensuring precise and uniform film growth, even when growing films on surfaces with complex morphologies [62].

The technique is known to have been first used in the 1970s, when he was searching for a technique that would allow him to fabricate thin films of zinc oxide for electroluminescent devices. At that time, its initial application was primarily in the production of thin film displays. Toward the end of the 1990s and the beginning of the new millennium, the technique became popular in the semiconductor industry, offering a level of thickness control unattainable with other deposition techniques [63]. Since then, ALD has become an indispensable engineering technique for producing advanced materials. One of the main reasons why ALD is gaining popularity in advanced materials is that, unlike methods such as CVD and PVD, which offer high deposition rates but with less control over uniformity and thickness, ALD synthesis is characterized by its ability to achieve angstrom growth, i.e., layer-by-layer film control. The self-limiting nature of the reactions allows the film to grow uniformly and continuously, even on irregularly shaped surfaces. This ensures that coatings maintain their conformity even when applied to complex three-dimensional structures, something that is more difficult to guarantee with CVD or PVD. However, this level of precision also means that the process is comparatively slower [64]. This technique is very useful in the semiconductor industry, as it is widely used in the manufacture of advanced transistors, such as high-constant dielectric coatings. In the energy sector, it has also been widely used in the synthesis of solar cells (anti-reflective and absorbent layers are obtained) and in lithium-ion batteries, where ALD coatings improve the stability and conductivity of the electrodes [65]. Furthermore, it is widely implemented in the field of biomedicine, where biocompatible coatings obtained through ALD are used in medical implants to increase their durability. Likewise, also widely used in emerging areas such as photonics and optoelectronics, the ALD technique is used to fine-tune the optical properties of synthesized films [66]. Chemical techniques, compared to physical techniques, offer substantial advantages in terms of uniformity in three-dimensional structures and greater integration with microfabrication processes, although they typically require rigorous control of reaction parameters, temperature, and precursor purity.

Electroplating

Another technique for chemically synthesizing thin films is electrodeposition (electrolytic deposition). This technique has its roots in the nineteenth century, when it began to be applied in metallurgy to obtain protective and decorative coatings on metals such as copper, nickel, or the commonly used gold. Initially, its development was linked to the search for methods to improve the appearance and durability of metallic objects in the jewelry industry. Over time, electrodeposition evolved from a process for decorative purposes to a valuable tool in the manufacture of functional devices, such as in the electronics and microelectronics industries [67, 68].

Some popular applications, such as in the energy sector, for example, are used to obtain nanoporous electrodes in lithium-ion batteries and supercapacitors. In microelectronics, this technique allows the manufacture of high-precision metallic interconnections, which are necessary for the functioning of integrated circuits. In addition, in the production of microchips, where millions of transistors must be connected to each other by extremely small metallic channels [70].

Chemical Bath Deposition

The chemical bath deposition (CBD) method is a thin film synthesis technique based on chemical precipitation reactions that occur in an aqueous solution. The precursors are ions, which are present in a solution. Unlike other deposition processes, CBD is characterized by its low cost and simplicity. Furthermore, it does not require expensive equipment and does not require high-vacuum equipment, making it a highly accessible alternative for thin film synthesis, very popular in the production of semiconductor, dielectric, or conductive materials [71].

The chemical bath method works through a slow and controlled precipitation of an insoluble compound directly on a surface that has been placed in the reaction solution. The control is achieved by using a complexing agent, which releases metal ions little by little so that the film can form gradually. One of the key roles of this agent is to make the coating grow evenly and adhere well to the substrate, instead of causing an instant precipitation in the bulk of the liquid. Because of that, the material deposits slowly on the solid surface.

The overall mechanism can be imagined in three main parts: first, the ionic precursors dissolve in the bath; then stable complexes are formed, releasing the ions little by little; finally, the material nucleates and grows on the surface of the substrate. With this process it is possible to obtain very uniform films of several materials, including metal sulfides such as CdS, ZnS, or PbS, and metal oxides like ZnO, CuO, or TiO_2. Some composite materials with optical or electronic functions can also be prepared in this way [72]. This technique does not need vacuum systems, very high temperatures, or expensive setups, which makes it more accessible and easier to handle in many labs. While the operating conditions of the chemical bath are simpler, as it operates at relatively low temperatures (60 and 90 °C), it does have limitations, such as the control of thickness at the nanometer level and material purity, attributes that the ALD technique offers [73].

The technique is very popular in materials synthesis and is widely used in the synthesis of semiconductor materials with applications in the energy industry, for example, in second—and third-generation solar cells, and buffer and window layer materials such as CdS, ZnS, and ZnO. As well as in the synthesis of gas sensors and optoelectronic devices, where the metal sulfides obtained using this technique demonstrate high sensitivity and stability [74].

Spray Pyrolysis

The chemical technique of spray pyrolysis began to be used in the mid-twentieth century as a simple and economical alternative for depositing films of semiconductor materials. For example, it was successfully applied in the growth of metal oxides (ZnO, SnO_2, TiO_2) with optoelectronic properties. With the rise of renewable energy, the technique gained greater importance in the production of absorber films for second-generation solar cells (especially CdS and CdTe). Currently, the technique has expanded its applications, being used in fields such as nanotechnology, catalysis, flexible electronics, and the biomaterials production industry [75]. It is worth noting that its main advantages include the use of inexpensive precursors, the absence of vacuum requirements, and the lack of expensive equipment. Furthermore, it is scalable, making it suitable for mass production, as well as capable of obtaining homogeneous coatings [76].

The method for synthesizing thin films and nanoparticles is based on the atomization of a precursor solution in the form of microdroplets, which subsequently pass through a high-temperature zone where evaporation, pyrolysis, and chemical decomposition processes occur. It is important to mention here that the microdroplets act as microreactors in which solid particles or coatings are formed on a substrate. Unlike more complex techniques, the spray pyrolysis technique is characterized by being a continuous, versatile, and scalable process, capable of producing nanoparticles and films with a wide range of compositions and morphologies. Its rationale is as follows, which includes a series of stages: first, the precursor is atomized into fine droplets using pneumatic, ultrasonic, or piezoelectric nozzles [77]. This is followed by the transport of the droplets to a reaction zone (a heated substrate). Subsequently, the solvent evaporates and the dissolved salts/metals thermally decompose, leading to nucleation and, consequently, the growth of solid particles, resulting in the formation of a thin layer on the substrate. The morphology of the material is controlled, as its parameters, such as precursor concentration, reaction temperature, carrier gas flow rate, and the distance between the nozzle and the substrate, can be adjusted [78]. When compared to other techniques, such as CVD or ALD, spray pyrolysis is distinguished by the fact that it does not require a vacuum or sophisticated plasma systems. Thus, it is more economical and its adaptability allows it to be used with a variety of soluble precursors. On the other hand, it is also important to mention that thickness and crystallinity control are usually lower, which means that a subsequent heat treatment is required to achieve the desired crystallinity. Other limitations include limited control of nanometric thickness, potential surface roughness (porosity in the films), critical dependence on temperature and atomization quality, and therefore the need for sintering to improve structural properties [79].

Spin Coating

Spin coating became popular in the 1980s, it was used in microelectronics thanks to the deposition of photoresists for lithography processes. Its application extended to the fields of nanotechnology and materials science, as it is a low-cost and highly reproducible method for obtaining high-quality coatings on substrates as varied as glass, silicon, metals, and polymers [80].

Spin coating is a coating method used to obtain homogeneous thin films. Its principle is as follows: it consists of depositing a small amount of precursor solution (e.g., polymer solutions, sol–gels, nanocomposites, or colloidal suspensions) on the surface of a substrate. Subsequently, centrifugal force disperses the solution toward the edges while the solvent evaporates, creating a uniform film. In practice, the final thickness of the layer changes for several reasons. It depends a lot on how fast the sample spins, but also on how long it keeps turning. When the rotation is faster, the liquid spreads more and the coating ends up thinner. If the spinning goes on for a while, the layer tends to level out better. The viscosity matters too, a dense solution gives a thicker film and the kind of solvent also affects it, because fast, evaporating solvents usually leave a thinner coating [81].

The spin coating process itself happens step by step. First, a few drops of the mixture are placed in the middle of the surface. As soon as it starts rotating, the liquid moves outward because of the centrifugal force. While the sample keeps spinning, the layer slowly adjusts its thickness. At the end, when the solvent evaporates, a solid and fairly uniform film remains. The main reason people like this method is that it's quick and simple. It doesn't need special equipment and, compared with other techniques, it gives quite consistent results [82].

For example, compared to dip coating, it is faster and consumes less solution, although it is limited to flat surfaces. Compared to spray coating, it offers better uniformity on flat substrates, although it does not adapt well to irregular surfaces. And, unlike CVD or ALD, it lacks control at the atomic level, but it is much more economical and does not require vacuum conditions [83].

Dip Coating

The dip coating technique is considered simple and inexpensive. It is used to obtain coatings on different types of substrates (e.g., ceramics, polymers, and fibrous materials). The steps are as follows: immerse the substrate in a liquid precursor solution, completely impregnating the substrate. It is then removed from the bath at a controlled rate, and the solvent begins to evaporate, leaving a uniform coating. The process seems simple, although it actually involves complex physical and chemical parameters. The thickness and final morphology of the film depend on factors such as immersion time and extraction speed from the substrate, the number of immersion cycles, the coating properties (viscosity, density, surface tension, and solids concentration), and even environmental conditions (temperature and humidity during solvent evaporation) [84]. It is important to mention that the Landau–Levich model describes the final thickness as a function of the capillary number, relating viscosity, extraction speed, surface tension, and solution density. Also, capillary action phenomena dominate at low speeds, while at high speeds, drainage prevails, giving rise to a characteristic V-shaped curve for film thickness [85].

Compared to other techniques, such as CVD, PVD, or ALD, dip coating has the following advantages: it does not require sophisticated equipment or vacuum conditions, making it a more accessible and scalable process. Compared to spray pyrolysis, dip coating offers greater uniformity on flat surfaces but less adaptability to complex geometries. Compared to spin coating, it has the advantage of being able to be applied to fibers, yarns, and three-dimensional substrates, although with less control over thickness [86].

Among its applications, it stands out in various areas, including functional textiles (self-cleaning surfaces based on TiO_2 with photocatalytic properties and hydrophobic fibers to repel water and dust). It also separates phases, for example, in coated fiber membranes for oil-water separation, with applications in wastewater treatment. In flexible electronics, fibers coated with conductive nanomaterials (graphene, CNTs, polypyrrole) are used to manufacture smart textiles, sensors, and energy-activated clothing. In the energy industry, it is used in supercapacitors and flexible battery electrodes using fibers coated with CNTs, graphene, or MnO_2, and in oxide membranes for fuel cells and solid electrolytes. However, it also has limitations, such as limited thickness control at the nanometer level and the potential for defects such as the "coffee-ring effect" in low-speed processes. Coatings are less homogeneous than vacuum methods on non-fibrous substrates. It is important to mention that a universal model, i.e., a general equation that accurately describes the process in non-Newtonian fluids such as sol–gel systems, is also lacking [87].

Inkjet Printing

Inkjet printing is a process that allows the deposit of materials. This process creates thin films by the controlled ejection of microdroplets of a precursor solution directly onto a substrate. It is a digital, contactless process carried out under ambient conditions, making it simpler compared to vacuum techniques (e.g., CVD, ALD, or sputtering). The mechanism is as follows: first, droplet generation occurs; that is, ink droplets with picoliter volumes are ejected through piezoelectric or thermal nozzles. This is followed by transport and deposition: the droplets follow controlled trajectories and are deposited in predetermined positions on the substrate. Finally, drying or sintering takes place; that is, the solvent evaporates and the active material forms the functional film [88].

Inkjet printing began to be used in graphic printing in the 1950s, but it began to be investigated as a technique for manufacturing electronic devices in the 1990s. Thus, it has been used in the deposition of conductive, semiconductor, and dielectric inks, becoming one of the pillars of so-called printed electronics. In recent years, it has gained importance due to the increased production of thin film transistors (TFTs), solar cells, LEDs, biosensors, and flexible devices, among others [89]. Its advantages are that virtually 100% of the material is recyclable. It also requires no vacuum conditions, masks, or low cost, and is scalable to industrial processes and compatible with flexible substrates. It also allows for digital printing and complex software-programmed patterns. Among its disadvantages is that the technique is prone to defects such as the "coffee-ring effect" and the materials obtained have electrical properties that are still inferior to devices manufactured using vacuum techniques [90].

3.4 Conclusions

When looking at the different ways thin films can be made, it becomes clear that there isn't really a single method that works for everything. Each technique has its own strengths and weak points, and the best choice usually depends on the material and what the film is going to be used for. The diversity of techniques responds to the specific needs our society faces, in terms of thickness, composition, uniformity, and functionality.

We know that physical methods (PVD, CVD, ALD, and variants) offer greater structural control and purity, being capable of generating superior-quality coatings, even at the atomic level. However, some limitations are that these processes require expensive equipment, vacuum conditions, and energy consumption, which can be detrimental to industrial scale.

On the other hand, chemical methods (such as sol–gel, electrodeposition, chemical bath, spray pyrolysis, spin coating, dip coating, and inkjet printing, to name a few examples) stand out for their experimental simplicity, versatility, and low costs. This allows them to work under ambient conditions. These techniques are especially useful for applications in renewable energy, sensors, catalysis, and functional coatings. Although it should be noted that they present limitations in angulation control and film homogeneity.

Finally, hybrid techniques are positioning themselves as an emerging field, as they combine the best of both methods (both physical and chemical). We look at examples such as the integration of spin coating with liquid-phase epitaxy and we know that films generated from MOFs demonstrate that it is possible to achieve uniformity, good crystalline control, and advanced properties. These combinations represent a trend toward technological convergence, where performance and cost can be optimized through a combination of strategies. Based on the points explained above, it is concluded that progress in thin film manufacturing does not depend on perfecting a single method, but rather, it is necessary to articulate chemical and physical techniques in a complementary manner. Different specific needs will arise according to the needs of each area. Having a comprehensive view allows for the design of increasingly sophisticated coatings, with a direct impact on strategic areas such as nanotechnology, microelectronics, optoelectronics, energy, and biomedicine.

References

1. Lim J, Kim D, Lee NH, Kim YG, Yu HK, Choi J-Y, et al. Optimizing electrostatic chuck performance through ZrO_2/Al_2O_3 ratio and doping components (SiO_2 and Y_2O_3). Ceramics International. 2025;51(9):11432–6. https://doi.org/10.1016/j.ceramint.2024.12.562
2. Sathish M, Radhika N, Saleh B. Current status, challenges, and future prospects of thin film coating techniques and coating structures. Journal of Bio-and Tribo-Corrosion. 2023;9(2):35. https://doi.org/10.1007/s40735-023-00754-9
3. Budida J, Srinivasan K. Review of thin film deposition and techniques. Materials today: proceedings. 2023;92:1030–3. https://doi.org/10.1016/j.matpr.2023.05.004
4. Jilani A, Abdel-Wahab MS, Hammad AH. Advance deposition techniques for thin film and coating. Modern technologies for creating the thin-film systems and coatings. 2017;2(3):137–49. https://doi.org/10.5772/65702
5. Zhang Y, Xia X, Ma K, Xia G, Wu M, Cheung YH, et al. Functional textiles with smart properties: their fabrications and sustainable applications. Advanced functional materials. 2023;33(33):2301607. https://doi.org/10.1002/adfm.202301607
6. Sakthinathan S, Meenakshi GA, Vinothini S, Yu C-L, Chen C-L, Chiu T-W, et al. A review of thin-film growth, properties, applications, and future prospects. Processes. 2025;13(2):587. https://doi.org/10.3390/pr13020587
7. Rahayu I, Insawang M, Muntini MS, Pramono YH, Pham ATT, Park S, et al. Thermoelectric properties of MoS_2 thin films prepared by RF magnetron sputtering. Vacuum. 2026;243:114817. https://doi.org/10.1088/1742-6596/3034/1/012008
8. Erken O, Sakin N, Gunes M, Ulutas C, Ozkendir OM, Gumus C. Role of Anionic Solution Temperature on the Electronic, Linear/Nonlinear Optical Properties and Dispersion Parameters of Iron (III) Oxide Thin Films. Optical Materials. 2025:117607. https://doi.org/10.1016/j.optmat.2025.117607
9. Sun D, Yang Y, Liu Q, Li D, Zhao Z, Li H. Controlling magnetic order via strain in epitaxial $CuCr_2O_4$ thin films: the critical role of film thickness. Journal of Magnetism and Magnetic Materials. 2025:173599. https://doi.org/10.1016/j.jmmm.2025.173599
10. Zhang Z, Cao L, Luo F, Zhou H, Ma K, Wang S, et al. Flexible In_2O_3 thin film with columnar structure for enhanced bending durability. Optical Materials. 2025;168:117449. https://doi.org/10.1016/j.optmat.2025.117449
11. Wang Y, Xu L, Ju Y, Yu W, Xu L, Hao J. Enduring macroscopic lubrication of thin graphene films achieved by pre-gelation with polymer nanofibers. Carbon. 2025;244:120655. https://doi.org/https://doi.org/10.1016/j.carbon.2025.120655
12. Abdullah M, Hosain MM, Hassan Parvez MM, Haque Motayed MS. Prospects and challenges of thin film coating materials and their applications. Inorganic Chemistry Communications. 2025;175:114117. https://doi.org/10.1016/j.inoche.2025.114117
13. Bachour Junior B, Batistuti Sawazaki MR, Guidelli ÉJ, Mulato M. Optimizing gold nanoparticles for thin film formation: insights into synthesis and surface functionalization with APTES. Colloids and Surfaces A: Physicochemical and Engineering Aspects. 2025;722:137180. https://doi.org/10.1016/j.colsurfa.2025.137180
14. Potočnik J, Latas N, Milošević M, Janakiev T, Stamenković T, Srejić I. Copper-enhanced TiO2 nanostructured thin films: Structural, chemical, and antibacterial properties. Applied Surface Science. 2025;708:163747. https://doi.org/10.1016/j.apsusc.2025.163747
15. Lin S-H, Wu Y-L, Yu R-S, Liu J-H, Kaneko S, Miao H-Y. Characterization of semiconductor properties and applications of Mg-doped $ZnCo_2O_4$ spinel-structured thin films. Materials Chemistry and Physics. 2025;342:130943. https://doi.org/10.1016/j.matchemphys.2025.130943
16. Ozdemir YB, Okudur OO, Gonzalez M, Merckling C. Physics-informed deep learning approach for nanoindentation-based thin film analysis. Microelectronics Reliability. 2025;173:115875. https://doi.org/10.1016/j.microrel.2025.115875

17. Jahanbakhsh A, Firuznia R, Badkoobeh Hezaveh S, Borzooei M, Ghasemi H. Physics-informed neural network based topology optimization for thin-film evaporation in hierarchical structures. International Journal of Heat and Mass Transfer. 2026;255:127902. https://doi.org/10.1016/j.ijheatmasstransfer.2025.127902
18. Jahanbakhsh A, Firuznia R, Nazifi S, Ghasemi H. Physics-informed neural network for thin-film evaporation in hierarchical structures. International Journal of Heat and Mass Transfer. 2024;224:125296. https://doi.org/10.1016/j.ijheatmasstransfer.2024.125296
19. Cho D-H, Hwang T-H, Chung Y-D, Kim R, Park J, Cho M-H, et al. Thickness-dependent structural evolution and quantum transport properties of Bi_2Se_3 thin films grown by thermal evaporation. Applied Surface Science Advances. 2025;30:100867. https://doi.org/10.1016/j.apsadv.2025.100867
20. Isik M, Surucu O, Bektas T, Parlak M. Cu doping of Sb_2Se_3 thin films via thermal evaporation: Tailoring structural and optical properties for enhanced photovoltaic performance. Optical Materials. 2025;167:117362. https://doi.org/10.1016/j.optmat.2025.117362
21. Rasmi T, Chandra Bose RS, T.S V, Aarya P, Srikanth P, V.K H, et al. Enhanced thermoelectric performance of n-type Bi_2Te_3 thin films via thermal evaporation and optimized post-annealing. Journal of Alloys and Compounds. 2025;1034:181378. https://doi.org/10.1016/j.jallcom.2025.181378
22. Nayak S, Prakash A, R G, G MM. Phase transition and switching behavior in thermally evaporated In_2Se_3 thin films for memory applications. Optical Materials. 2025;167:117353. https://doi.org/10.1016/j.optmat.2025.117353
23. Idoudi M, Akkari FC, Gallas B, Kanzari M. Optical parameters of obliquely thermally vacuum evaporated $CuSbS_2$ thin films. Materials Chemistry and Physics. 2025;345:131233. https://doi.org/10.1016/j.matchemphys.2025.131233
24. Khanlary MR, Rahmani E, Pasalari Z, Reyhani A. Post-annealing effect on the structural and optical properties of tungsten trioxide thin films prepared by thermal evaporation. Materials Chemistry and Physics. 2025;338:130670. https://doi.org/10.1016/j.matchemphys.2025.130670
25. Li D, Zeng L, Zi G, Hong R. Rapid preparation of Cu $(In,Ga)Se_2$ thin films by thermal evaporation. Surfaces and Interfaces. 2025;60:106064. https://doi.org/10.1016/j.surfin.2025.106064
26. Hsu J-C, Chen H-L, Wu H-Y, Sun Y-S. Optical and microstructural properties of gadolinium fluoride thin films fabricated by thermal evaporation with ion assisted deposition using SF_6 as working gas. Materials Today Communications. 2025;43:111841. https://doi.org/10.1016/j.mtcomm.2025.111841
27. Grayeli A, Matos RS, da Fonseca Filho HD, Ferreira NS, Arman A, Rezaee S, et al. Thickness dependent crystallinity, hydrophilicity, and surface microtexture in CaF_2 thin films deposited via thermal evaporation method. Ceramics International. 2025;51(1):267–78. https://doi.org/10.1016/j.ceramint.2024.10.455
28. Kumar A, Kumar P, Dhaliwal AS. Phase evolution and surface analysis of electron beam evaporated calcium and yttria-stabilized zirconia thin films. Thin Solid Films. 2024;808:140575. https://doi.org/10.1016/j.tsf.2024.140575
29. Rampai MM, Mtshali CB, Nemukula E, Seroka NS, Khotseng L. Characterization of ZrC-V-Ti-ZrC multilayer hydrogen storage thin films prepared by e-beam evaporator. International Journal of Hydrogen Energy. 2025;124:18–27. https://doi.org/10.1016/j.ijhydene.2025.04.004
30. Kander NS, Gajar B, Islam S, Moulick S, Guchhait S, Das AK. Evidence of topological surface states in Bi_2Te_3 thin film grown by electron beam evaporator through co-deposition technique. Materials Chemistry and Physics. 2024;326:129790. https://doi.org/10.1016/j.matchemphys.2024.129790
31. Yadav M, Shankar A. Impact of deposition rate on structural, morphological and optical properties of E-beam evaporated chromium thin film. Materials Today: Proceedings. 2024. https://doi.org/10.1016/j.matpr.2024.04.098
32. Das CR, Rangwala M, Ghosh A. Influence of substrate bias voltage on microstructure and mechanical characteristics of TiAlSiN coating deposited by High Power Impulse Magnetron

Sputtering (HiPIMS). Surface and Coatings Technology. 2023;458:129351. https://doi.org/10.1016/j.surfcoat.2023.129351
33. Martínez-González CJ, Bravo-Barcenas D, Pérez-Alvarez J, Jiménez O, Flores M. Mechanical and electrochemical properties of CrAl-CrAlN and CrAl-CrAlN-(a-CNx) multilayer coatings obtained from Cr/Al sectioned targets deposited by HIPIMS. Surface and Coatings Technology. 2025;516:132706. https://doi.org/10.1016/j.surfcoat.2025.132706
34. He Y, Gao K, Yang H, Pang X, Volinsky AA. Nitrogen effects on structure, mechanical and thermal fracture properties of CrN films. Ceramics International. 2021;47(21):30729–40. https://doi.org/10.1016/j.ceramint.2021.07.252
35. Herman J, Hrbek T, Kúš P, Matolínová I. Performance and degradation analysis of sputter deposited thin-film platinum cathodes in PEM water electrolyzers analyzed by electrochemical impedance spectroscopy. Journal of Power Sources. 2025;655:237947. https://doi.org/10.1016/j.jpowsour.2025.237947
36. Schrader JS, Huguenin-Love JL, Soukup RJ, Ianno NJ, Exstrom CL, Darveau SA, et al. Thin films of GeC deposited using a unique hollow cathode sputtering technique. Solar Energy Materials and Solar Cells. 2006;90(15):2338–45. https://doi.org/10.1016/j.solmat.2006.03.007
37. Chen L, Zhu Y, Huang H, Zhou D, Peng K, Wang Y. Process-surface topography-tribological performance of surface micro-textures by nanosecond laser ablation with low-melting alloy coatings. Tribology International. 2026;214:111167. https://doi.org/10.1016/j.triboint.2025.111167
38. Liu J, Liu C, Gao L, Liu S. Energy dissipation mechanism of femtosecond laser ablation of DLC/Ti/TiN multilayer heterogeneous films. International Communications in Heat and Mass Transfer. 2025;169:109822. https://doi.org/10.1016/j.icheatmasstransfer.2025.109822
39. Feng Q, Zhang M, Li Q, Xu W, Wang L, Shi X, et al. Femtosecond laser ablation behavior of diamond/SiC composites in different environments: air and dynamic water. Composites Part A: Applied Science and Manufacturing. 2026;200:109346. https://doi.org/10.1016/j.compositesa.2025.109346
40. O'Connor M, Lackey TG, Tesema N, Johnson AM, Amin MR. KTP Laser Ablation of Benign Vocal Fold Lesions in Performers—Assessing Patient Outcomes. Journal of Voice. 2025. https://doi.org/10.1016/j.jvoice.2025.02.037
41. Giulioni C, Pirola GM, Maggi M, Brocca C, Tramanzoli P, Stramucci S, et al. Current Evidence on Utility, Outcomes, and Limitations of Endoscopic Laser Ablation for Localized Upper Urinary Tract Urothelial Carcinoma: Results from a Scoping Review. European Urology Open Science. 2024;59:7–17. https://dx.doi.org/https://doi.org/10.1016/j.euros.2023.11.005
42. Morimoto A, Asada K, Minamikawa T, Yonezawa Y, Shimizu T. Effect of hot filament on preparation of YBCO superconducting films by pulsed laser ablation in nitrous oxide gas. Thin Solid Films. 2001;395(1):51–4. https://doi.org/10.1016/S0040-6090(01)01206-8
43. Dowding C. 19 - Laser ablation. In: Lawrence J, Pou J, Low DKY, Toyserkani E, editors. Advances in Laser Materials Processing: Woodhead Publishing; 2010. p. 575-628. https://doi.org/10.1533/9781845699819.7.575
44. Lee J, Liu Y, Malhado C, Xu M, Wang K, Qiu Z, et al. MetaMode Deposition of Niobium Oxide Thin Films Using a Broad-Beam Ion Source. Journal of Alloys and Compounds. 2025:184505. https://doi.org/10.1016/j.jallcom.2025.184505
45. Matias V, Hammond RH. Ion beam induced crystalline texturing during thin film deposition. Surface and Coatings Technology. 2015;264:1–8. https://doi.org/10.1016/j.surfcoat.2014.12.018
46. Ma S, Ma S, Ran G. Ion beam irradiation techniques: Enabling the synthesis, modification and enhancement catalytic applications in nanostructured materials. Materials Today Chemistry. 2025;43:102529. https://doi.org/10.1016/j.mtchem.2025.102529
47. Maharana B, Chatterjee S. Ion beam engineering of nanostructured materials for superior performance in supercapacitor applications: Recent progress and future perspectives. Journal of Energy Storage. 2025;109:115039. https://doi.org/10.1016/j.est.2024.115039
48. Gordon S, Rodriguez-Suarez T, Roa JJ, Jiménez-Piqué E, Llanes L. Three-dimensional microstructural characterization of a polycrystalline cubic boron nitride composite by means

of focused ion beam (FIB) tomography. International Journal of Refractory Metals and Hard Materials. 2025;128:107045. https://doi.org/10.1016/j.ijrmhm.2025.107045
49. Jambur V, Wang Z, Sunderland J, Im S, Hu X, Akinyemi S, et al. Ion beam assisted deposition of a thin film metallic glass. Thin Solid Films. 2025;812:140612. https://doi.org/10.1016/j.tsf.2025.140612
50. Sun Y, Liang S, Li Y, Miao J, Liu S, Chen D, et al. Fiber-based anode for lithium metal battery: Ion deposition behavior, interface stabilization mechanisms, and advanced characterization. Journal of Alloys and Compounds. 2025;1010:178118. https://doi.org/10.1016/j.jallcom.2024.178118
51. Geng Y, Xu Y, Li X, Wang X, Wu H, Liu C. Molecular beam epitaxy of flexible InN thin films on fluorophlogopite mica. Journal of Crystal Growth. 2025;671:128355. https://doi.org/10.1016/j.jcrysgro.2025.128355
52. Zhou H, Li J, Wang X, Chen Y, Liu C. Antimony background in molecular beam epitaxy growth of antimonide superlattices. Journal of Crystal Growth. 2025;668:128312. https://doi.org/10.1016/j.jcrysgro.2025.128312
53. Tran M-AN, Nguyen VQ, Nguyen CK, Cho S. Strain modulation in 2D GaSe epitaxial films by substrate engineering via molecular beam epitaxy (MBE). Journal of Crystal Growth. 2025;671:128369. https://doi.org/10.1016/j.jcrysgro.2025.128369
54. Akutsu K, Suzuno M, Kawakami H, Suemasu T. Reduction of carrier concentrations of β-$FeSi_2$ films by atomic hydrogen-assisted molecular beam epitaxy. Physics Procedia. 2011;11:19–22. https://doi.org/10.1016/j.phpro.2011.01.028
55. Majchrzak D, Grodzicki M, Moszak K, Zdanowicz E, Serafińczuk J, Pucicki D, et al. Influence of pulsed Al deposition on quality of Al-rich Al(Ga)N structures grown by molecular beam epitaxy. Surfaces and Interfaces. 2021;27:101560. https://doi.org/10.1016/j.surfin.2021.101560
56. Kaspi R, Lu C. Monolayer oscillations of the Sb desorption rate during molecular beam epitaxy of GaSb. Journal of Crystal Growth. 2021;568–569:126179. https://doi.org/10.1016/j.jcrysgro.2021.126179
57. Zaera F. Mechanisms of surface reactions in thin solid film chemical deposition processes. Coordination Chemistry Reviews. 2013;257(23):3177–91. https://doi.org/10.1016/j.ccr.2013.04.006
58. Akamatsu K, Suzuki J, Nakao S-i, Wang X-l. Trimethylmethoxysilane-derived silica membranes with high hydrogen permeance and high hydrogen/nitrogen selectivity prepared by chemical vapor deposition. International Journal of Hydrogen Energy. 2025;183:151876. https://doi.org/10.1016/j.ijhydene.2025.151876
59. Winter CH, Upadhyay A, Overbeek M, Hollin J, Cwik S. 5.14 - Group 7 and 8 Compounds for Chemical Vapor Deposition. In: Constable EC, Parkin G, Que Jr L, editors. Comprehensive Coordination Chemistry III. Oxford: Elsevier; 2021. p. 824–41. https://doi.org/10.1016/B978-0-12-409547-2.14951-0
60. Marumo NM, Selahle SK, Kabir A, Nomngongo PN. Innovative sol-gel titania and silica-based membranes for fabric phase sorptive extraction of sulfonamides in environmental samples. Microchemical Journal. 2025;217:115106. https://doi.org/10.1016/j.microc.2025.115106
61. Ma Y, Li W, Qi W, Zhong Y, Liu Y, Zhang H, et al. A refined method for assessing the sodium silicate sol-gel rate during silica aerogel preparation: The influence of Ca_2+ as an application example. Colloids and Surfaces A: Physicochemical and Engineering Aspects. 2025;727:138131. https://doi.org/10.1016/j.colsurfa.2025.138131
62. Wang B, Cheng Z, Xi F, Li S, Lu J, Chen X, et al. Recent advances in interfacial engineering of lithium battery anodes by atomic layer deposition technology. Chinese Chemical Letters. 2025:111973. https://doi.org/10.1016/j.cclet.2025.111973
63. Leskelä M, Ritala M. Atomic layer deposition (ALD): from precursors to thin film structures. Thin Solid Films. 2002;409(1):138-46. https://doi.org/10.1016/S0040-6090(02)00117-7
64. Justin Kunene T, Kwanda Tartibu L, Ukoba K, Jen T-C. Review of atomic layer deposition process, application and modeling tools. Materials Today: Proceedings. 2022;62:S95–S109. https://doi.org/10.1016/j.matpr.2022.02.094

65. Oliveira MP, Silva RM, Yasakau KA, Bastos A, Kallip S, Zheludkevich ML, et al. Atomic layer deposition of nanometric alumina for corrosion protection of heterogeneous metallic surfaces – The case of aeronautical grade aluminium alloy 2024-T3. Corrosion Science. 2022;209:110773. https://doi.org/10.1016/j.corsci.2022.110773
66. Kaindl R, Homola T, Rastelli A, Schwarz A, Tarre A, Kopp D, et al. Atomic layer deposition of oxide coatings on porous metal and polymer structures fabricated by additive manufacturing methods (laser-based powder bed fusion, material extrusion, material jetting). Surfaces and Interfaces. 2022;34:102361. https://doi.org/10.1016/j.surfin.2022.102361
67. Huang Y, Li T, He X, Lang Y, Ning Y, Dong H. Multi-Objective optimization of electroplating parameters for low-passive intermodulation radio frequency connectors. Results in Engineering. 2025;28:107509. https://doi.org/10.1016/j.rineng.2025.107509
68. Cheng L, Sun Y, Deng H, Kang X. A convenient electroplating method for emitter wetting of indium FEEP. Materials Today Communications. 2025;49:113770.
69. Giurlani W, Zangari G, Gambinossi F, Passaponti M, Salvietti E, di Benedetto F, et al. Electroplating for Decorative Applications: Recent Trends in Research and Development. Coatings. 2018. https://doi.org/10.1016/j.mtcomm.2025.113770
70. Rytlewski P, Augustyn P, Malinowski R, Budner B, Antończak A. Surface metallisation of acrylonitrile–butadiene–styrene (ABS) composites via a novel electroplating technique. Applied Surface Science. 2025;712:164146. https://doi.org/10.1016/j.apsusc.2025.164146
71. Arepalli VK, Yang E, Patil AA, Wi J-S, Park JS, Lee J-M, et al. ZnO nanowire broadband ultra-wide-angle optical diffusers grown by aqueous chemical bath deposition. Journal of Alloys and Compounds. 2024;1008:176660. https://doi.org/10.1016/j.jallcom.2024.176660
72. Dong R, Tan J, Ji Y, Qi Z, Chen Z, Chen Y. Synthesis of zinc hydroxyl acetate nanosheet-assembled honeycomb-like and flower-like structures by chemical bath deposition method for the construction of hydrophobic surfaces. Thin Solid Films. 2024;808:140569. https://doi.org/10.1016/j.tsf.2024.140569
73. Sengupta S, Grover R. Chemical bath deposition-based synthesis of tin sulfide on silicon substrate and its application as a photodetector. Physica B: Condensed Matter. 2024;679:415789. https://doi.org/10.1016/j.physb.2024.415789
74. Rahmawati AI, Wirzal MDH, Sufian S, Abdul Aziz MS, Salleh MS, Widayatno T, et al. Iron dopant in zinc oxide nanorods-based photoanode using chemical bath deposition method for photoelectrochemical water-splitting. International Journal of Hydrogen Energy. 2024;72:422–36. https://doi.org/10.1016/j.ijhydene.2024.05.296
75. Siyu G, Qihao G, Bingguo L, Guangxiong J, Wang C, Peng L, et al. Synergistic spray pelletizing-microwave pyrolysis for morphology-controlled ruthenium powder synthesis: Kinetics and precursor engineering. Journal of Analytical and Applied Pyrolysis. 2025;192:107338. https://doi.org/10.1016/j.jaap.2025.107338
76. Chauhan AS, Dipak P, Gahlaut UPS, Goswami YC. Unveiling the gas sensing potential of copper-doped zinc oxide thin films deposition by spray pyrolysis. Sustainable Chemistry One World. 2025;8:100123. https://doi.org/10.1016/j.scowo.2025.100123
77. LÜ X, Hu J, Hou J, Wang J, Bao W, Chang L. Study on the synthesis of CeO_2-Co_3O_4 catalyst by flame spray pyrolysis method and its catalytic oxidation performance for ethane. Journal of Fuel Chemistry and Technology. 2025;53(10):1470–81. https://doi.org/10.1016/S1872-5813(25)60548-2
78. Zala P, Panchal K, Shah V, Zala D, Ray A. Pyrolysis temperature-controlled structure and properties of spray deposited n-type Sb_2S_3 thin films for photoelectrochemical applications. Physica B: Condensed Matter. 2025;716:417710. https://doi.org/10.1016/j.physb.2025.417710
79. Abed Al-Jubbori M, Sinjar MS, saleh A. Chemical spray pyrolysis of cadmium oxide thin films: UVC-induced structural and optical modifications for enhanced photodetector performance. Optical Materials. 2025;168:117430. https://doi.org/10.1016/j.optmat.2025.117430
80. Boon-on P, Raksa P, Ponhan W, Chaiworn P, Tubtimtae A. Effect of Sn concentration on antimony oxide thin films via using spin coating method: Analysis of structural and optical properties. Materials Letters. 2026;404:139579. https://doi.org/10.1016/j.matlet.2025.139579

81. Wang Y, Qi M, Xu H, Huang M, Liang J, Yang L, et al. Structural color patterns derived from photonic crystals with high saturation and brightness using spin coating method. Optics Communications. 2025;596:132380. https://doi.org/10.1016/j.optcom.2025.132380
82. El Berjali W, Colas V, Ould Saad Hamady S, Ng SS, Hamzah NA, Boulet P, et al. Copper oxide by spin-coating: Cost-effective deposition and post-annealing process for thin films with modulated Cu_2O/CuO phase ratio. Thin Solid Films. 2025;826:140762. https://doi.org/10.1016/j.tsf.2025.140762
83. Ţălu Ş. Morphological study of Cu_2CdSnS_4 nanostructures synthesized via spin coating method at different temperatures. Materials Today Communications. 2025;48:113696. https://doi.org/10.1016/j.mtcomm.2025.113696
84. Wang S, Xie N, Ma X, Chen S, Liu H, Ma R, et al. Study on the self-healing behavior and corrosion mechanism of hot-dipping Zn-6Al-3Mg alloy coatings in corrosive environments. Surfaces and Interfaces. 2025;74:107714. https://doi.org/10.1016/j.surfin.2025.107714
85. Yu Y, Shi Q, Qiao D, Du X, Zhao X, Wang J, et al. Effect of steel substrate roughness on microstructure and properties of hot-dip Zn–Al–Mg coating. Journal of Materials Research and Technology. 2025;39:1877–93. https://doi.org/10.1016/j.jmrt.2025.09.233
86. Sowmya S, Sivabharathy M, Padmaja S, Babu SM. A comprehensive guide to enhance elemental properties of $AgGa0.5In0.5Se_2$ thin films via dip coating for solar cell applications. Surfaces and Interfaces. 2025;72:107454. https://doi.org/10.1016/j.surfin.2025.107454
87. Hmud K, Kannan AM, Bootchanont A, Pavasupree S, Ponchio C. Advanced automated dip-coating strategy for high-performance $WO_3/BiVO_4$ photoanodes in durable photoelectrocatalytic water splitting and organic dye degradation. International Journal of Hydrogen Energy. 2025;150:150143. https://doi.org/10.1016/j.ijhydene.2025.150143
88. Ahn J-H, Roh M-J, Jeon E-J, Lee C-Y. Development of high-sensitivity multi-strain sensors using EHD inkjet printing for structural health monitoring in advanced air mobility. Sensors and Actuators A: Physical. 2025;395:116987. https://doi.org/10.1016/j.sna.2025.116987
89. Wang S, Shen H, Ren Y, Zhao M, Jin Z, Wang C. Numerical investigation of deposition formation in inkjet printing of phase change material droplets. International Journal of Heat and Fluid Flow. 2025;116:109940. https://doi.org/10.1016/j.ijheatfluidflow.2025.109940
90. Tekoglu E, Liao S, Kutschke Z, O'Brien AD, Lettiere B, Li J, et al. Rapid exploration of nanoparticle-modified alloys in metal additive manufacturing by combining inkjet printing and laser powder bed fusion. Additive Manufacturing Letters. 2025;14:100315. https://doi.org/10.1016/j.addlet.2025.100315

Chapter 4
Magnetoelectric Coupling in Thin Films

Pawan Kumar

4.1 Introduction

Magnetoelectric (ME) coupling, the phenomenon in which magnetic properties can be controlled by electric fields and electric polarization can be tuned by magnetic fields, has emerged as one of the most fascinating topics in condensed matter physics and materials science. The importance of ME coupling extends beyond mere curiosity; it also has significant implications for developing multitasking gadgets such as sensors, actuators, spintronic components, and next-generation memories [1, 2]. Over the past twenty years, thin-film designs have been central to this field because they provide specialized platforms to investigate, control, and benefit from ME coupling at the nanoscale [3]. The study of ME effects began in the mid-twentieth century. Shortly, after Astrov [5] confirmed Dzyaloshinskii's ([4] theoretical prediction of ME coupling in certain antiferromagnetic materials experimentally—specifically in chromium oxide (Cr_2O_3)—these groundbreaking works laid the foundation for systematic research into the ME effect. However, the weak coupling coefficients in most bulk materials limit their technological potential. For many years, the ME effect was mainly regarded as a scientific curiosity with limited practical application due to material constraints [6].

The renaissance of the field occurred with the discovery and exploration of multiferroics, i.e., materials that exhibit more than one ferroic order parameter such as ferroelectricity, ferromagnetism, and ferroelasticity within a single phase [7]. The coexistence of these orderings creates the possibility of coupling between electric and magnetic degrees of freedom, offering an intrinsic pathway for ME effects. Among single-phase multiferroics, $BiFeO_3$ has attracted particular attention due to its robust ferroelectric polarization (~90 $\mu C/cm^2$) and antiferromagnetic ordering above room temperature [8]. Intrinsic single-phase multiferroics, however, are rather uncommon

P. Kumar (✉)
School of Basic and Applied Sciences, K. R. Mangalam University, Gurugram, Haryana, India

S. Sharma (ed.), *Multifunctional Heterostructure Thin Films Handbook*,
https://doi.org/10.1007/978-3-032-18692-8_4

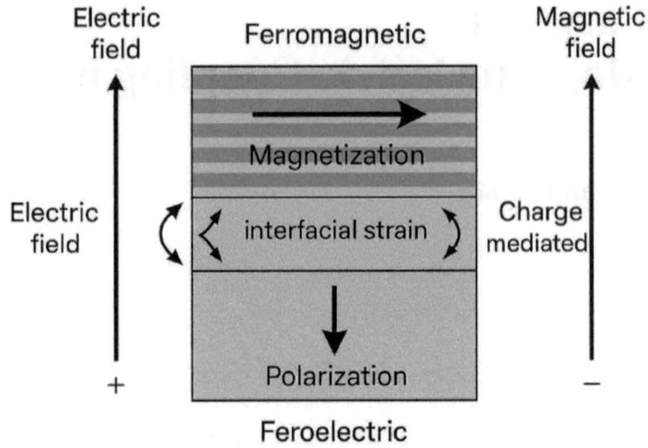

Fig. 4.1 Magnetoelectric (ME) coupling in thin film heterostructures

and frequently have trade-offs like low ordering temperatures or weak coupling despite these developments.

Thin films and engineered heterostructures became the main focus to address these problems. There are several benefits of thin films: They allow accurate strain tuning via epitaxial growth, enable atomic-level control over interfaces, and allow integration with semiconductor technologies [9]. Moreover, composite thin films, combining ferroelectric and ferromagnetic constituents, can generate strong ME effects via strain-mediated or charge-mediated coupling across interfaces [10]. These heterostructures circumvent the intrinsic limitations of single-phase materials and provide design flexibility for enhancing ME coupling as shown in Fig. 4.1.

The fabrication of high-quality thin films has been made possible by advanced deposition techniques such as pulsed laser deposition (PLD), molecular beam epitaxy (MBE), sputtering, and atomic layer deposition (ALD). These methods not only enable precise stoichiometric control but also allow the exploration of novel phases stabilized only in thin-film form [11]. Epitaxial strain, for instance, can stabilize ferroelectricity in otherwise non-ferroelectric perovskites, therefore creating opportunities for finding new qualities in thin films [12].

Technologically, ME thin films show promise for low-power spintronic systems, especially in the creation of magnetoelectric random access memory (ME-RAM). Such gadgets use ferroelectric switching [13], which combines the non-volatility of magnetic memory with its low energy cost, where information is written electronically but read magnetically. The possibility of such methods and the great device-oriented ME thin film research were fueled by a landmark demonstration of deterministic, electric-field-driven switching of ferromagnetism in a $BiFeO_3$-based heterostructure [14].

ME thin films are also being used in nanoscale sensors and transducers, not just as memory aids. For instance, ME composites have shown remarkable sensitivity to magnetic fields, so allowing the development of tiny magnetic sensors running without the need for big biasing magnets [10]. In layered heterostructures, the union of piezoelectric and magnetostrictive thin films offers effective routes for switching between electric and magnetic signals, which is much sought-after for energy gathering techniques and wireless communication.

Theoretically, our knowledge of ME coupling has greatly grown. Early phenomenological models gave a macroscopic account of coupling mechanisms derived from Landau–Ginzburg–Devonshire (LGD) theory [12]. Concurrent with these macroscopic models, microscopic ones have found that ME phenomena depend mostly on spin–orbit coupling [15], exchange striction [16], and charge/orbital hybridization [9]. In thin films, these observations are especially pertinent since interfaces, decreased dimensionality, and lattice strain offer fresh degrees of freedom that may be used to improve coupling.

New innovations have also stressed how topological effects and non-collinear spin textures such as skyrmions help to create original ME reactions [17]. These developments emphasize the growing breadth of ME thin film research, which now crosses over other advanced issues including oxide electronics, 2D materials, and quantum technologies.

Many obstacles still exist despite quick advancement. Achieving room temperature operation with strong and reversible ME coupling is still an ongoing pursuit. Reliable device integration depends on defect control, interfacial chemistry, and leakage currents in thin films. Furthermore, scalability and compatibility with CMOS processes are crucial for translating laboratory-scale demonstrations into commercial products.

In summary, magnetoelectric coupling in thin films represents a vibrant research frontier that bridges fundamental physics with technological applications. The thin-film approach provides unparalleled opportunities for engineering ME effects by exploiting strain, interfaces, and dimensionality. With ongoing advances in synthesis, characterization, and theoretical modeling, ME thin films are poised to play a transformative role in future multifunctional devices. This chapter aims to provide a comprehensive overview of the field, beginning with fundamental concepts and mechanisms, followed by thin film growth strategies, characterization techniques, and finally, potential applications in emerging technologies. Timeline of key developments in magnetoelectric coupling shown in Fig. 4.2.

4.2 Fundamentals of Magnetoelectric Coupling

Magnetoelectric (ME) coupling denotes the physical linkage between electric and magnetic order parameters: An applied electric field modifies magnetic order and, conversely, an applied magnetic field modifies electric polarization. In its simplest linear form the effect is expressed by the relations

$$P_i = \alpha_{ij} H_j, \quad M_i = \alpha_{ij} E_j,$$

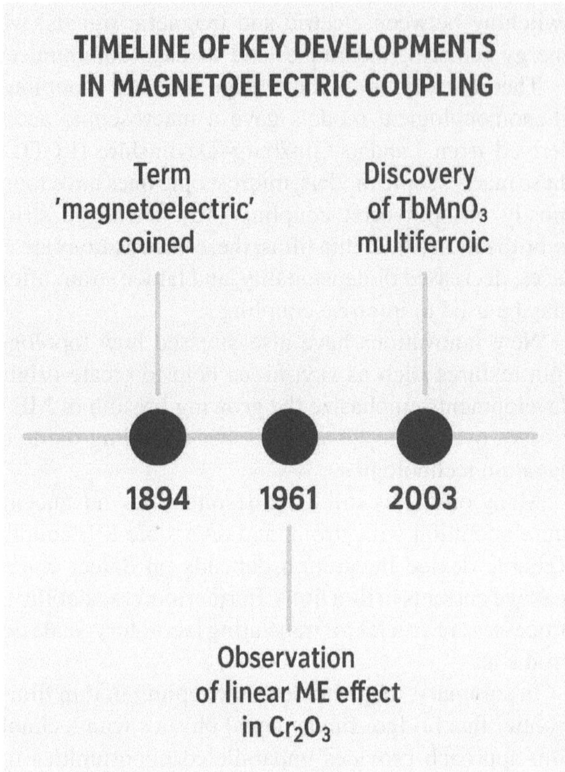

Fig. 4.2 Timeline of key developments in magnetoelectric coupling

where P is polarization, M magnetization, E electric field, H magnetic field, and α_{ij} the magnetoelectric tensor (units: SI-s·m^{-1}). The tensorial character of α encodes crystallographic and magnetic symmetry and therefore determines which components of the electric and magnetic fields can couple [1, 3].

4.2.1 Linear Versus Nonlinear Effects; Direct and Converse ME Responses

ME coupling can be linear (first-order in fields) or nonlinear (higher-order terms). The free-energy expansion of a magnetoelectric medium helps clarify this:

$$F = F_0 - P_i E_i - M_j H_j - \alpha_{ij} E_i H_j - \frac{1}{2}\beta_{ijk} E_i H_j H_k - \frac{1}{2}\beta_{ijk} H_i E_j E_k + \ldots,$$

Linear coupling (α) yields direct and reversible control and is preferred for devices. Nonlinear terms (β,) become relevant in many composite and interfacial systems, particularly where symmetry-allowed linear coupling is weak or absent [2].

4.2.2 Symmetry Constraints and Selection Rules

Symmetry is the principal filter for magnetoelectricity. A material must break space-inversion symmetry (to permit a polar axis) and time-reversal symmetry (to permit magnetic order). Historically, Dzyaloshinskii predicted such cross-coupling on symmetry grounds [4] and Astrov provided the first experimental confirmation in Cr_2O_3 [5]. The magnetic point group of a crystal determines which α_{ij} components are nonzero; Neumann's principle implies that physical tensors must be invariant under the symmetry operations of the crystal's point group [1]. In thin films, symmetry can be further reduced by epitaxial strain, interface reconstructions, and structural distortions, thereby enabling ME responses that are forbidden in the corresponding bulk crystals [12].

4.2.3 Microscopic Mechanisms

At the microscopic level, several mechanisms give rise to ME coupling; these often coexist and can be tuned in thin films:

Spin–orbit coupling (SOC): SOC links the spin orientation to lattice and orbital degrees of freedom. In non-centrosymmetric systems, SOC can translate electric-field-induced modifications in orbital occupation into changes of magnetic anisotropy [15].

Exchange striction: Here, magnetic exchange interactions depend on interatomic separations. A change in magnetic order can cause lattice displacements and polarization; similarly, an applied electric field that moves ions alters exchange constants [16].

Non-collinear spin textures (spiral or cycloidal orders) create polarization via anti-symmetric exchange; this mechanism underpins multiferroicity in some manganites [15].

Polarization-bound charge at boundaries between a ferroelectric and a correlated magnetic oxide modifies carrier density and band occupation in the magnetic layer, therefore altering exchange interactions and magnetization, hence generating charge-mediated (electronic reconstruction). In oxide heterostructures with small screening lengths [18], this process is particularly strong.

Electric-field-induced strain in a piezoelectric layer is conveyed to a magnetostrictive layer in composites and heterostructures, so changing its magnetization. Engineered thin films greatly use this extrinsic path to get high effective ME coefficients [10].

Deliberately enhancing specific processes is made possible by thin films; hence, they have particular benefits. For example, epitaxial strain can amplify lattice distortions that couple strongly to magnetism; engineered polar discontinuities at interfaces can maximize charge-mediated effects; and vertical nanocomposites can increase the interfacial area for strain transfer [8, 10].

4.2.4 Phenomenological and Microscopic Models

Phenomenological approaches such as Landau–Ginzburg–Devonshire (LGD) theory are commonly used to model ME coupling in thin films. LGD free energies include order-parameter terms (polarization P, magnetization M), elastic energy, gradient terms for domain walls, and explicit $P{-}M$ coupling terms; boundary conditions (mechanical) and clamping, depolarization fields must be employed to capture thin-film behavior [12]. First-principles density functional theory (DFT) complements LGD by providing microscopic parameters—e.g., the dependence of exchange integrals or anisotropy on lattice distortions and electronic occupancy—and has been essential in predicting strain-stabilized phases and interfacial magnetism [3].

4.2.5 Figures of Merit and Measurement Approaches

The key figure of merit is the ME tensor α_{ij}. Experimental conventions vary: Some report the magnetoelectric voltage coefficient (e.g., mV cm^{-1} Oe^{-1}), while SI-consistent literature reports α in s·m^{-1} [2]. Measurement techniques include direct methods (apply H, measure induced P or charge), converse methods (apply E, detect induced magnetization or magnetic anisotropy), dynamic lock-in approaches (to improve signal-to-noise), magnetocapacitance (change of dielectric constant with H), and optical probes for ultra-fast dynamics. In thin films, small sample volume and substrate contributions complicate measurements; hence, sensitive methods such as second-harmonic generation, X-ray magnetic circular dichroism (XMCD) under applied E, and polarized neutron reflectometry are often employed to obtain element-specific and depth-resolved information [2, 18].

4.2.6 Thin-Film Specific Considerations

Thin films introduce phenomena absent in bulk: Substrate clamping suppresses or modifies strain relaxation pathways; depolarization fields and finite screening reduce effective polarization unless compensated by electrodes or interfacial charge; oxygen vacancies and interfacial chemical mixing can drastically alter both charge-mediated and strain mechanisms. Domain structures and domain-wall contributions become pronounced at the nanoscale, and domain walls themselves may carry enhanced conductivity or magnetoelectric activity [8]. Furthermore, thickness-dependent phase transitions (e.g., strain-driven morphotropic-like boundaries) offer a powerful knob to tune ME responses but also require precise thickness and stoichiometry control [8, 12].

4.2.7 Summary

The fundamentals of magnetoelectric coupling weave together symmetry, thermodynamics, microscopic exchange and orbital physics, and mesoscale elasticity. Thin films enrich this tapestry by providing levers—strain, interfaces, dimensional confinement—to strengthen, tune, or even create ME coupling that is weak or absent in the bulk. A close interplay between phenomenological modeling, first-principles calculations, and advanced characterization is essential to both understand and engineer ME effects in nanoscale heterostructures. Figure 4.3 shows a conceptual illustration of linear and higher-order magnetoelectric coupling and the role of symmetry in determining allowed alpha components.

4.3 Thin Film Growth and Fabrication Techniques

Thin film synthesis is one of the most critical aspects of magnetoelectric (ME) research because the growth route strongly determines crystallinity, stoichiometry, interfacial quality, and defect concentration, all of which influence coupling mechanisms. The fabrication of thin films allows control over dimensional confinement, epitaxial strain, and chemical composition—levers that can either enhance or suppress ME responses. This section surveys the most widely employed deposition techniques, growth modes, and structural challenges, while highlighting their relevance for ME coupling.

4.3.1 Growth Modes and Thermodynamic Considerations

Thin films typically grow following one of three classical modes: (i) Frank–van der Merwe (layer-by-layer), producing atomically smooth films with sharp interfaces; (ii) Volmer–Weber (island growth), resulting in discontinuous films that may require post-deposition annealing; and (iii) Stranski–Krastanov (layer plus island), a hybrid that initiates with smooth layers but relaxes via island formation due to strain accumulation [19].

Particularly perovskite oxides like $BiFeO_3$, ME thin films are very sensitive to growth method since ferroelectric and magnetic order parameters are intimately

Fig. 4.3 Conceptual illustration of linear and higher-order magnetoelectric coupling and the role of symmetry in determining allowed α_{ij} components

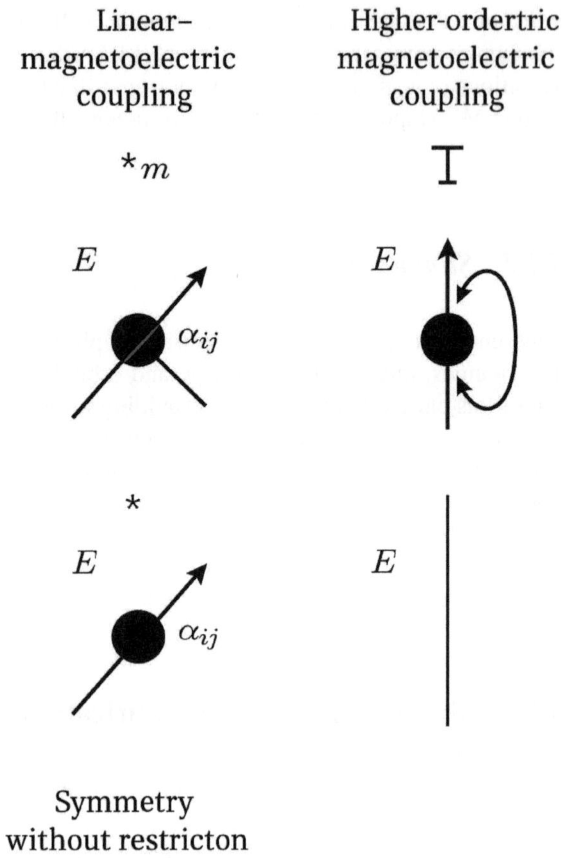

related to strain and defect levels. For instance, while islanded development can produce roughness that diminishes order [8], coherent epitaxial growth on lattice-mismatched substrates can stabilize metastable phases with increased polarization or magnetization.

Nucleation kinetics and defect densities are determined by thermodynamic factors including deposition rate, oxygen partial pressure, and substrate temperature. In perovskite oxides, oxygen stoichiometry is crucial: Oxygen vacancies can screen polarization, alter cation valence states, and in turn modify magnetic order [20].

4.3.2 Deposition Techniques for ME Thin Films

Pulsed Laser Deposition (PLD)

PLD is among the most commonly used methods for complex oxide ME films. In PLD, a high-energy laser ablates a stoichiometric target, generating a plasma plume that deposits material on a heated substrate. PLD's advantages include stoichiometric transfer from target to film, relatively fast growth, and versatility for multi-component oxides [21].

For $BiFeO_3$, PLD has been instrumental in achieving epitaxial films with robust ferroelectric polarization and weak ferromagnetism [8]. However, plume dynamics and energetic species can induce non-equilibrium defects, and uniformity over large areas remains challenging.

Molecular Beam Epitaxy (MBE)

With ultra-high vacuum conditions, MBE offers atomic-layer control over growth, therefore suitable for creating precisely calibrated interfaces. Traditional MBE (for metallic ferromagnets) and oxide MBE (using oxygen plasma or ozone) are both used. MBE allows fine-tuned cation stoichiometry and doping, which is essential for customizing magnetoelectric properties.

La-substituted $BiFeO_3$ films made by oxide MBE have shown, for instance, regulated strain states and decreased leakage currents [22]. Limitations include slow deposition rates and high system costs, but the unparalleled control of MBE has made it invaluable for fundamental studies of ME interfaces.

Sputtering

RF magnetron sputtering and off-axis sputtering are attractive for large-area growth, scalable device fabrication, and industrial compatibility. In sputtering, energetic ions (often Ar^+) dislodge atoms from a target, which then deposit onto a substrate. While stoichiometry control in multi-component oxides is more difficult than in PLD, sputtering excels in producing uniform, large-area films [23].

ME heterostructures such as PZT/$CoFe_2O_4$ composites have been fabricated by sputtering, showing strain-mediated coupling [10]. Optimization of substrate bias, oxygen pressure, and post-deposition annealing is critical to minimize defects.

Chemical Solution Deposition (CSD)

CSD, including sol–gel spin coating and chemical vapor deposition (CVD), offers cost-effectiveness and compositional versatility. Metal-organic precursors are spun, dried, and annealed in sol–gel techniques to produce crystalline films. Though achieving epitaxial quality is more challenging [24], CSD is well suited for compositional tuning and multilayer architectures.

For magnetoelectric composites, CSD offers a scalable route to multilayer PZT/ $NiFe_2O_4$ systems [25]. CVD variations like metal-organic CVD (MOCVD) enable high-quality epitaxial oxide films but call for meticulous precursor design.

Atomic Layer Deposition (ALD)

ALD employs self-limiting surface reactions, yielding conformal films with sub-nanometer thickness precision. This is particularly advantageous for ultra-thin ME films and 3D nanostructured templates. ALD has been applied to deposit dielectric ferroelectric layers (e.g., HfO_2-based ferroelectrics) integrated with ferromagnetic layers [26]. The main challenge is the limited availability of volatile precursors for complex oxides.

4.3.3 Substrate Selection and Strain Engineering

The choice of substrate profoundly influences ME thin films. Lattice mismatch imposes biaxial strain, which can stabilize otherwise unstable phases [12]. For instance, epitaxial strain in $BiFeO_3$ films on $SrTiO_3$ substrates enhances polarization and alters magnetic anisotropy [8]. Substrate symmetry also matters: Perovskite substrates ($SrTiO_3$, $LaAlO_3$, $DyScO_3$) transmit different strain states and impose orientation constraints.

Buffer layers are often introduced to mediate mismatch or suppress interdiffusion. Conducting oxides like $SrRuO_3$ serve as bottom electrodes while maintaining epitaxial continuity.

4.3.4 Interfaces, Defects, and Stoichiometry

Interfaces define ME heterostructures. Sharp, chemically abrupt interfaces maximize charge-mediated coupling, while intermixing or roughness reduces it. Defects such as oxygen vacancies, dislocations, and cation intermixing act as pinning centers, degrade ferroelectricity, and alter exchange interactions [20].

Advanced techniques such as reflection high-energy electron diffraction (RHEED), scanning transmission electron microscopy (STEM), and atom probe tomography are indispensable for monitoring interface quality.

4.3.5 *Nanostructuring Approaches: Vertical Heterostructures and Composites*

Beyond planar bilayers, vertical nanocomposites (e.g., $CoFe_2O_4$ nanopillars embedded in a $BaTiO_3$ matrix) have been developed using self-assembly during PLD or sputtering. These structures maximize interfacial area for strain coupling and circumvent substrate clamping [10]. Similarly, multilayer stacks fabricated by repeated deposition cycles allow systematic tuning of ME coupling.

These architectures illustrate the creative potential of thin-film growth: By designing geometry at the nanoscale, researchers can amplify specific coupling mechanisms Table 4.1.

Table 4.1 Comparative analysis of thin film fabrication techniques

Technique	Principle	Advantages	Limitations	Typical applications
Pulsed laser deposition (PLD)	Laser ablation of target material onto substrate	High-quality epitaxial films, precise stoichiometry	Limited uniformity over large areas, expensive	Multiferroic thin films, ME heterostructures
Molecular beam epitaxy (MBE)	Atom-by-atom deposition in ultra-high vacuum	Atomic-level control, high purity	Slow growth rate, costly, UHV requirement	Fundamental studies, quantum ME interfaces
Sputtering	Target atoms ejected by ion bombardment	Large-area deposition, scalable	Poor stoichiometry control, lower crystallinity	Sensors, device prototypes
Chemical solution deposition	Solution-based precursor deposited and annealed	Low cost, simple setup, flexible substrates	Film thickness control, crystallinity variability	Flexible devices, lab-scale ME films
Atomic layer deposition (ALD)	Sequential surface reactions for monolayer growth	Excellent thickness control, conformal coating	Slow deposition, limited material options	Nanoscale devices, ultrathin barriers
Sol–gel technique	Hydrolysis and condensation of metal alkoxides	Low cost, compositional flexibility	Cracking during drying, annealing required	Lab-scale ME composites, spintronics prototypes

4.3.6 *Summary*

Thin-film growth is both an art and a science: The choice of technique, substrate, and growth conditions determines the balance between crystallinity, strain, stoichiometry, and interfacial integrity. For ME thin films, growth control is tantamount to property control. By refining deposition parameters and leveraging nanostructuring strategies, researchers have created heterostructures that exhibit coupling strengths far exceeding bulk values. Growth science thus remains central to advancing magnetoelectric devices.

4.4 Interface Engineering and Strain Effects in Thin Films

4.4.1 *Introduction*

In thin film systems, the interface is often more influential than the bulk material in determining functional properties. The broken symmetry, lattice mismatch, and chemical discontinuities at interfaces can profoundly impact both ferroelectric and magnetic order parameters. Similarly, strain induced by lattice mismatch or external stress plays a decisive role in tuning the strength of magnetoelectric (ME) coupling. These effects are particularly relevant in multiferroic heterostructures where coupling mechanisms—such as strain-mediated, charge-mediated, and exchange-mediated interactions—emerge primarily at the interface [27].

Thus, understanding and engineering interfacial and strain effects are critical for optimizing device performance. This section explores the physics of interface phenomena, epitaxial strain, defect chemistry, and novel approaches to strain engineering.

4.4.2 *Interface Engineering in Magnetoelectric Thin Films*

Electronic Reconstruction at Interfaces

At oxide interfaces, polar discontinuities can give rise to electronic reconstruction, where charge carriers redistribute to minimize electrostatic potential. This phenomenon can create conductive two-dimensional electron gases (2DEGs) at otherwise insulating interfaces, such as $LaAlO_3/SrTiO_3$ [28]. In ME thin films, such charge rearrangements modulate both ferroelectric polarization and magnetic ordering, thereby enhancing cross-coupling.

Exchange Bias at Ferromagnet/Antiferromagnet Interfaces

Coupling between ferromagnetic and antiferromagnetic layers can induce exchange bias, which manifests as a shift in the magnetic hysteresis loop. In composite ME thin films, ferroelectric polarization can modulate exchange bias via interface charge transfer or structural distortion [29].

Chemical and Structural Compatibility

High-quality ME heterostructures require chemical stability at the interface. Interdiffusion of cations, oxygen vacancy migration, or secondary phase formation can degrade properties. Interface engineering strategies such as inserting buffer layers, controlling oxygen pressure during growth, and atomic-layer epitaxy have been successfully employed to mitigate such challenges [30].

4.4.3 Strain Engineering and its Role in ME Coupling

Strain-Induced Ferroelectricity

In certain perovskite oxides, epitaxial strain can stabilize ferroelectric phases not present in the bulk. For example, $SrTiO_3$, normally paraelectric at room temperature, becomes ferroelectric under compressive strain due to softening of polar phonon modes [31]. Such strain-induced ferroelectricity provides additional pathways for ME coupling.

Modulation of Magnetic Anisotropy

Strain directly influences magnetic anisotropy by altering bond lengths and angles in transition-metal oxides. Tensile strain can enhance in-plane magnetic anisotropy, while compressive strain favors out-of-plane magnetization. Coupled with ferroelectric distortion, this tunability allows for strong ME coupling in heterostructures [6].

Strain-Mediated Magnetoelectric Coupling

In artificial multiferroic composites, strain is the dominant coupling mechanism. Electric-field-induced piezoelectric strain in the ferroelectric layer is transferred to the magnetostrictive layer, modulating its magnetic state. This mechanism underpins many device prototypes, including strain-mediated magnetic memory and voltage-tunable microwave devices [10].

4.4.4 Approaches to Interface and Strain Engineering

Substrate Selection: Choice of substrate with appropriate lattice parameters is critical. For instance, $DyScO_3$ and $LaAlO_3$ substrates induce distinct strain states in $BiFeO_3$ films, modifying their ferroelectric domain structure [32].

Superlattice Design: Alternating ferroelectric and ferromagnetic layers in superlattices enable exact control of interfacial stress and charge. Customized periodicity can improve ME responses [13].

Application of external stress, hydrostatic pressure, or flexible substrates helps dynamic strain tuning. Reversible electric-field management of magnetization [14] has been accomplished with this approach.

Controlled addition of oxygen vacancies or cation substitution changes local strain fields and interfacial polarization, therefore affecting ME behavior.

4.4.5 Case Studies

$BiFeO_3$ Thin Films: Epitaxial $BiFeO_3$ on $SrTiO_3$ shows enhanced polarization and ferromagnetic order due to compressive strain, making it a model system for strain-engineered ME coupling [8].

$CoFe_2O_4/BaTiO_3$ Heterostructures: Strong strain-mediated coupling between the piezoelectric $BaTiO_3$ substrate and the magnetostrictive $CoFe_2O_4$ layer allows electric-field control of magnetization [33].

$La_{0.7}Sr_{0.3}MnO_3/BaTiO_3$ Systems: Magnetoresistance modulation via ferroelectric strain demonstrates potential for spintronic devices [34].

4.4.6 Challenges and Future Directions

While interface and strain engineering has enabled breakthroughs in ME thin films, challenges remain:

- Maintaining chemical sharpness at interfaces during high-temperature growth.
- Balancing strain to avoid relaxation or dislocation formation in thicker films.
- Scaling up epitaxial strain strategies for large-area device fabrication.

Future advances may involve adaptive heterostructures where strain and interfacial properties are dynamically tunable, possibly aided by machine learning-guided materials design [35] (Fig. 4.4).

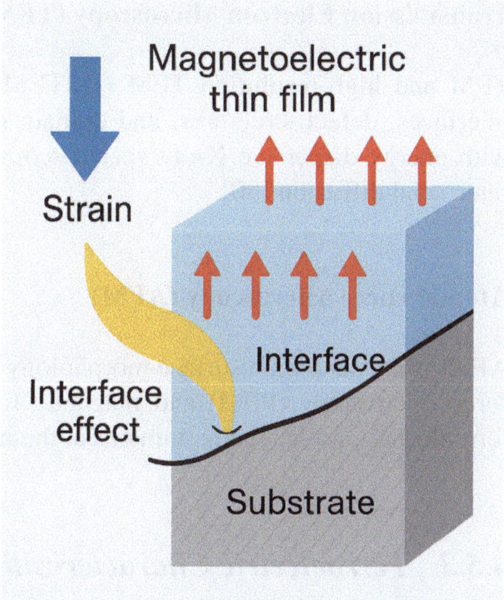

Fig. 4.4 Schematic representation of interface and strain effects in magnetoelectric thin films

4.5 Characterization Techniques for Magnetoelectric Thin Films

4.5.1 Introduction

Characterizing magnetoelectric (ME) thin films requires techniques that probe ferroelectric, magnetic, structural, and coupled responses. Since ME coupling often arises from interfacial or nanoscale effects, high-resolution and multifunctional characterization is essential to capture both intrinsic and extrinsic contributions. A combination of structural, electrical, magnetic, and spectroscopic tools provides the most comprehensive understanding [36].

4.5.2 Structural Characterization

X-Ray Diffraction (XRD) and Reciprocal Space Mapping

XRD is fundamental for determining crystallinity, phase purity, lattice constants, and epitaxial strain. High-resolution reciprocal space maps reveal strain states and relaxation behavior, crucial for interpreting strain-mediated ME effects [3].

Transmission Electron Microscopy (TEM)

TEM and high-resolution TEM (HRTEM) allow direct imaging of atomic-scale interfaces, defect structures, and domain configurations. Scanning TEM (STEM) with energy-dispersive X-ray spectroscopy (EDS) provides chemical mapping of interfacial diffusion [30].

Atomic Force Microscopy (AFM)

AFM offers nanoscale surface morphology analysis. Variants such as piezoresponse force microscopy (PFM) and magnetic force microscopy (MFM) directly probe ferroelectric and magnetic domains at the nanoscale [37].

4.5.3 *Ferroelectric Characterization*

Polarization–Electric Field (P–E) Hysteresis

The P–E loop is the primary indicator of ferroelectric switching. In ME thin films, distortions in hysteresis can indicate coupling to magnetic states [36].

Piezoresponse Force Microscopy (PFM)

PFM enables visualization and manipulation of ferroelectric domains. Strain-mediated ME effects can be quantified by correlating domain wall movement with applied magnetic fields [38].

4.5.4 *Magnetic Characterization*

Magnetization–Field (M–H) Hysteresis

Using vibrating sample magnetometry (VSM) or superconducting quantum interference device (SQUID) magnetometry, the M–H loop provides insights into magnetic anisotropy and coercivity, which may be tunable by electric fields [10].

Magnetic Force Microscopy (MFM)

MFM maps stray fields from domain structures, allowing visualization of magnetization reversal under applied electric bias.

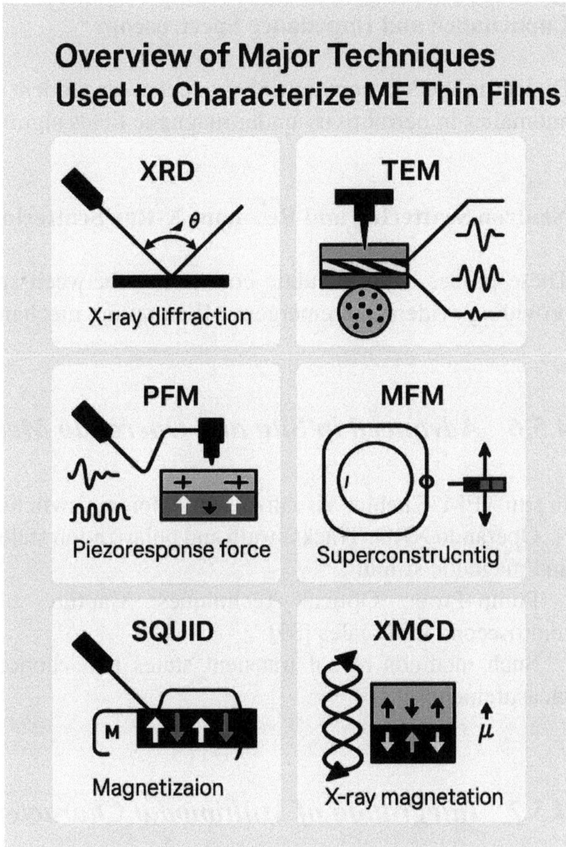

Fig. 4.5 Schematic overview of major techniques used to characterize ME thin films (XRD, TEM, PFM, MFM, SQUID, XMCD)

X-Ray Magnetic Circular Dichroism (XMCD)

Synchrotron-based XMCD offers element-specific probing of magnetic moments. This is critical for heterostructures where interface magnetism differs from bulk contributions [14]. Figure 4.5 shows schematic overview of major techniques used to characterize ME thin films (XRD, TEM, PFM, MFM, SQUID, XMCD).

4.5.5 Magnetoelectric Coupling Characterization

Direct Measurement of ME Coefficient

The ME coupling coefficient (α) is determined by applying an electric field and measuring induced magnetization (or vice versa). Lock-in techniques improve sensitivity, especially in thin films where signals are weak [10].

Capacitance and Impedance Spectroscopy

Dielectric spectroscopy reveals frequency-dependent coupling effects. For instance, anomalies in permittivity under magnetic fields signify ME interactions [8].

Neutron Scattering and Resonant X-Ray Scattering

These probes capture subtle correlations between spin, lattice, and orbital order, providing evidence of emergent ME coupling mechanisms at the atomic scale [2].

4.5.6 Advanced in Situ and Operando Methods

In situ TEM: Enables visualization of domain switching under applied fields.

Operando XRD: Tracks strain and polarization states during simultaneous electric and magnetic stimuli.

Pump–Probe Optical Techniques: Capture ultra-fast ME dynamics on femtosecond timescales [39].

Such methods reveal transient states that cannot be accessed in equilibrium measurements.

4.5.7 Integration of Multimodal Characterization

Given the complexity of ME coupling, no one approach will do. Correlative approaches combining structural, ferroelectric, and magnetic probes are therefore vital. Real-space correlation between polarization and magnetization domains, for example coupling PFM with MFM provides; by mixing XRD with XMCD, interface-driven coupling is seen [27].

Additionally rising is machine learning-aided data analysis, which helps to decipher vast multimodal datasets and speeds the finding of fresh ME materials [35].

4.5.8 Challenges and Outlook

Challenges in characterization include the following:

- Sensitivity limitations for ultrathin films (<10 nm).
- Separating intrinsic ME effects from extrinsic contributions such as leakage currents.

- Need for in situ multimodal measurements to capture coupled dynamics.

Future progress may involve synchrotron and neutron facilities combined with nanoscale scanning probes and computational modeling to provide a holistic picture of ME coupling in thin films.

4.6 Device Architectures and Applications of Magnetoelectric Thin Films

4.6.1 Introduction

The advancement of magnetoelectric (ME) thin films has shifted the research focus from fundamental phenomena to practical device implementations. ME thin films provide unique opportunities for electric-field control of magnetism and magnetic-field control of polarization, enabling low power and multifunctional devices. Their integration with existing semiconductor technology makes them attractive for spintronics, non-volatile memories, energy-efficient sensors, and energy harvesting systems [6, 13]. This section explores different architectures and application domains, emphasizing the interplay between materials science and device engineering.

4.6.2 Spintronics Applications

ME-Based Spin Valves and Tunnel Junctions

ME thin films have been incorporated into magnetic tunnel junctions (MTJs), where ferroelectric layers serve as barriers or gate electrodes to control spin transport. Voltage-driven modulation of tunneling magnetoresistance reduces Joule heating and enhances device scalability [40].

Voltage-Controlled Magnetic Anisotropy (VCMA)

One of the key advantages of ME coupling is enabling electric-field control of magnetic anisotropy, which is crucial for spintronic memory devices. VCMA allows non-volatile, ultra-low-power magnetic switching without the need for large current densities, thereby offering superior energy efficiency compared to conventional spin-transfer torque mechanisms [41].

4.6.3 *Memory Devices*

Magnetoelectric Random Access Memory (MERAM)

MERAM exploits electric-field-driven magnetic switching for non-volatile data storage. Unlike traditional MRAM, which relies on current-induced spin torque, MERAM reduces power consumption by eliminating Joule heating. Recent studies on $BiFeO_3$-based thin films demonstrate stable room-temperature switching, showing promise for commercialization [14].

Hybrid Ferroelectric–Magnetic Memory

Composite thin films combining ferroelectric and ferromagnetic layers have been used to design four-state memory devices (two polarization states $\times$ two magnetic states). Such devices offer enhanced storage density and multifunctional read/write capabilities [42].

4.6.4 *Sensors*

Magnetic Field Sensors

ME thin film sensors can detect ultra-weak magnetic fields by converting them into measurable electric signals. These sensors are lightweight, compact, and capable of operating without cryogenic cooling, making them competitive alternatives to SQUID sensors for biomedical applications such as magnetocardiography and magnetoencephalography [43].

Pressure and Acoustic Sensors

ME heterostructures sensitive to mechanical strain are also used in pressure and acoustic sensing. Thin film architectures allow for flexible and wearable sensor platforms, with high sensitivity arising from strain-mediated ME coupling [44].

4.6.5 *Energy Harvesting*

Vibration Energy Harvesters

Strain-mediated ME thin films have been designed for vibration-to-electricity conversion, where mechanical vibrations induce magnetostrictive strain, which is transferred to the piezoelectric component to generate voltage. Such harvesters are ideal for powering low-energy devices in the Internet of Things (IoT) [10].

Wireless Power Transfer and Antenna Applications

The ME coupling in thin film devices has also been exploited in resonant antennas and wireless power transfer systems. ME antennas provide compact form factors with tunable frequency responses, essential for next-generation communication devices [45].

4.6.6 *Neuromorphic and Emerging Computing Architectures*

Beyond conventional applications, ME thin films are being explored in neuromorphic computing, where coupled ferroelectric–ferromagnetic states mimic synaptic plasticity. Their multifunctional switching capability enables artificial neurons and synapses that are energy-efficient and adaptive [46].

4.6.7 *Challenges and Outlook*

Despite significant progress, several challenges remain in translating ME thin film architectures into large-scale applications:

Scalability: Keeping excellent ME coupling in nanoscale devices.

Reliability: managing thermal stability, leak currents, and exhaustion.

Integration: Compatibility with CMOS procedures meant for industry adoption.

Hybrid architectures combining 2D materials with ME thin films, as well as flexible and wearable electronics, should be future studies. Development of energy-autonomous, multipurpose devices could place ME thin films at the center of next-generation technology. Figure 4.6 shows representation of ME thin film applications, highlighting spintronics, MERAM, sensors, and energy harvesters.

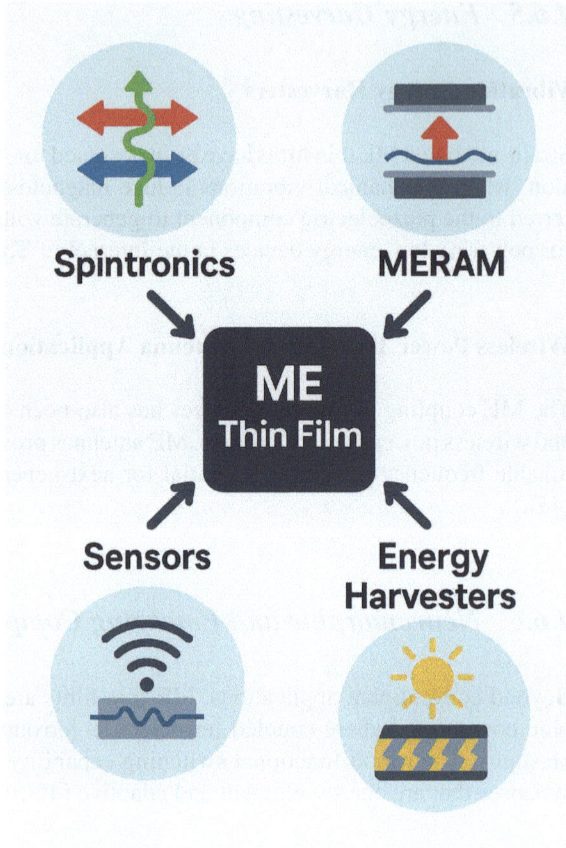

Fig. 4.6 Schematic representation of ME thin film applications, highlighting spintronics, MERAM, sensors, and energy harvesters

4.7 Emerging Trends and Future Directions in Magnetoelectric Thin Films

4.7.1 Introduction

Magnetoelectric (ME) thin films have already established themselves as a versatile platform for multifunctional devices in spintronics, sensors, and energy harvesting. However, the field remains highly dynamic, with new material systems, fabrication strategies, and device concepts emerging rapidly. These advancements are motivated by the need for higher efficiency, reduced energy consumption, and compatibility with future computing and communication paradigms [7, 47]. This section highlights major emerging trends and provides insights into where research on ME thin films is heading in the next decade.

4.7.2 *Integration with Two-Dimensional (2D) Materials*

The emergence of 2D materials such as graphene, transition-metal dichalcogenides (TMDs), and hexagonal boron nitride has opened new avenues for ME coupling research. 2D materials offer atomically sharp interfaces and improved charge transport when combined with ferroelectric or ferromagnetic thin films. Graphene combined with ferroelectric thin films, for instance, allows voltage-controlled modulating of spin transport with hardly any energy loss [48]. TMDs coupled with ME layers can create new spin–orbit effects, thereby clearing the route for very-scalable spintronic systems.

4.7.3 *Wearable and Flexible ME Thin Films*

The development of flexible, lightweight ME thin films for wearable and biomedical uses is a huge frontier. Depositing ME heterostructures on polymer substrates has made bendable sensors and energy harvesters [49] possible. Such devices could power next-generation implantable gadgets, wearable portable communication devices, and health monitors. Maintaining great ME coupling while guaranteeing mechanical resilience under repeated flexing is the difficulty here.

4.7.4 *Quantum Materials and Topological Effects*

The discovery of topological insulators and Weyl semimetals has influenced the direction of ME research. Incorporating ME thin films with topological materials enables new coupling pathways mediated by spin–orbit interactions and Berry curvature effects [50]. These hybrid systems could unlock room-temperature quantum ME devices, including topological spin transistors and quantum neuromorphic networks.

4.7.5 *Neuromorphic and Brain-Inspired Computing*

Neuromorphic computing aims to mimic the structure and function of biological brains. ME thin films, with their multistate switching capability, are highly promising for artificial synapses and neurons [46]. For instance, ferroelectric polarization states can represent synaptic weights, while magnetic domain configurations can emulate short-term and long-term plasticity. The low power requirements and multifunctional behavior of ME devices could accelerate the development of brain-inspired architectures.

4.7.6 Energy-Efficient and Green Electronics

As global energy demands rise, energy efficiency in electronics has become a central challenge. ME thin films inherently provide voltage-driven control of magnetism, drastically lowering energy consumption compared to current-driven devices. Furthermore, sustainable fabrication approaches—such as chemical solution deposition, inkjet printing, and additive manufacturing—are gaining traction [51]. These methods reduce material waste and processing costs, aligning ME device development with green technology goals.

4.7.7 Multifunctional and Hybrid Devices

Emerging devices are moving beyond single functionality toward multimodal platforms. For example, ME thin films integrated with photonic materials could enable light-controllable spintronic devices. Similarly, hybrid systems combining ferroelectric, ferromagnetic, and ionic conductors can deliver electrical, magnetic, and ionic control within a single device [52]. This multifunctionality is crucial for adaptive electronics and autonomous systems.

4.7.8 Challenges Ahead

Despite remarkable progress, several critical challenges remain:

Scalability: Maintaining robust ME coupling at sub-10 nm thicknesses remains non-trivial.

Fatigue and Stability: Repeated switching can degrade interfaces, leading to reduced device lifetimes.

Integration: CMOS compatibility must be further optimized for industrial adoption.

Standardization—Lack of standardized characterization protocols hampers cross-laboratory comparisons.

Addressing these bottlenecks will require collaboration between materials scientists, device engineers, and industry stakeholders.

4.7.9 Outlook

The next decade promises a paradigm shift in how ME thin films are understood and applied. Research is increasingly interdisciplinary, spanning condensed matter physics, nanotechnology, and quantum information science. By integrating 2D materials, quantum materials, and flexible substrates, ME thin films are poised to transform sectors ranging from biomedicine to neuromorphic computing. The pursuit of multifunctional, energy-efficient, and eco-friendly device architectures ensures that ME thin films will remain at the forefront of material science and technology development. Figure 4.7 shows emerging directions in ME thin films, highlighting integration with 2D materials, flexible devices, neuromorphic computing, and quantum materials.

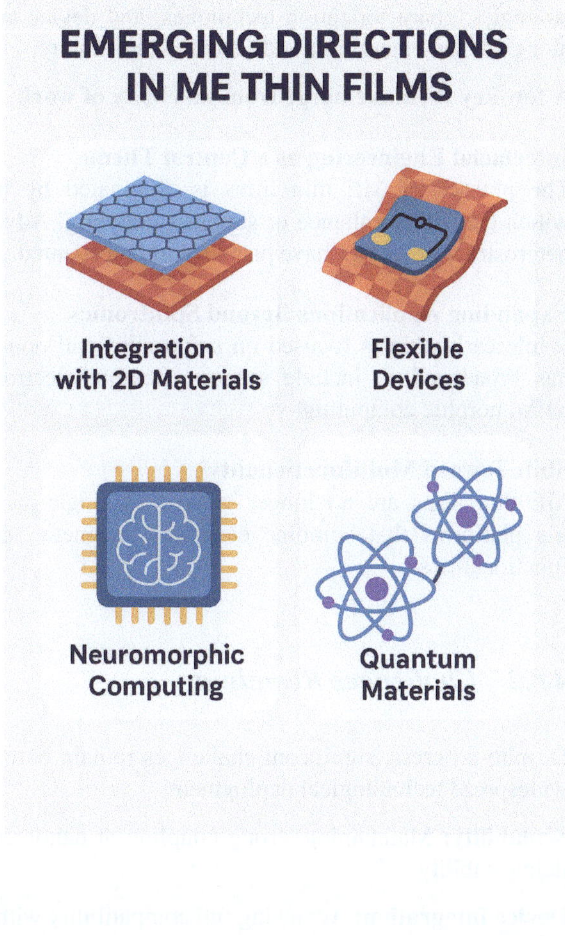

Fig. 4.7 Emerging directions in ME thin films, highlighting integration with 2D materials, flexible devices, neuromorphic computing, and quantum materials

4.8 Conclusion and Outlook

4.8.1 Conclusion

Research on magnetoelectric (ME) coupling in thin films has progressed from a fundamental curiosity to a dynamic, application-driven field. The unique ability of ME thin films to enable reciprocal control of electric and magnetic order parameters has positioned them as a cornerstone of next-generation materials and devices. The preceding sections of this chapter outlined the fundamental principles, fabrication strategies, characterization techniques, and device applications of ME thin films, alongside the emerging trends shaping their future.

A few key themes emerge from this body of work:

Interfacial Engineering as a Central Theme
The physics of ME thin films is dominated by interfacial and strain effects, which can either enhance or suppress coupling. Advances in epitaxial growth and heterostructure design have provided unprecedented control at the atomic scale.

Expanding Applications Beyond Spintronics
While early studies focused on spintronics and non-volatile memories, the scope has broadened to include sensors, flexible electronics, energy harvesters, and neuromorphic computing.

Shift Toward Multifunctionality
ME thin films are no longer viewed as single-purpose materials but as versatile platforms that combine electrical, magnetic, mechanical, and even optical functionalities.

4.8.2 Challenges Remaining

Despite progress, significant challenges remain before ME thin films can achieve widespread technological deployment:

Scalability: Maintaining strong coupling at nanometer dimensions without sacrificing stability.

Device Integration: Achieving full compatibility with CMOS technology for large-scale adoption.

Reliability: Overcoming issues of fatigue, leakage, and thermal stability under continuous cycling.

Standardization: Lack of uniform testing protocols makes cross-study comparison difficult.

These limitations represent opportunities for future innovation.

4.8.3 Future Outlook

The outlook for ME thin films is promising and multidimensional. Future research will likely focus on the following directions:

Integration with 2D and Quantum Materials
Heterostructures that incorporate graphene, TMDs, or topological insulators can open new quantum pathways for ME interactions.

Flexible and Bio-Integrated Electronics
Wearable and implantable devices based on flexible ME thin films could revolutionize healthcare and personalized monitoring.

Energy-Efficient Computing
By leveraging voltage-driven switching, ME thin films could enable ultra-low-power logic, neuromorphic computing, and possibly quantum-inspired architectures.

Sustainable Fabrication
Adoption of additive manufacturing and solution-based processing can lower costs and environmental impact, making ME devices more accessible.

System-Level Demonstrations
Translating laboratory-scale results into full prototype systems—such as ME memory arrays, biomedical sensors, or ME antennas—will determine the readiness of this technology for commercialization.

In summary, ME thin films represent a highly interdisciplinary field where materials innovation, device engineering, and system integration converge. Their unique tunability and multifunctionality make them central to future technologies that demand energy efficiency, scalability, and adaptability.

4.8.4 Summary Table

See Table 4.2.

4.8.5 Final Remarks

The field of magnetoelectric thin films embodies the spirit of modern materials science—a synergy of theory, synthesis, characterization, and application. ME thin

Table 4.2 Current status and future directions of magnetoelectric thin films

Domain	Current achievements	Future directions
Materials	$BiFeO_3$, composite heterostructures, strain-engineered films	2D/quantum materials, additive manufacturing
Fundamentals	Interface-driven coupling, strain-mediated effects	Operando studies, quantum coupling, AI/ML predictive modeling
Characterization	PFM, MFM, TEM, Raman, XMCD, direct ME measurements	In situ multimodal techniques, nanoscale operando probes
Applications	Spintronics, MERAM, sensors, energy harvesters	Neuromorphic computing, bioelectronics, multifunctional wearables, ME antennas
Challenges	Stability, scalability, CMOS integration, fatigue	Standardized protocols, long-term reliability, device upscaling

films are set not only as academic curiosities but also as enablers of technological change as energy-efficient and multifunctional electronics become a necessity in the twenty-first century.

Should the path of invention continue, ME thin films will develop from lab prototypes to the basis of practical gadgets covering spintronics, medicine, green energy, and quantum computing. Their potential comes from their capacity to fit with developing technologies as well as from their inherent features.

References

1. Spaldin, N. A., & Fiebig, M. (2005). The renaissance of magnetoelectric multiferroics. Science, 309(5733), 391–392.
2. Fiebig, M., Lottermoser, T., Meier, D., & Trassin, M. (2016). The evolution of multiferroics. Nature Reviews Materials, 1(8), 16046.
3. Ramesh, R., & Spaldin, N. A. (2007). Multiferroics: Progress and prospects in thin films. Nature Materials, 6(1), 21–29.
4. Dzyaloshinskii, I. (1960). On the magneto-electrical effect in antiferromagnets. Soviet Physics JETP, 10, 628–629.
5. Astrov, D. N. (1960). The magnetoelectric effect in antiferromagnetics. Soviet Physics JETP, 11, 708–709.
6. Eerenstein, W., Mathur, N. D., & Scott, J. F. (2006). Multiferroic and magnetoelectric materials. Nature, 442(7104), 759–765.
7. Spaldin, N. A., & Ramesh, R. (2019). Advances in magnetoelectric multiferroics. Nature Materials, 18(3), 203–212.
8. Wang, J., Neaton, J. B., Zheng, H., Nagarajan, V., Ogale, S. B., Liu, B., Viehland, D., Vaithyanathan, V., Schlom, D. G., Waghmare, U. V., Spaldin, N. A., Rabe, K. M., Wuttig, M., & Ramesh, R. (2003). Epitaxial $BiFeO_3$ multiferroic thin film heterostructures. Science, 299(5613), 1719–1722.
9. Chakhalian, J., Freeland, J. W., Millis, A. J., Panagopoulos, C., & Rondinelli, J. M. (2014). Colloquium: Emergent properties in plane view: Strong correlations at oxide interfaces. Reviews of Modern Physics, 86(4), 1189.

10. Nan, C. W., Bichurin, M. I., Dong, S., Viehland, D., & Srinivasan, G. (2008). Multiferroic magnetoelectric composites: Historical perspective, status, and future directions. Journal of Applied Physics, 103(3), 031101.
11. Hu, J. M., Li, Z., & Nan, C. W. (2021). Multiferroic heterostructures for novel device applications. Advanced Materials, 33(32), 2005063.
12. Pertsev, N. A., Zembilgotov, A. G., & Tagantsev, A. K. (1998). Effect of mechanical boundary conditions on phase diagrams of epitaxial ferroelectric thin films. Physical Review Letters, 80(9), 1988–1991.
13. Bibes, M., & Barthélémy, A. (2008). Multiferroics: Towards a magnetoelectric memory. Nature Materials, 7(6), 425–426.
14. Heron, J. T., Bosse, J. L., He, Q., Gao, Y., Trassin, M., Ye, L., Clarkson, J. D., Wang, C., et al. (2014). Deterministic switching of ferromagnetism at room temperature using an electric field. Nature, 516(7531), 370–373.
15. Mostovoy, M. (2006). Ferroelectricity in spiral magnets. Physical Review Letters, 96(6), 067601.
16. Sergienko, I. A., & Dagotto, E. (2006). Role of the Dzyaloshinskii-Moriya interaction in multiferroic perovskites. Physical Review B, 73(9), 094434.
17. Tokura, Y., & Kanazawa, N. (2021). Magnetic skyrmion materials. Chemical Reviews, 121(5), 2857–2897.
18. Vaz, C. A. F. (2012). Electric field control of magnetism in multiferroic heterostructures. Journal of Physics: Condensed Matter, 24(33), 333201.
19. Venables, J. A., Spiller, G. D. T., & Hanbücken, M. (1984). Nucleation and growth of thin films. Reports on Progress in Physics, 47(4), 399.
20. Kalinin, S. V., et al. (2007). Defect-mediated polarization switching in ferroelectrics and related materials. Reports on Progress in Physics, 71(5), 056502.
21. Chrisey, D. B., & Hubler, G. K. (Eds.). (1994). Pulsed laser deposition of thin films. Wiley.
22. Qi, X., et al. (2005). Epitaxial $BiFeO_3$ thin films on $SrTiO_3$ substrates: Growth, structure, and properties. Applied Physics Letters, 86(7), 071913.
23. Sreenivas, K., & Mansingh, A. (1985). RF sputtered $BaTiO_3$ thin films for integrated optics. Journal of Applied Physics, 57(12), 5239–5245.
24. Schlom, D. G., Chen, L.-Q., Pan, X., Schmehl, A., & Zurbuchen, M. A. (2008). A thin film approach to engineering functionality into oxides. Journal of the American Ceramic Society, 91(8), 2429–2454.
25. Dong, S., et al. (2004). Magnetoelectric effect in PZT/$NiFe_2O_4$ multilayer thin films. Applied Physics Letters, 83(11), 2265–2267.
26. Boschker, H., & Mannhart, J. (2017). Quantum-matter heterostructures. Annual Review of Condensed Matter Physics, 8, 145–164.
27. Zubko, P., Gariglio, S., Gabay, M., Ghosez, P., & Triscone, J. M. (2011). Interface physics in complex oxide heterostructures. Annual Review of Condensed Matter Physics, 2, 141–165.
28. Ohtomo, A., & Hwang, H. Y. (2004). A high-mobility electron gas at the $LaAlO_3$/$SrTiO_3$ heterointerface. Nature, 427(6973), 423–426.
29. Martin, L. W., Chu, Y. H., & Ramesh, R. (2008). Advances in the growth and characterization of magnetic, ferroelectric, and multiferroic oxide thin films. Materials Science and Engineering: R: Reports, 68(4–6), 89–133.
30. Yu, P., Chu, Y. H., & Ramesh, R. (2012). Emergent phenomena at multiferroic heterointerfaces. Philosophical Transactions of the Royal Society A, 370(1976), 4856–4873.
31. Haeni, J. H., Irvin, P., Chang, W., Uecker, R., Reiche, P., Li, Y. L.,... & Schlom, D. G. (2004). Room-temperature ferroelectricity in strained $SrTiO_3$. Nature, 430(7001), 758–761.
32. Chu, Y. H., Martin, L. W., Holcomb, M. B., Gajek, M., Han, S. J., He, Q.,... & Ramesh, R. (2007). Electric-field control of local ferromagnetism using a multiferroic. Nature Materials, 7(6), 478–482.
33. Zheng, H., Wang, J., Lofland, S. E., Ma, Z., Mohaddes-Ardabili, L., Zhao, T.,... & Ramesh, R. (2004). Multiferroic $BaTiO_3$–$CoFe_2O_4$ nanostructures. Science, 303(5658), 661–663.

34. Zavaliche, F., Shafer, P., Ramesh, R., et al. (2005). Electric field-induced magnetization switching in epitaxial multiferroic heterostructures. Nano Letters, 5(9), 1793–1796.
35. Xie, Y., Chen, J., & Zhang, W. (2020). Data-driven design of multiferroic heterostructures. npj Computational Materials, 6(1), 98.
36. Scott, J. F. (2007). Applications of modern ferroelectrics. Science, 315(5814), 954–959.
37. Gruverman, A., & Kholkin, A. (2006). Nanoscale ferroelectrics: Processing, characterization and future trends. Reports on Progress in Physics, 69(8), 2443–2474.
38. Kalinin, S. V., & Gruverman, A. (2007). Scanning Probe Microscopy of Functional Materials: Nanoscale Imaging and Spectroscopy. Springer.
39. Mikhaylovskiy, R. V., Hendry, E., Kruglyak, V. V., et al. (2015). Ultrafast optical control of magnetization dynamics in multiferroics. Nature Communications, 6, 8190.
40. Tsymbal, E. Y., & Kohlstedt, H. (2006). Tunneling across a ferroelectric. Science, 313(5784), 181–183.
41. Nozaki, T., Shiota, Y., Miwa, S., Murakami, S., Bonell, F., Ishibashi, S.,... & Suzuki, Y. (2012). Voltage-induced magnetic anisotropy change in magnetic tunnel junctions. Nature Physics, 8(6), 491–496.
42. Velev, J. P., Jaswal, S. S., & Tsymbal, E. Y. (2011). Multi-ferroic and magnetoelectric materials and interfaces. Philosophical Transactions of the Royal Society A, 369(1948), 3069–3097.
43. Li, M., Dong, S., & Viehland, D. (2017). Research updates: Magnetoelectric sensors: A review. APL Materials, 5(4), 042902.
44. Dong, S., Zhai, J., Li, J., Viehland, D., & Sun, N. X. (2006). Magnetic field sensors using multiferroic composites. Applied Physics Letters, 89(25), 252904.
45. Zhu, M., Nan, C. W., & Wang, Y. (2012). Magnetoelectric composite antennas: A review. Journal of Electroceramics, 29(1), 9–15.
46. Manipatruni, S., Nikonov, D. E., Lin, C. C., Gosavi, T. A., Liu, H., Prasad, B.,... & Young, I. A. (2019). Scalable energy-efficient magnetoelectric spin–orbit logic. Nature, 565(7737), 35–42.
47. Nan, T., Ma, J., Lin, H., & Sun, N. X. (2021). Recent progress in magnetoelectric materials and devices: Beyond multiferroics. Advanced Functional Materials, 31(7), 2007615.
48. Zheng, Y., Cheng, R., & Yu, G. (2018). Integration of two-dimensional materials with multiferroics: Opportunities and challenges. Nano Letters, 18(5), 2935–2942.
49. Li, J., Wang, Y., & Dong, S. (2020). Flexible magnetoelectric thin film devices: Opportunities and challenges. Applied Physics Reviews, 7(4), 041310.
50. Tokura, Y., Kawasaki, M., & Nagaosa, N. (2019). Emergent functions of quantum materials. Nature Physics, 15(11), 1056–1068.
51. Roy, S., Datta, S., & Bhattacharya, S. (2020). Additive manufacturing of multiferroic materials: A review. Journal of Materials Chemistry C, 8(15), 4923–4940.
52. Kim, T. H., Lee, S., & Park, J. H. (2021). Ionic control of multifunctional magnetoelectric devices. Advanced Materials, 33(10), 2007279.

Chapter 5
Magnetic Thin Films for Spintronics

Parminder Singh

5.1 Introduction

The theoretical foundation of two-dimensional (2D) magnetism was established in 1944 with Onsager's exact solution of the 2D Ising model, which revealed the possibility of a finite-temperature phase transition leading to long-range magnetic order [1]. Due to persistent experimental limitations, the practical deployment of 2D magnets remained unrealized for decades, resulting in the continued reliance on traditional 3D magnetic materials for major technological applications (Fig. 5.1) such as spin-based quantum dots, spin-orbit torque (SOT) devices, magnetic tunnel junctions (MTJs), and spin field-effect transistors (Spin-FETs).

The landscape of low-dimensional magnetism has significantly advanced with the realization of magnetic van der Waals (vdW) crystals, which maintain intrinsic magnetic order even at the atomic scale [3, 4]. These advances have enabled unprecedented exploration of low-dimensional magnetism, leading to breakthroughs in spintronic phenomena and paving the way for emerging applications in chemical sensing and memory technologies [5]. A fundamental requirement for both 2D and three-dimensional (3D) magnetic applications is the growth of long-range ferromagnetic (FM) order below a critical temperature (T_c), where spontaneous symmetry breaking enables the material to maintain a net magnetization without external field [6–8]. While magnetite (Fe_3O_4) and other bulk FM materials have served technological and navigational roles for millennia, the concept of achieving intrinsic magnetic order in atomically thin layers has gained momentum only in recent years [9, 10]. The analytical resolution of the 2D Ising model by Onsager marked a foundational advance, establishing the possibility of spontaneous magnetic ordering, albeit under the constraint of spin states limited to binary orientations along one crystallographic direction [11]. Unlike the Ising model, the Heisenberg model offers a more realistic

P. Singh (✉)
Department of Physics, School of Sciences, IILM University, Greater Noida, India
e-mail: pagi.pagi20@gmail.com

S. Sharma (ed.), *Multifunctional Heterostructure Thin Films Handbook*,
https://doi.org/10.1007/978-3-032-18692-8_5

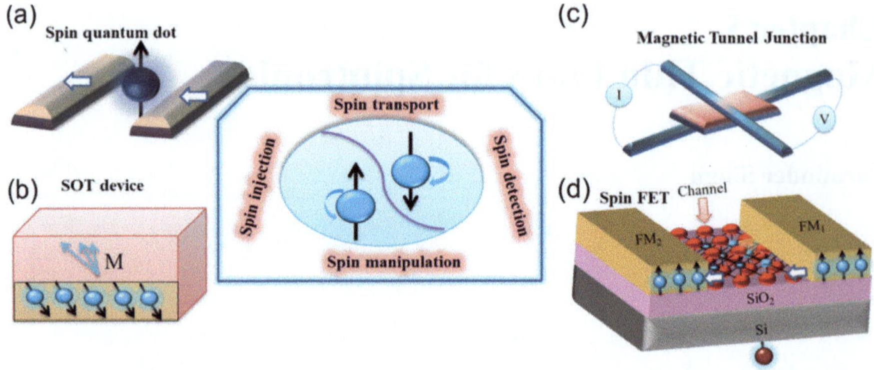

Fig. 5.1 Schematic illustrations depict diverse applications, including a spin-based quantum dots, b SOT devices, c MTJs, and d Spin-FETs, each enabling unique control over spin degrees of freedom for next-generation information technologies [2]. Copyright 2022. Reproduced with permission from Elsevier.

description of magnetism in most materials, allowing spin vectors to orient freely in three dimensions [12, 13]. The Mermin-Wagner theorem later established that continuous rotational symmetry in 2D Heisenberg systems prohibits long-range order at finite temperatures due to divergent spin-wave (magnon) fluctuations [14]. However, this restriction does not extend to anisotropic systems such as Ising magnets, where spin orientation constraints circumvent the limitations imposed by the theorem, preserving the feasibility of 2D magnetic ordering. Consequently, the modern realization of 2D magnets with robust anisotropy has validated long-standing theoretical predictions and paved the way for advancements in exploring basic science and developing cutting-edge technologies.

This chapter presents a comprehensive overview of magnetic thin films and their pivotal role in contemporary spintronic architectures. Section 5.2 introduces the foundational aspects of magnetic thin films, outlining their structural, electronic, and magnetic characteristics relevant to spin-dependent transport. Section 5.3 traces the evolution of magnetic thin films toward the emergence of 2D magnetic materials, emphasizing both intrinsic and proximity-induced magnetism within vdW systems. The classification and properties of distinct magnetic material families are elaborated in Sect. 5.4, providing insight into their exchange mechanisms and anisotropic behaviors. Section 5.5 delves into interface-driven phenomena, such as spin-orbit coupling (SOC), exchange bias, and various magnetoresistance effects that underpin device functionality and tunability. Section 5.6 details the synthesis routes employed for fabricating high-quality thin films and heterostructures, while Sect. 5.7 discusses the characterization techniques essential for probing magnetic, structural, and interfacial properties. Finally, Sect. 5.8 addresses the practical integration challenges and strategic considerations for realizing scalable, high-performance spintronic devices based on these advanced magnetic thin films.

5.2 2D Magnetic Films for Spintronics

The integration of thin film magnetism with spintronic principles has catalyzed significant breakthroughs in data processing, advanced memory architectures, and emerging non-traditional computing technologies. Owing to their tunable magnetic properties, interfacial engineering potential, and seamless integration with low-dimensional systems, thin magnetic films and vdW materials have become pivotal in spin-based device architectures [15]. Spintronic device architectures from magnetic tunnel junctions (MTJs) to spin valves rely heavily on magnetic thin films for their efficient spin-transfer torque (STT) property [16]. Although 2D MTJs exhibit promising magnetoresistance characteristics, their performance remains limited by interface contamination, weak interlayer exchange coupling, and spin-flip scattering at vdW interfaces. These devices operate through the controlled manipulation of spin currents and magnetization dynamics within multilayer thin film heterostructures comprising ferromagnetic (FM) layers and non-magnetic spacers, where interface quality critically governs spin transport efficiency and overall device performance [17]. Key phenomena such as giant magnetoresistance and tunneling magnetoresistance (TMR) arise from spin-polarized transport in these structures, leading to their widespread application in hard drives and magnetic random-access memory (MRAM) [18]. The functionality of these devices can be tailored by tuning magnetic anisotropy, spin polarization, and damping constants through material composition, thickness, and interfacial engineering. For instance, systems exhibiting PMA, such as CoFeB/MgO multilayers, have enabled reduced switching current densities and enhanced thermal stability in memory applications [19]. Moreover, incorporation of spin Hall materials like Pt and W allows for non-contact magnetization switching through SOTs, minimizing wear on magnetic layers and promoting energy-efficient operation in next-generation memory technologies [20].

The integration of antiferromagnetic (AFM) or ferrimagnetic layers further expands the design space. Anti-ferromagnet based spintronics offers ultrafast dynamics and insensitivity to external magnetic fields, while ferrimagnets such as GdFeCo enable ultrafast all-optical switching, bridging photonics and magnetism [21]. Thin film spintronics is now recognized as a scalable and versatile framework, enabling not only conventional memory and logic devices but also next-generation neuromorphic and quantum systems, where spin degrees of freedom facilitate non-Boolean computation and quantum state manipulation.

5.2.1 Neuromorphic Computing

Neuromorphic computing aims to imitate the architecture and functionality of operations of the human brain using non-von Neumann models of computation [22]. The inherent stochasticity, non-volatility, and reconfigurability of thin film and 2D spintronic elements closely mirror the probabilistic and adaptive nature of biological synapses, thereby offering an excellent physical substrate for hardware-efficient neuromorphic computing. These physical properties enable localized state retention, energy-efficient switching, and adaptive modulation of conductance, all of which are essential for realizing synaptic plasticity and spike-timing-dependent plasticity in neuromorphic systems [22].

In this context, topologically protected nanoscale spin configurations known as magnetic skyrmions have shown great potential for use in neuromorphic and quantum computing systems. Their compact size, ultra-low current-driven mobility, and topologically robust behavior make them highly suitable for energy-efficient, densely packed synaptic architectures. Moreover, the interplay between skyrmions and proximity-induced superconductivity in 2D magnetic heterostructures opens up new avenues to explore topological superconductivity, a critical ingredient in fault-tolerant quantum computing. As illustrated in Fig. 5.2a, a skyrmion-based synapse can be realized in a 2D ferromagnet/heavy metal bilayer, where SOTs driven by bidirectional spin currents can modulate skyrmion behavior to emulate synaptic functionality. Further, reservoir computing architectures leveraging the nonlinear dynamics of skyrmion motion offer a compelling approach to transforming low-dimensional temporal inputs into high-dimensional magnetic responses, as shown in Fig. 5.2b. This approach exploits the intrinsic memory and nonlinearity of skyrmion dynamics, facilitating temporal pattern recognition and real-time signal processing with minimal training overhead. MTJs act as artificial synapses by modulating conductance via spin-transfer torque or voltage-controlled magnetic anisotropy (VCMA), enabling analog synaptic weight tuning. Patterned magnetic nanowires support domain wall motion that emulates spiking neuron behavior, facilitating spintronic implementations of leaky-integrate-and-fire models. 2D spintronic materials, such as $CrGeTe_3$, offer compact, low-power alternatives with stochastic switching and gate-tunable magnetism ideal for probabilistic neuromorphic computing [24]. Integrating 2D magnets with ferroelectrics or phase-change materials enables the formation of multiferroic heterostructures exhibiting magnetoelectric coupling, which allows voltage-driven modulation of synaptic weights in neuromorphic circuits. These advances offer a pathway toward energy-efficient and scalable hardware capable of mimicking biologically inspired learning dynamics. However, despite their potential for ultra-low-power, non-volatile information processing, practical deployment of neuromorphic spintronic devices faces significant challenges. Endurance is a major concern, as repeated synaptic updates induce cumulative stress on magnetic tunnel junctions. Furthermore, Stochastic switching, while advantageous for probabilistic operations, introduces challenges for deterministic learning, and heterogeneities in anisotropy, interfacial quality, and thermal stability can reduce array-level reliability. Robust materials engineering alongside adaptive circuit-level strategies, such as stochastic resonance control, variation-aware algorithms, and hybrid CMOS-spintronic designs, is essential to ensure reproducible and durable neuromorphic devices.

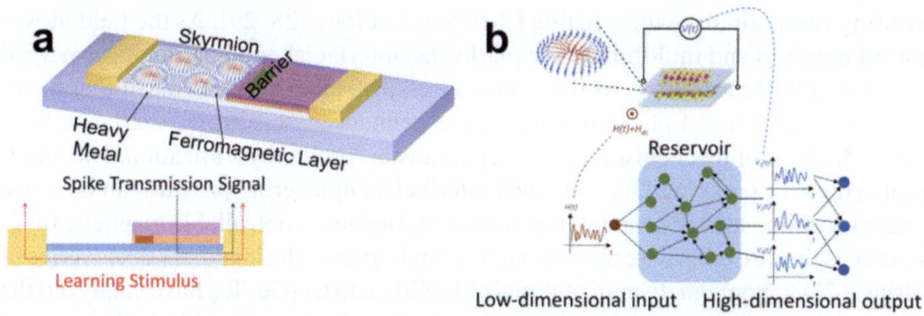

Fig. 5.2 Magnetic skyrmions enable neuromorphic and quantum computing and facilitate topological superconductivity exploration in 2D heterostructures. **a** Skyrmionic synapse driven by spin currents in 2D FM/heavy metal bilayers. **b** Reservoir computing using skyrmion dynamics for input-output transformation [23]. Copyright 2024. Reproduced with permission under a creative commons CC BY 4.0 license from Ref. [23]

5.2.2 *Quantum Computing*

In quantum computing, 2D magnetic materials offer compelling opportunities for realizing coherent spin qubits, quantum transducers, and topologically robust devices [23]. Unlike bulk systems, atomically thin magnets suppress decoherence by minimizing spin scattering pathways, facilitating long-lived spin coherence. Their proximity coupling with 2D semiconductors (e.g., $MoTe_2$, WS_2) enables gate-tunable spin splitting via magnetic exchange interactions, forming the basis of spin-selective quantum architectures [23]. The integration of 2D magnetic materials with superconductors has unveiled signatures of topological superconductivity and emergent Majorana zero modes, pivotal for achieving fault-tolerant quantum computing through non-Abelian statistics.

Furthermore, the emergence of chiral spin textures, including skyrmions and non-collinear states, in 2D magnets offers robust, topologically protected carriers of quantum information, manipulable via electric fields or SOTs [25]. These features open pathways to braiding-based quantum gates. Complementarily, optical control using magneto-optical Kerr effect (MOKE), exciton-spin interactions, and photo-induced switching enables coherent spin-photon coupling, advancing quantum transduction strategies essential for scalable, hybrid quantum networks.

5.3 Evolution of Thin Film Magnets

Thin film magnets play a foundational role in advancing scalable and energy-efficient spintronic devices, primarily by offering precise control over magnetic properties at the nanoscale [26, 27]. Their inherent tunability makes them ideally suited for integration into next-generation memory, logic, and neuromorphic architectures, while ensuring compatibility with existing CMOS technology [28, 29]. As the field moves toward ultrathin and multifunctional platforms, interfacial engineering has emerged as a critical design strategy. At the atomic level, interfaces between dissimilar materials give rise to novel phenomenon, such as, enhanced anisotropy, exchange bias, and SOC through mechanisms like charge transfer, orbital hybridization, and structural symmetry breaking [30, 31]. Such interfacial engineering provides an effective strategy to tailor spin-dependent interactions and achieve controllable magnetic functionalities that are otherwise absent in bulk counterparts. The landmark discoveries of intrinsic 2D ferromagnetism in materials like CrI_3 and $Cr_2Ge_2Te_6$ have catalyzed the exploration of vdW magnets, establishing a new frontier for probing low-dimensional magnetism. These layered magnetic crystals offer exfoliatable, chemically tunable, and heterostructure-compatible platforms, making them ideal for both fundamental investigations and device-level implementations. Building upon these foundational discoveries, a growing family of 2D vdW magnetic materials has emerged. These include: transition metal halides (MX_3, MX_2; M = V, Cr; X = Cl, Br, I), binary transition metal chalcogenides (e.g., MX, M_2X_3, M_5X_8, MX_2; M = Cr, V; X = S, Se, Te), transition metal phosphorus chalcogenides (MPX_3; M = Fe, Co, Ni, Mn; X = S, Se), including $CrPS_4$, Mn(Sb)-Bi-Te family, $Fe_3Ge(Ga)Te_2$ family, and ternary compounds (MXY; M = Cr, V; X = O, S, Se, Te; Y = Cl, Br, I), as systematically illustrated in Fig. 5.3.

These material systems not only exhibit diverse magnetic ground states (e.g., FM, AFM, and spin glass materials) but also host a range of interfacial and quantum effects. Importantly, emergent interfacial effects enable the realization of functionalities not accessible in bulk systems, thereby unlocking new pathways for device innovation in low-dimensional magnetism and spin-based computing [32, 33]. In particular, SOC at interfaces involving heavy metals such as Pt, W, or Ta and adjacent FM layers (e.g., Co, CoFeB) induces efficient spin-charge interconversion via the spin Hall effect, enabling the generation of spin currents that can manipulate magnetization through SOT [34, 35]. The magnitude and symmetry of SOC can be systematically engineered through heterostructure design, interface crystallinity, stacking sequence, and interlayer spacing. Incorporation of heavy metals or atomically thin non-magnetic spacers (e.g., Bi_2Se_3, MoS_2) enhances SOC via orbital hybridization, improving spin transparency and charge–spin conversion efficiency. Additionally, electric field gating provides dynamic tunability of SOC, offering a versatile route for voltage-controlled spin manipulation and energy-efficient spintronic device operation.

Magnetic anisotropy, a key parameter dictating the stability and directionality of magnetization, is highly sensitive to both epitaxial strain and layer thickness [37, 38]. In ultrathin magnetic films, interfacial perpendicular magnetic anisotropy (PMA) arises from broken inversion symmetry and SOC-induced orbital reconstruction at the interface [39, 40]. Systems such as CoFeB/MgO and Fe_3GeTe_2 exhibit strong PMA that is tunable via film thickness, enabling voltage-controlled magnetic switching and low-power magnetoresistive devices [41, 42].

Moreover, epitaxial strain induced during film growth on lattice-mismatched substrates modulates the crystal field environment and spin exchange pathways,

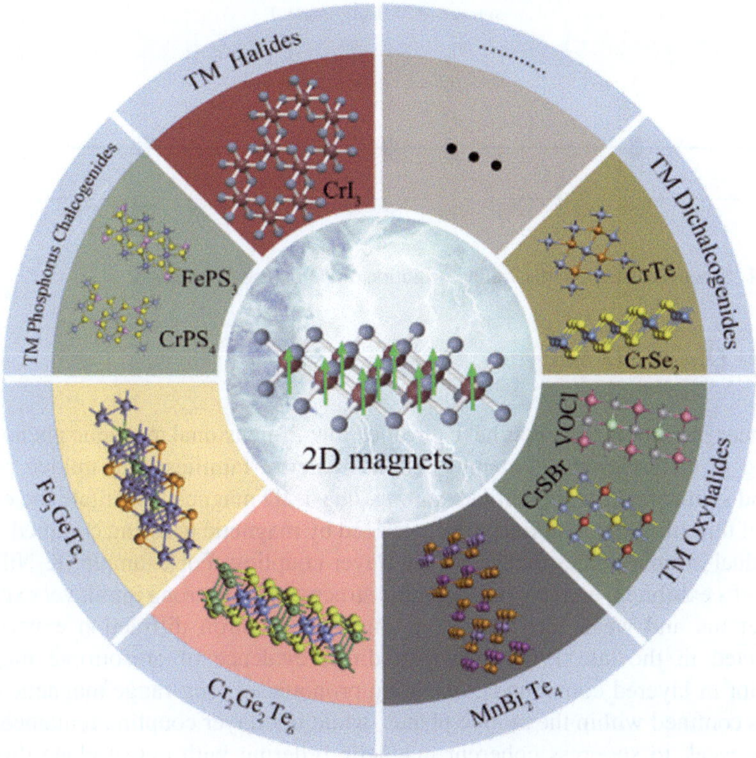

Fig. 5.3 Representative group of intrinsic 2D magnets [36]. Copyright 2023. Reproduced with permission under a creative commons CC BY 4.0 license from Ref. [36]

thereby affecting magnetic anisotropy and spin transport [43, 44]. For example, tensile strain can stabilize out-of-plane magnetization in vdW magnets or Heusler alloys, while compressive strain may enhance in-plane spin coherence or induce spin-reorientation transitions. Therefore, the interplay of interfacial chemistry, spin-orbit interactions, and elastic strain offers a powerful toolbox for optimizing thin film magnetic behavior.

5.4 2D Magnetic Families

Figure 5.4 illustrates the chronological evolution and material expansion of 2D magnetic systems, which can be broadly categorized into seven distinct material families. Each of these material families exhibits distinct crystallographic architectures, magnetic exchange mechanisms, and degrees of chemical or electronic tunability, which are examined in detail in the following sections.

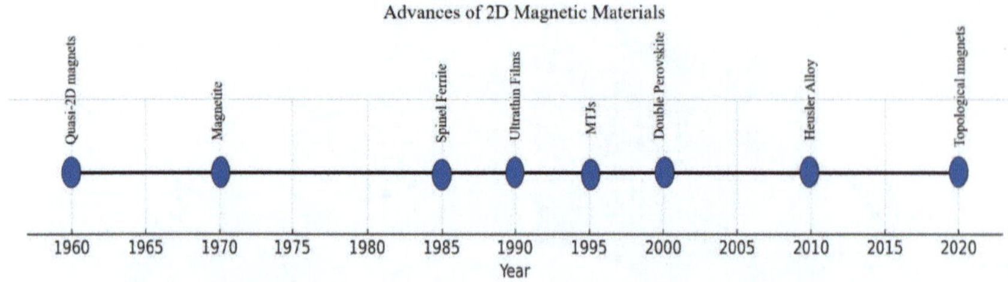

Fig. 5.4 Schematic timeline illustrating evolution of 2D magnetic materials

5.4.1 *Quasi 2D Magnets*

Advances in thin film magnets have unveiled low-dimensional magnetic phenomena, leading to their classification into distinct material families with unique properties and spintronic functionalities. Among, quasi-2D magnets constitute the earliest form of thin film-like materials, characterized by magnetic ordering confined within individual atomic planes and weak interlayer coupling. For example, K_2NiF_4 and Rb_2MnF_4 exhibit layered crystallographic structures with strong intralayer exchange interactions and weak interlayer coupling [45]. Neutron diffraction experiments conducted in the late 1960s provided direct evidence of anisotropic magnetic behavior in layered compounds, revealing pronounced long-range magnetic correlations confined within the atomic planes, while interlayer coupling remained sufficiently weak to suppress coherent magnetic ordering with out-of-plane direction above Neel temperature [46]. These findings established the foundational understanding of magnetic dimensionality and inspired subsequent research on ultrathin films and true 2D magnets.

5.4.2 *Magnetic Oxide Thin Films*

Magnetic oxide thin films (MOTF) stand out for their ability to couple spin polarization with strongly correlated electronic states and multifunctional coupling effects, features rarely observed in conventional FM metals [47]. Unlike metallic counterparts, oxides such as manganites, ferrites, and cobaltites harness the synergy of spin, charge, orbital, and lattice interactions that result in emergent phenomena including colossal magnetoresistance, charge ordering, and multiferroicity. These intricate couplings provide an effective platform to modulate magnetic and electronic states via electric fields, strain, or interface engineering. A prime example is $La_{1-x}Sr_xMnO_3$ (LSMO), a colossal magnetoresistive (CMR) manganite that demonstrates nearly 100% spin polarization at the fermi level, making it a prototypical half-metallic FM material [48, 49]. Its T_c can be engineered to surpass room temperature (RT) by varying doping concentration, making it a viable contender for RT spintronics [50]. LSMO has been extensively used as an electrode in oxide-based

MTJs, particularly with $SrTiO_3$ as a tunnel barrier, where TMR effects exceeding 180% have been observed at low temperatures [51]. Such configurations underscore the importance of interface engineering in optimizing spin polarization and tunneling efficiency. Spinel ferrites like Fe_3O_4 and $CoFe_2O_4$ also contribute significantly to the spintronics landscape. These materials are ferrimagnetic insulators with high magnetic ordering temperatures and sizable spin-filtering efficiency, making them ideal for spin injection and barrier layers in spin-filter tunnel junctions [52, 53]. Their compatibility with various substrates and chemical stability further enhances their appeal in device fabrication. Moreover, the use of perovskite and double perovskite oxides has led to the exploration of emergent properties such as multiferroicity and magnetoelectric coupling, which are instrumental in realizing electric field-controlled spintronic devices [54, 55]. The interfacial coupling between magnetic oxides and non-magnetic layers allows for manipulation of magnetic anisotropy and spin dynamics, enabling low-power switching mechanisms crucial for memory and logic applications.

Despite these advances, challenges remain, particularly in achieving high-quality interfaces, managing oxygen vacancies, and integrating oxide thin films with conventional semiconductor platforms. However, advancements in pulsed laser deposition (PLD), molecular beam epitaxy (MBE), and atomic layer deposition (ALD) have significantly improved the control over film stoichiometry and epitaxial growth, addressing many of these concerns.

5.4.3 Heusler Alloys

Heusler alloys have emerged as pivotal materials in the advancement of spintronic technologies due to their high spin polarization, structural tunability, and compatibility with standard thin film fabrication techniques. These intermetallic compounds, primarily consisting of full-Heusler (X_2YZ) and half-Heusler (XYZ) phases, display range of electronic and magnetic characteristics that can be tailored through compositional engineering and atomic ordering [56]. Moreover, these are particularly valued for their half-metallic nature, which yields nearly complete spin polarization at Fermi level. As a result, they are ideally suited for making efficient spin injection in MTJs, spin valves, and giant magnetoresistive sensors [57, 58].

Among the extensively studied systems, Co-based full Heusler alloys such as Co_2MnSi, and Co_2FeSi stand out for their high T_c (> 900 K), robust ferromagnetism, and low magnetic damping, which are crucial for efficient spin-transfer torque and low-power operation in next-generation devices [59, 60]. Furthermore, the ability to grow these materials as epitaxial thin films with controlled B2 or $L2_1$ ordering on substrates such as MgO or GaAs enhances their crystalline quality and spintronic performance [61, 62]. Precise control of atomic-scale ordering is essential, as deviations can significantly suppress spin polarization and thereby limit device efficiency. The emergence of Mn_3Ga and Mn_2Ru_xGa as compensated Heusler systems introduces a pathway toward ultrafast, stray-field-free spintronics, benefiting spin–orbit torque applications [63, 64]. Moreover, their integration into spin-filter devices and PMA-based memory cells demonstrates their versatility in diverse spintronic architectures. Overall, Heusler alloy thin films represent a cornerstone of modern spintronics, offering scalable and multifunctional platforms for both fundamental research and practical device development.

5.4.4 *Magnetic Insulator Thin Films*

Magnetic insulator thin films have emerged as material in the advances of spintronics because of their ability to support pure spin currents without accompanying charge transport, thereby minimizing Joule heating [65]. Among these, yttrium iron garnet (YIG) stands out as the prototypical magnetic insulator, calibrated for minimal magnetic damping and long magnon propagation lengths, which can exceed several millimeters [66]. Recent developments in epitaxial growth methods have made it possible to fabricate fine YIG films exhibiting PMA, further broadening their utility in field-free switching scenarios driven by SOTs [67].

Beyond garnets, alternative insulators such as spinel ferrites (e.g., $NiFe_2O_4$, $CoFe_2O_4$) and hexaferrites have garnered attention due to their tunable magnetic anisotropies and compatibility with a broader range of substrates [68]. These materials support efficient spin injection and detection when integrated with heavy metals such as Pt, which act as spin Hall sources and detectors [69]. The exchange proximity effect (PE) at interface between magnetic insulators and metals plays crucial role in enabling phenomena like spin pumping and inverse spin Hall effect (ISHE), central to modern spintronic operation [70]. Importantly, the insulating nature of these materials eliminates parasitic currents, making them ideal for energy-efficient, high-frequency spintronic architectures.

5.4.5 *Synthetic Antiferromagnets (SAFs)*

SAFs have emerged as essential building blocks in modern spintronics due to their engineered AFM coupling between FM layers via non-magnetic spacers [71]. Unlike natural antiferromagnets, SAFs offer tunable interlayer exchange through Ruderman-Kittel-Kasuya-Yosida (RKKY) interactions, typically mediated by metallic spacers such as Ru, Ir, or Cr [72, 73]. This artificial coupling enables the realization of a zero net magnetic moment configuration, which significantly reduces stray fields and enhances magnetic stability. The unique advantage of SAFs lies in their ability to retain fast switching dynamics and low-power consumption while offering enhanced thermal stability and robustness against external magnetic perturbations. Recent developments have demonstrated that SAFs can support effective SOT switching even in compensated magnetic states, enabling ultrafast and deterministic magnetization control without any external field [74]. This opens up pathways for advanced non-volatile memory architectures such as SOT-MRAM. Overall, SAFs represent a compelling materials platform that merges the benefits of both FM and AFM orders, paving the way for highly tunable and robust spintronic technologies.

5.4.6 Van der Wall Magnetic Materials

The advent of vdW magnetic materials has introduced a paradigm shift in the realm of thin film spintronics. Exfoliation breakthroughs led to the discovery of monolayer vdW magnets like CrI_3 and Fe_3GeTe_2 with intrinsic long-range magnetic order [75]. This remarkable property defies the conventional understanding framed by the Mermin-Wagner theorem, which prohibits spontaneous symmetry breaking in low-dimensional isotropic Heisenberg magnets at finite temperatures [76]. The key lies in the substantial magnetic anisotropy and SOC inherent to these materials, which suppress magnonic excitations and stabilize ferromagnetism even in atomically thin geometries [77]. Crucially, the precise modulation of interfacial properties in vdW heterostructures has been shown to significantly influence spin-related phenomena, including interlayer FM coupling and enhanced SOC effects.

As illustrated in Fig. 5.5, recent advancements have strategically refined these interfaces and PE to engineer critical magnetic parameters, including T_c, robust exchange bias, spin-lifetime anisotropy, and efficient SOTs.

Furthermore, vdW magnets offer exceptional tunability through external electric fields, strain modulation, electrostatic gating, and heterostructure engineering, enabling dynamic control over magnetic phases and anisotropy [75, 79]. Their atomically clean and dangling-bond-free surfaces facilitate seamless stacking with other 2D materials, allowing the construction of designer spintronic heterostructures with tailored interfacial interactions [80]. Such architectures pave the way for novel phenomena such as gate-tunable tunneling magnetoresistance, proximity-induced magnetism in adjacent layers, and electrically reconfigurable spin textures.

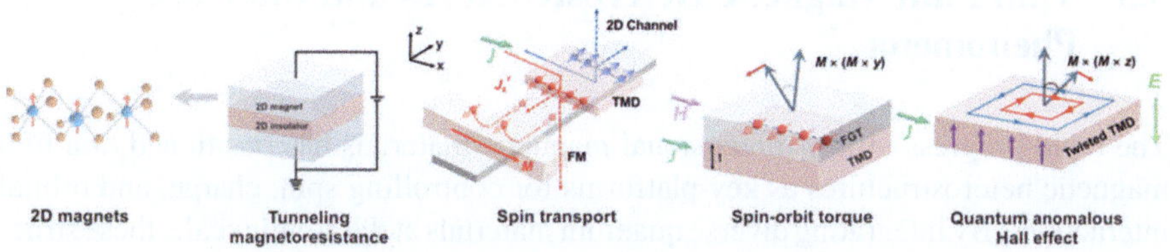

Fig. 5.5 Exploration of vdW heterostructures for advanced spintronic device engineering across multiple architectures [78]. Copyright 2025. Reproduced with permission under a creative commons CC BY 4.0 license from Ref. [78]

These characteristics render vdW magnets ideal candidates for next-generation, ultra-compact, low-power spintronic devices, including memory cells, magnetic sensors, and neuromorphic logic units.

5.4.7 Topological Magnetic Thin Films

Topological magnetic thin films have recently attracted significant interest in spintronics for hosting symmetry-protected quantum states, enabling robust and energy-efficient spin transport [81]. These systems, which typically combine magnetic order with strong SOC, give rise to emergent phenomena such as the quantum anomalous Hall effect (QAHE), topological surface states, and chiral edge modes that offers dissipation less spin transport critical for next-generation low-power spintronic devices [81, 82]. One of the most prominent examples is the intrinsic AFM topological insulator $MnBi_2Te_4$, which exhibits QAHE in thin film form at relatively elevated temperatures due to its layered vdW structure and robust magnetic exchange interactions [82, 83]. These materials allow precise control over thickness, doping, and interface properties, enabling engineered manipulation of the topological band structure and magnetic domains. Moreover, when integrated into magnetic topological insulator (TI) heterostructures, thin films such as Cr-or V-doped $(Bi, Sb)_2Te_3$ display large magneto-transport signatures and enhanced QAHE due to interfacial PE [84]. The interplay of magnetism and topology in such heterostructures leads to tunable Berry curvature and spin-momentum locking, thereby supporting efficient spin current generation and robust edge conduction [85, 86]. Notably, the synthesis of these topological thin films via molecular beam epitaxy (MBE) or pulsed laser deposition (PLD) enables atomically smooth interfaces and high crystalline quality, which is essential for device-grade implementation. As a result, topological magnetic thin films serve as a potential platform for enabling integrating non-volatile memory, logic, and quantum functionalities within a unified spintronic platform.

5.5 Thin Film Magnetic Heterostructures and Interface Phenomena

The rapid progress in low-dimensional magnetic materials has positioned thin film magnetic heterostructures as key platforms for controlling spin, charge, and orbital interactions. By integrating diverse quantum materials at the atomic scale, these structures enable strong interfacial coupling, giving rise to tunable magnetic anisotropy, topological states, and efficient spin-charge conversion. The interface characterized by symmetry breaking and electronic band reconstruction serves as a critical region for the emergence of phenomena such as exchange bias, PMA, and proximity-induced magnetism. Two core motivations drive interest in 2D magnetic heterostructures, i.e., enhancing functional capabilities for spintronic applications and leveraging PE to engineer novel magnetic properties. Particularly, interfacing ferromagnetic layers with heavy metals (Pt, W, Ta) induces robust SOC, generating spin Hall currents that enable SOT-based switching [87]. As depicted in Fig. 5.6, such interface-driven magnetism reveals richer physical behavior than conventional charge-based systems, offering promising pathways for next-generation quantum and spintronic technologies. In particular, vdW heterostructures formed by stacking 2D magnetic layers with non-magnetic 2D crystals offer unprecedented flexibility for tailoring interlayer coupling, magnetic phase behavior, and spin transport characteristics. These heterostructures exploit weak interlayer forces to assemble clean, lattice-mismatched interfaces without the constraints of conventional epitaxy. For instance, CrI_3-based heterostructures with graphene or WSe_2 exhibit gate-tunable interlayer exchange coupling, allowing magnetic phase transitions to be electrically modulated at low power [88]. Additionally, heterostructures involving Fe_3GeTe_2 have demonstrated gate-controlled modulation of T_c and magnetic anisotropy, highlighting the tunability of magnetism at interfaces [89]. Its inherent ferromagnetism originates from strong intralayer exchange interactions within the Fe-Ge-Te framework and can be effectively modulated through electrostatic gating, chemical substitution (e.g., Co or Ni doping), or strain engineering. Fe_3GeTe_2-based heterostructures exhibit pronounced PMA and a sizable anomalous Hall effect, rendering them highly suitable for voltage-controlled spintronic and SOT device architectures. Recent demonstrations of deterministic magnetization switching in Fe_3GeTe_2/Pt bilayers further underscore the promise of Fe-based magnets for scalable, room temperature spintronic technologies.

AFM and ferrimagnetic material interfaces also play a critical role in spintronic heterostructures. When coupled to FM layers, AFM materials such as $MnPS_3$ or NiO induce interfacial exchange bias, stabilizing the magnetization direction and enhancing thermal resistance [90]. Moreover, ferrimagnetic heterostructures combining rare-earth and transition metal thin films (e.g., Gd/Co multilayers) have demonstrated ultrafast magnetization dynamics, driven by sub-picosecond angular momentum transfer across interfaces. These effects are central to the realization of ultrafast spintronic and opto-magnetic devices.

Importantly, interfaces can also host topologically nontrivial spin textures such as skyrmions, which arise from Dzyaloshinskii-Moriya interactions (DMI) at asymmetrical boundaries [91]. Skyrmion-based racetrack memories rely on engineered multilayers, where magnetic and non-magnetic layers alternate to amplify interfacial DMI, enabling the stabilization and current-driven motion of nanoscale spin vortices [91].

The continued development of thin film magnetic heterostructures is expected to unlock new device paradigms that combine ultrathin geometry, reconfigurable magnetism, and efficient spin control. Strategic control over interfacial interactions unlocks novel quantum states and spin configurations, driving progress in low-energy memory, synaptic emulation, and topological quantum systems.

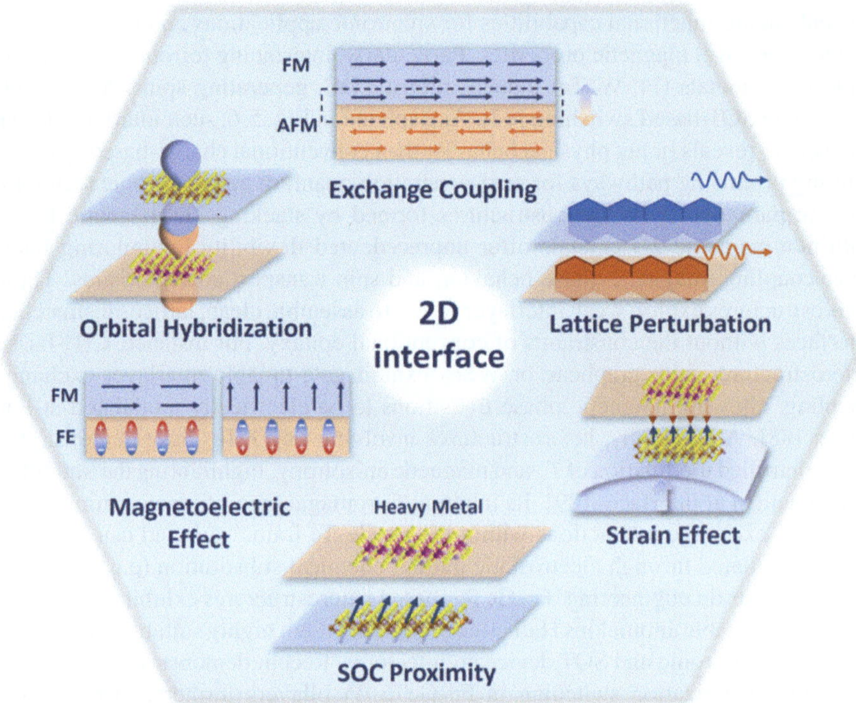

Fig. 5.6 Illustration of interfacial phenomena in 2D magnetic heterostructures, encompassing orbital hybridization, interlayer exchange coupling, lattice perturbation, strain modulation, SOC PE, and magnetoelectric coupling mechanisms [23]. Copyright 2024. Reproduced with permission under a creative commons CC BY 4.0 license from Ref. [23]

5.6 Synthesis Techniques

The controlled synthesis of high-quality 2D magnetic materials is central to the exploration and realization of spin-dependent phenomena at the atomic limit. Given the sensitivity of magnetic order to crystalline symmetry, layer thickness, chemical purity and the choice of synthesis route is pivotal in shaping both the magnetic ground states and the functionality of the device. Several distinct synthesis strategies have emerged and each is systematically analyzed to understand its contribution to structural and functional characteristics.

5.6.1 Mechanical and Liquid Exfoliation

Mechanical exfoliation has been the most accessible and widely used technique for isolating monolayer and few-layer crystals from bulk vdW magnets. This method involves cleaving layered magnetic materials, such as CrI_3 or Fe_3GeTe_2, using adhesive tapes to obtain atomically thin flakes [36]. While it produces pristine crystals with minimal defects, this technique lacks uniformity and scalability, making it more suitable for fundamental studies than industrial-scale device fabrication.

Liquid-phase exfoliation, on the other hand, disperses bulk crystals in solvents under ultrasonication to achieve thickness reduction. Though it offers higher throughput, the process often introduces structural disorder and yields polydisperse flakes, limiting control over flake size, shape, and magnetic uniformity.

5.6.2 Chemical Vapor Deposition

Chemical vapor deposition (CVD) has become a cornerstone technique for the scalable synthesis of 2D magnetic materials. It involves decomposing gaseous precursors (e.g., metal chlorides, organometallics, or chalcogenides) at elevated temperatures onto substrates such as sapphire or SiO_2/Si [92]. The process parameters include substrate temperature, precursor concentration, and carrier gas flow that can be precisely tuned to control layer thickness, grain size, and crystal orientation.

For instance, CVD growth of monolayer VSe_2 and $CrTe_2$ has demonstrated promising control over stoichiometry and magnetic phase stabilization [92]. Moreover, recent advancements in low-pressure CVD and plasma-enhanced CVD have improved crystallinity while minimizing defect formation, thereby enhancing the long-range magnetic ordering in these atomically thin films.

5.6.3 Molecular Beam Epitaxy (MBE)

MBE is regarded as the most precise method for fabricating 2D magnetic heterostructures and ultrathin films with atomic layer control. Conducted in ultra-high vacuum (UHV), MBE enables the co-deposition of magnetic and non-magnetic elements with sub-monolayer accuracy. The slow deposition rate allows for epitaxial alignment, smooth interface formation, and defect suppression that attributes that are critical for investigating interface-induced phenomena such as exchange bias, SOT, and proximity magnetism.

MBE has been successfully applied to synthesize high-purity films of $MnSe_2$, Fe_3GeTe_2, and Cr-doped topological insulators, often on lattice-matched substrates like GaAs or sapphire [92, 93]. The in situ characterization tools integrated within MBE chambers, such as, reflection high-energy electron diffraction (RHEED), which further allow real-time monitoring of growth dynamics.

5.6.4 *Intercalation and Ion Implantation*

Beyond direct growth, intercalation methods introduce foreign atoms or molecules between vdW layers to induce or tune magnetic properties. Alkali metal or 3D-transition metal intercalation can modify charge distribution, introduce magnetic exchange interactions, or even convert non-magnetic hosts into FM systems. Similarly, ion implantation or focused ion beam techniques allow for site-specific doping and defect engineering, providing a route to locally modulate magnetic anisotropy or induce magnetic domains in otherwise non-magnetic matrices.

5.7 Characterization Techniques

Comprehensive characterization of 2D magnetic materials is vital to assess their structural integrity, magnetic ordering, and spintronic functionalities. Given the atomically thin nature and low magnetic moment of such systems, advanced and highly sensitive characterization techniques are required. Spanning various experimental domains, these tools facilitate detailed analysis of morphological features, crystal structure, electronic configurations, magnetic responses, and dynamic spin phenomena, each contributing critical insights into the physics of 2D magnets.

5.7.1 *Structural Characterization*

At the core of thin film analysis lies atomic force microscopy, which offers high-resolution topography images of exfoliated or synthesized flakes, revealing thickness uniformity, grain boundaries, and step heights down to the sub-nanometer scale. For monolayers and heterostructures, scanning tunneling microscopy (STM) not only resolves atomic-scale surface features but, when spin-polarized (SP-STM), can image local magnetic domains and spin textures directly.

X-ray diffraction is employed to analyze the crystal structure, phase purity, and epitaxial orientation, especially in films grown via CVD or MBE. Selected area electron diffraction (SAED) and high-resolution transmission electron microscopy (HRTEM) further elucidate interlayer stacking, interface sharpness, and lattice strain in heterostructures with sub-angstrom precision.

5.7.2 *Magnetic Property Measurements*

MOKE microscopy is a highly sensitive, non-destructive technique for probing magnetization switching and hysteresis in ultrathin magnetic films. It detects changes

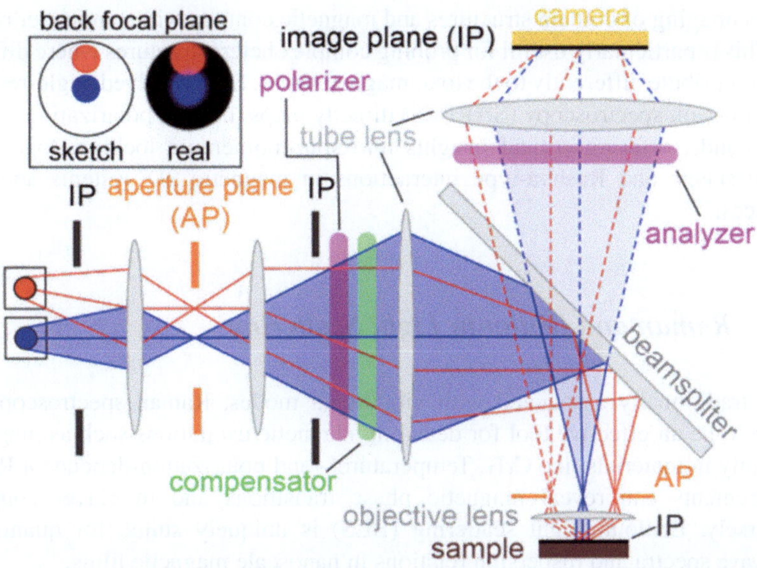

Fig. 5.7 Schematic representation of the MOKE microscope setup, with the inset displaying a photographic image of the experimental apparatus [94]. Copyright 2021. Reproduced with permission from Elsevier

in the polarization of reflected light, which are directly linked to the sample's magnetization.

A linearly polarized beam uniformly illuminates the sample (Fig. 5.7), and the reflected light is analyzed through a compensator analyzer setup, with spatial information captured by a CCD camera. This enables magnetic domain imaging after background subtraction. MOKE is particularly advantageous for its compatibility with cryogenic conditions, external fields, and ultrafast dynamics, including spin precession and damping via time-resolved measurements. Superconducting quantum interference device (SQUID) magnetometry and vibrating sample magnetometry (VSM) offer extremely high sensitivity for measuring net magnetization, magnetic susceptibility, and T_c, albeit with limited spatial resolution. These techniques are essential for verifying the presence of long-range magnetic order and determining anisotropy fields.

5.7.3 Spectroscopic and Element-Specific Probes

X-ray magnetic circular dichroism (XMCD) provides element-specific information on spin and orbital magnetic moments through synchrotron-based absorption spectroscopy. When coupled with photoemission electron microscopy (PEEM), XMCD enables imaging of domain structures and magnetic contrast with nanometer resolution. This is particularly useful for probing complex heterostructures where different layers contribute differently to the total magnetization. Spin-resolved angle-resolved photoemission spectroscopy (SARPES) directly maps the spin polarization of electronic bands, offering critical insights into spin-momentum locking, topological surface states, and Rashba-type interactions in magnetic 2D systems and their interfaces.

5.7.4 Raman and Brillouin Light Scattering

While traditionally associated with vibrational modes, Raman spectroscopy has proven to be an effective tool for detecting magnetic excitations such as magnons, especially in materials like CrI_3. Temperature—and polarization-dependent Raman measurements can reveal magnetic phase transitions and interlayer coupling. Conversely, Brillouin light scattering (BLS) is uniquely suited for quantifying spin-wave spectra and dispersion relations in nanoscale magnetic films.

5.8 Integration Challenges and Strategies

The integration of spintronic devices into mainstream electronic platforms requires meticulous attention to CMOS compatibility, thermal stability, and scaling feasibility. To ensure successful deployment alongside conventional semiconductor technologies, spintronic components must conform to back-end-of-line processing conditions, typically capped below 400 °C, to prevent thermal damage to prefabricated circuitry [95]. Materials such as CoFeB, MgO, and certain Heusler alloys have demonstrated promising CMOS-compatible deposition, particularly through industry-adapted techniques like magnetron sputtering and atomic layer deposition (ALD) [95, 96]. ALD, with its atomic precision and conformal coating capabilities, proves particularly advantageous for scaling spintronic architectures into three-dimensional configurations. Beyond compatibility, thermal robustness of the magnetic layers is essential for ensuring reliable data retention and prolonged device lifetime, especially in applications like STT-MRAM. Magnetic anisotropy energy and T_c are fundamental parameters that determine the thermal stability of magnetic order, serving as key indicators for the robustness of spintronic functionality under operational conditions.

Table 5.1 represents the 2D magnetic materials with the details of their magnetic ordering, T_c, and synthesis methods. Unlike their bulk counterparts or conventional thin films, 2D magnetic systems often exhibit emergent properties arising from reduced dimensionality and enhanced quantum confinement. To improve thermal resilience while preserving fast magnetization switching, several engineering strategies have been adopted. These include interfacial engineering, strain modulation, and the introduction of heavy metal underlayers, which enhance SOC and thereby increase magnetic anisotropy. Such approaches not only stabilize magnetic phases at elevated temperatures but also maintain compatibility with high-speed spintronic applications. As the lateral dimensions of spintronic elements are reduced to meet the ever-tightening demands of miniaturization, scaling challenges become increasingly prominent. These include increased stochasticity in magnetization switching, edge-induced variability, and enhanced thermal noise, all of which threaten device reproducibility and energy efficiency. To address these limitations, materials exhibiting PMA, such as CoFeB/MgO and L10-FePt, are leveraged to sustain stability at reduced bit volumes, enabling sub-10 nm device footprints [98, 99]. Furthermore, VCMA has emerged as a promising approach to lower switching energy by modulating the energy barrier via electric-field application, thereby enabling ultra-low-power, high-speed operation. In parallel, the exploration of atomically thin 2D magnets and topologically protected spin textures, such as skyrmions that offers a radical alternative for ultra-scalable, thermally stable spintronic devices. These systems reduce the critical switching current while enabling high-density information storage through their nanoscale size and inherent robustness. Although still in the developmental stage, the integration of such materials with CMOS architectures opens up avenues for hybrid logic and memory platforms. In this evolving landscape, success hinges on the synergy between material innovation, precise thin film engineering, and interface control, all of which are indispensable for translating the rich physics of spintronics into viable, scalable, and thermally resilient technologies compatible with future computing paradigms.

5.9 Conclusion

The evolution of thin film magnetism, from early quasi-2D systems to modern vdW monolayers, has fundamentally reshaped our understanding of low-dimensional magnetic phenomena and their technological relevance. Magnetic thin films and heterostructures offer a highly tunable platform to engineer spin-dependent transport, interfacial coupling, and anisotropy core mechanisms that underpin spintronic functionality. The emergence of intrinsic magnetic films further extends the frontiers of spin-based devices, enabling atomically thin systems that exhibit long-range order, gate-tunability, and compatibility with flexible architectures. These advances are not only enhancing conventional spintronic applications, such as non-volatile memory and magnetic logic, but also unlocking new possibilities in neuromorphic and quantum computing. The ability of thin film and 2D systems to emulate synaptic plasticity, support stochastic switching, and exhibit topologically protected spin states makes them promising candidates for brain-inspired and quantum information processing. As material synthesis, interface engineering, and heterostructure integration continue to mature, thin film and 2D spintronics are poised to become foundational elements of next-generation computing technologies.

Table 5.1 T_c of distinct 2D magnetic materials [97]

Material	Magnetic ordering	T_c (K)	Synthesis technique
CrI_3	FM	45	Exfoliation
$CrCl_3$	FM/AFM	17 K	Exfoliation
$CrBr_3$	FM	47	Exfoliation
β-MnSe	FM	42.3	CVD
1 T-$MnSe_2$	FM	300	MBE
T-FeTe	AFM	70	CVD
H-FeTe	FM	220	CVD
Fe_5Se_8	FM	300 K	CVD
$CrTe_2$	FM	305–320	Exfoliation
CrTe	FM	350	CVD/exfoliation
VSe_2	FM	300	MBE/exfoliation
Cr_2S_3	FM	120	CVD
ε-Fe_2O_3	FM	291	CVD
γ-Fe_2O_3	FM	950	CVD
$CoFe_2O_4$	FM	390	CVD
Fe_3O_4	FM	350	CVD
Fe_3GeTe_2	FM	203	Exfoliation
$CrSiTe_3$	FM	17–33	Exfoliation
$Cr_2Ge_2Te_6$	FM	30–68	Exfoliation
$MnBi_2Te_4$	AFM	20.6–21.6	Exfoliation

References

1. Onsager L (1944). Crystal Statistics. I. A Two-Dimensional Model with an Order-Disorder Transition. Phys Rev 65:117–149. https://doi.org/10.1103/PhysRev.65.117
2. Elahi E, Dastgeer G, Nazir G, et al (2022). A review on two-dimensional (2D) magnetic materials and their potential applications in spintronics and spin-caloritronic. Comput Mater Sci 213:111670. https://doi.org/10.1016/j.commatsci.2022.111670
3. Patel K, Zhang J, Ren S (2018). Rare-earth-free high energy product manganese-based magnetic materials. Nanoscale 10:11701–11718. https://doi.org/10.1039/C8NR01847B
4. Arnold DP (2007). Review of Microscale Magnetic Power Generation. IEEE Trans Magn 43:3940–3951. https://doi.org/10.1109/TMAG.2007.906150
5. Burch KS, Mandrus D, Park J-G (2018). Magnetism in two-dimensional van der Waals materials. Nature 563:47–52. https://doi.org/10.1038/s41586-018-0631-z
6. Gangopadhyay AK, Roy SB, Majumdar AK (1986). Phase transition to long-range ferromagnetism in $Au_{82}Fe_{18}$ and the associated critical exponents. Phys Rev B 33:5010–5015. https://doi.org/10.1103/PhysRevB.33.5010
7. Ren J, Wang Z, Chen W, You W-L (2022). Long-range order and quantum criticality in antiferromagnetic chains with long-range staggered interactions. Phys Rev E 105:034128. https://doi.org/10.1103/PhysRevE.105.034128
8. Cannas SA, Tamarit FA (1996). Long-range interactions and nonextensivity in ferromagnetic spin models. Phys Rev B 54:R12661–R12664. https://doi.org/10.1103/PhysRevB.54.R12661

9. Nguyen MD, Tran H-V, Xu S, Lee TR (2021). Fe_3O_4 Nanoparticles: Structures, Synthesis, Magnetic Properties, Surface Functionalization, and Emerging Applications. Appl Sci 11:11301. https://doi.org/10.3390/app112311301
10. Liu M, Ye Y, Ye J, et al (2023). Recent Advances of Magnetite (Fe_3O_4)-Based Magnetic Materials in Catalytic Applications. Magnetochemistry 9:110. https://doi.org/10.3390/magnetochemistry9040110
11. BM McCoy (1973). CHAPTER V. Specific Heat of Onsager's Lattice in the Absence of a Magnetic Field. In: The Two-Dimensional Ising Model. Harvard University Press, pp. 76–112
12. Zhang L, Fan J, Li L, et al (2010). Critical properties of the 3D-Heisenberg ferromagnet $CdCr_2Se_4$. EPL (Europhysics Lett) 91:57001. https://doi.org/10.1209/0295-5075/91/57001
13. Katsura S (1962). Statistical Mechanics of the Anisotropic Linear Heisenberg Model. Phys Rev 127:1508–1518. https://doi.org/10.1103/PhysRev.127.1508
14. Halperin BI (2019). On the Hohenberg–Mermin–Wagner Theorem and Its Limitations. J Stat Phys 175:521–529. https://doi.org/10.1007/s10955-018-2202-y
15. Hirohata A, Yamada K, Nakatani Y, et al (2020). Review on spintronics: Principles and device applications. J Magn Magn Mater 509:166711. https://doi.org/10.1016/j.jmmm.2020.166711
16. Kawahara T, Ito K, Takemura R, Ohno H (2012). Spin-transfer torque RAM technology: Review and prospect. Microelectron Reliab 52:613–627. https://doi.org/10.1016/j.microrel.2011.09.028
17. Bauer GEW, Tserkovnyak Y, Brataas A, Kelly PJ (2007). Chapter Two Magnetic Nanostructures: Currents and Dynamics. pp 123–148
18. Nguyen VD, Rao S, Wostyn K, Couet S (2024). Recent progress in spin-orbit torque magnetic random-access memory. npj Spintron 2:48. https://doi.org/10.1038/s44306-024-00044-1
19. Sbiaa R, Meng H, Piramanayagam SN (2011). Materials with perpendicular magnetic anisotropy for magnetic random access memory. Phys status solidi—Rapid Res Lett 5:413–419. https://doi.org/10.1002/pssr.201105420
20. Manchon A, Železný J, Miron IM, et al (2019). Current-induced spin-orbit torques in ferromagnetic and antiferromagnetic systems. Rev Mod Phys 91:035004. https://doi.org/10.1103/RevModPhys.91.035004
21. Davies CS, Mentink JH, Kimel AV, et al (2022). Helicity-independent all-optical switching of magnetization in ferrimagnetic alloys. J Magn Magn Mater 563:169851. https://doi.org/10.1016/j.jmmm.2022.169851
22. Marrows CH, Barker J, Moore TA, Moorsom T (2024). Neuromorphic computing with spintronics. npj Spintron 2:12. https://doi.org/10.1038/s44306-024-00019-2
23. Zhang B, Lu P, Tabrizian R, et al (2024). 2D Magnetic heterostructures: spintronics and quantum future. npj Spintron 2:6. https://doi.org/10.1038/s44306-024-00011-w
24. Lin GT, Zhuang HL, Luo X, et al (2017). Tricritical behavior of the two-dimensional intrinsically ferromagnetic semiconductor $CrGeTe_3$. Phys Rev B 95:245212. https://doi.org/10.1103/PhysRevB.95.245212
25. Petrović AP, Psaroudaki C, Fischer P, et al (2025). Colloquium : Quantum properties and functionalities of magnetic skyrmions. Rev Mod Phys 97:031001. https://doi.org/10.1103/RevModPhys.97.031001
26. Majumdar S, Dijken S van (2014). Pulsed laser deposition of $La_{1-x}Sr_xMnO_3$: thin-film properties and spintronic applications. J Phys D Appl Phys 47:034010. https://doi.org/10.1088/0022-3727/47/3/034010
27. Parkes DE, Shelford LR, Wadley P, et al (2013). Magnetostrictive thin films for microwave spintronics. Sci Rep 3:2220. https://doi.org/10.1038/srep02220
28. Thompson D, Romankiw L, Mayadas A (1975). Thin film magnetoresistors in memory, storage, and related applications. IEEE Trans Magn 11:1039–1050. https://doi.org/10.1109/TMAG.1975.1058786
29. Lappalainen J, Mizsei J, Huotari M (2019). Neuromorphic thermal-electric circuits based on phase-change VO_2 thin-film memristor elements. J Appl Phys 125: https://doi.org/10.1063/1.5037990

30. Zollner K, Kurpas M, Gmitra M, Fabian J (2025). First-principles determination of spin–orbit coupling parameters in two-dimensional materials. Nat Rev Phys 7:255–269. https://doi.org/10.1038/s42254-025-00818-4
31. Soumyanarayanan A, Reyren N, Fert A, Panagopoulos C (2016). Emergent phenomena induced by spin–orbit coupling at surfaces and interfaces. Nature 539:509–517. https://doi.org/10.1038/nature19820
32. Falicov LM, Pierce DT, Bader SD, et al (1990). Surface, interface, and thin-film magnetism. J Mater Res 5:1299–1340. https://doi.org/10.1557/JMR.1990.1299
33. Zhao Y-P, Palasantzas G, Wang G-C, De Hosson JTM (1999). Surface/interface-roughness-induced demagnetizing effect in thin magnetic films. Phys Rev B 60:1216–1226. https://doi.org/10.1103/PhysRevB.60.1216
34. Husain S, Figueiredo-Prestes N, Fayet O, et al (2023). Origin of the anomalous Hall effect at the magnetic insulator/heavy metals interface. Appl Phys Lett 122: https://doi.org/10.1063/5.0132895
35. Meng KK, Miao J, Xu XG, et al (2016). Anomalous Hall effect in $Mn_{1.5}Ga$/Ta and $Mn_{1.5}Ga$/Pt bilayers: Modification from spin-orbit coupling of heavy metals. Phys Rev B 93:060406. https://doi.org/10.1103/PhysRevB.93.060406
36. Liu P, Zhang Y, Li K, et al (2023). Recent advances in 2D van der Waals magnets: Detection, modulation, and applications. iScience 26:107584. https://doi.org/10.1016/j.isci.2023.107584
37. Ranno L, Llobet A, Tiron R, Favre-Nicolin E (2002). Strain-induced magnetic anisotropy in epitaxial manganite films. Appl Surf Sci 188:170–175. https://doi.org/10.1016/S0169-4332(01)00730-9
38. Huang W, Zhu J, Zeng HZ, et al (2006). Strain induced magnetic anisotropy in highly epitaxial $CoFe_2O_4$ thin films. Appl Phys Lett 89: https://doi.org/10.1063/1.2424444
39. Heinrich B, Cochran JF (1993). Ultrathin metallic magnetic films: magnetic anisotropies and exchange interactions. Adv Phys 42:523–639. https://doi.org/10.1080/00018739300101524
40. Jaiswal S, Lee K, Langer J, et al (2019). Tuning of interfacial perpendicular magnetic anisotropy and domain structures in magnetic thin film multilayers. J Phys D Appl Phys 52:295002. https://doi.org/10.1088/1361-6463/ab1c42
41. Khoo KH, Wu G, Jhon MH, et al (2013). First-principles study of perpendicular magnetic anisotropy in CoFe/MgO and CoFe/$Mg_3B_2O_6$ interfaces. Phys Rev B 87:174403. https://doi.org/10.1103/PhysRevB.87.174403
42. Yu Z, Yang L, Guo K, et al (2025). Field-Free Switching of Perpendicular Magnetization in Low-Symmetry Materials. Adv Funct Mater. https://doi.org/10.1002/adfm.202505170
43. Rao RA, Lavric D, Nath TK, et al (1999). Effects of film thickness and lattice mismatch on strain states and magnetic properties of $La_{0.8}Ca_{0.2}MnO_3$ thin films. J Appl Phys 85:4794–4796. https://doi.org/10.1063/1.370484
44. Xie Y, Zhan Q, Shang T, et al (2017). Effect of epitaxial strain and lattice mismatch on magnetic and transport behaviors in metamagnetic FeRh thin films. AIP Adv 7: https://doi.org/10.1063/1.4976301
45. de Wijn HW, Walker LR, Walstedt RE (1973). Spin-Wave Analysis of the Quadratic-Layer Antiferromagnets K_2NiF_4, K_2MnF_4, and Rb_2MnF_4. Phys Rev B 8:285–299. https://doi.org/10.1103/PhysRevB.8.285
46. Birgeneau RJ, Guggenheim HJ, Shirane G (1970). Neutron Scattering Investigation of Phase Transitions and Magnetic Correlations in the Two-Dimensional Antiferromagnets K_2NiF_4, Rb_2MnF_4, Rb_2FeF_4. Phys Rev B 1:2211–2230. https://doi.org/10.1103/PhysRevB.1.2211
47. Trier F, Noël P, Kim J-V, et al (2021). Oxide spin-orbitronics: spin–charge interconversion and topological spin textures. Nat Rev Mater 7:258–274. https://doi.org/10.1038/s41578-021-00395-9
48. Kovaleva EA, Kuzubov AA, Avramov PV, et al (2016). Characterization of LSMO/C_{60} spinterface by first-principle calculations. Org Electron 37:55–60. https://doi.org/10.1016/j.orgel.2016.06.021
49. Li F, Song C, Gu YD, et al (2017). Chemical modulation of electronic structure at the excited state. Phys Rev B 96:245108. https://doi.org/10.1103/PhysRevB.96.245108

50. Sun T, Ji F, Liu Y, et al (2019). $La_{1-x}Sr_xMnO_3$:$Ag_{0.2}$ ($0.1 \leq x \leq 0.2$) ceramics with large room-temperature TCR for uncooled infrared bolometers. J Eur Ceram Soc 39:352–357. https://doi.org/10.1016/j.jeurceramsoc.2018.10.004
51. Bowen M, Bibes M, Barthélémy A, et al (2003). Nearly total spin polarization in $La_{2/3}Sr_{1/3}MnO_3$ from tunneling experiments. Appl Phys Lett 82:233–235. https://doi.org/10.1063/1.1534619
52. Wu H-C, Coileáin CÓ, Abid M, et al (2015). Spin-dependent transport properties of Fe_3O_4/MoS_2/Fe_3O_4 junctions. Sci Rep 5:15984. https://doi.org/10.1038/srep15984
53. Nonaka Y, Wakabayashi YK, Shibata G, et al (2021). Reduced magnetocrystalline anisotropy of $CoFe_2O_4$ thin films studied by angle-dependent x-ray magnetic circular dichroism. AIP Adv 11: https://doi.org/10.1063/5.0057576
54. Liu H, Yang X (2017). A brief review on perovskite multiferroics. Ferroelectrics 507:69–85. https://doi.org/10.1080/00150193.2017.1283171
55. Fernández-Posada CM, Castro A, Kiat J-M, et al (2016). A novel perovskite oxide chemically designed to show multiferroic phase boundary with room-temperature magnetoelectricity. Nat Commun 7:12772. https://doi.org/10.1038/ncomms12772
56. Sharma P, Kant R, Bhardwaj P (2024). Heusler Alloys: Structural, Electronic, Magnetic, Mechanical, and Thermoelectric Features - A Review. Int J Adv Sci Eng 11:4044–4054. https://doi.org/10.29294/IJASE.11.2.2024.4044-4054
57. Elphick K, Frost W, Samiepour M, et al (2021). Heusler alloys for spintronic devices: review on recent development and future perspectives. Sci Technol Adv Mater 22:235–271. https://doi.org/10.1080/14686996.2020.1812364
58. HIROHATA A, KIKUCHI M, TEZUKA N, et al (2006). Heusler alloy/semiconductor hybrid structures. Curr Opin Solid State Mater Sci 10:93–107. https://doi.org/10.1016/j.cossms.2006.11.006
59. Wurmehl S, Fecher GH, Kandpal HC, et al (2006). Investigation of Co_2FeSi: The Heusler compound with highest Curie temperature and magnetic moment. Appl Phys Lett 88: https://doi.org/10.1063/1.2166205
60. Abdallah I, Pradines B, Ratel-Ramond N, et al (2017). Evolution of magnetic properties and damping coefficient of Co_2MnSi Heusler alloy with Mn/Si and Co/Mn atomic disorder. J Phys D Appl Phys 50:035003. https://doi.org/10.1088/1361-6463/50/3/035003
61. Takamura Y, Suzuki T, Fujino Y, Nakagawa S (2014). Full-Heusler Co_2FeSi alloy thin films with perpendicular magnetic anisotropy induced by MgO-interfaces. J Appl Phys 115:. https://doi.org/10.1063/1.4865971
62. Chen J, Sakuraba Y, Yakushiji K, et al (2020). Fully epitaxial giant magnetoresistive devices with half-metallic Heusler alloy fabricated on poly-crystalline electrode using three-dimensional integration technology. Acta Mater 200:1038–1045. https://doi.org/10.1016/j.actamat.2020.04.002
63. Elliott P, Müller T, Dewhurst JK, et al (2016). Ultrafast laser induced local magnetization dynamics in Heusler compounds. Sci Rep 6:38911. https://doi.org/10.1038/srep38911
64. Banerjee C, Teichert N, Siewierska KE, et al (2020). Single pulse all-optical toggle switching of magnetization without gadolinium in the ferrimagnet Mn2RuxGa. Nat Commun 11:4444. https://doi.org/10.1038/s41467-020-18340-9
65. Emori S, Li P (2021). Ferrimagnetic insulators for spintronics: Beyond garnets. J Appl Phys 129: https://doi.org/10.1063/5.0033259
66. Dreyer R, Liebing N, Edwards ERJ, et al (2021). Spin-wave localization and guiding by magnon band structure engineering in yttrium iron garnet. Phys Rev Mater 5:064411. https://doi.org/10.1103/PhysRevMaterials.5.064411
67. Guo CY, Wan CH, Zhao MK, et al (2019). Spin-orbit torque switching in perpendicular $Y_3Fe_5O_{12}$/Pt bilayer. Appl Phys Lett 114: https://doi.org/10.1063/1.5098033
68. Fritsch D, Ederer C (2010). Epitaxial Strain Effects in the Spinel Ferrites $CoFe_2O_4$ and $NiFe_2O_4$ From First Principles. Phys Rev B 82:104117. https://doi.org/10.1103/PhysRevB.82.104117
69. Gray MT, Emori S, Gray BA, et al (2018). Spin-Current Generation in Low-Damping $Ni_{0.65}Zn_{0.35}Al_{0.8}Fe_{1.2}O_4$ Spinel Ferrite. Phys Rev Appl 9:064039. https://doi.org/10.1103/PhysRevApplied.9.064039

70. Amamou W, Pinchuk I V., Trout AH, et al (2018). Magnetic proximity effect in Pt / $CoFe_2O_4$ bilayers. Phys Rev Mater 2:011401. https://doi.org/10.1103/PhysRevMaterials.2.011401
71. Gomonay O, Jungwirth T, Sinova J (2017). Concepts of antiferromagnetic spintronics. Phys status solidi—Rapid Res Lett 11: https://doi.org/10.1002/pssr.201700022
72. Xu T, Liu J, Zhang X, et al (2022). Systematic Control of the Interlayer Exchange Coupling in Perpendicularly Magnetized Synthetic Antiferromagnets. Phys Rev Appl 18:054051. https://doi.org/10.1103/PhysRevApplied.18.054051
73. Liang S, Chen R, Cui Q, et al (2023). Ruderman–Kittel–Kasuya–Yosida-Type Interlayer Dzyaloshinskii–Moriya Interaction in Synthetic Magnets. Nano Lett 23:8690–8696. https://doi.org/10.1021/acs.nanolett.3c02607
74. Zhang PX, Liao LY, Shi GY, et al (2018). Spin-orbit torque in a completely compensated synthetic antiferromagnet. Phys Rev B 97:214403. https://doi.org/10.1103/PhysRevB.97.214403
75. Gibertini M, Koperski M, Morpurgo AF, Novoselov KS (2019). Magnetic 2D materials and heterostructures. Nat Nanotechnol 14:408–419. https://doi.org/10.1038/s41565-019-0438-6
76. Jenkins S, Rózsa L, Atxitia U, et al (2022). Breaking through the Mermin-Wagner limit in 2D van der Waals magnets. Nat Commun 13:6917. https://doi.org/10.1038/s41467-022-34389-0
77. Li Y, Jiang Z, Li J, et al (2019). Magnetic anisotropy of the two-dimensional ferromagnetic insulator $MnBi_2Te_4$. Phys Rev B 100:134438. https://doi.org/10.1103/PhysRevB.100.134438
78. Zhao Z, Lin Y, Avsar A (2025). Novel spintronic effects in two-dimensional van der Waals heterostructures. npj 2D Mater Appl 9: https://doi.org/10.1038/s41699-025-00546-4
79. Miao F, Liang S-J, Cheng B (2021). Straintronics with van der Waals materials. npj Quantum Mater 6:59. https://doi.org/10.1038/s41535-021-00360-3
80. Choi E, Sim KI, Burch KS, Lee YH (2022). Emergent Multifunctional Magnetic Proximity in van der Waals Layered Heterostructures. Adv Sci 9:. https://doi.org/10.1002/advs.202200186
81. Chang C-Z, Liu C-X, MacDonald AH (2023). Colloquium : Quantum anomalous Hall effect. Rev Mod Phys 95:011002. https://doi.org/10.1103/RevModPhys.95.011002
82. Liu J, Hesjedal T (2023). Magnetic Topological Insulator Heterostructures: A Review. Adv Mater 35: https://doi.org/10.1002/adma.202102427
83. Li J, Li Y, Du S, et al (2019). Intrinsic magnetic topological insulators in van der Waals layered $MnBi_2Te_4$ -family materials. Sci Adv 5:. https://doi.org/10.1126/sciadv.aaw5685
84. Yao Q, Ji Y, Chen P, et al (2021). Topological insulators-based magnetic heterostructures. Adv Phys X 6:. https://doi.org/10.1080/23746149.2020.1870560
85. Bac S-K, Le Mardelé F, Wang J, et al (2025). Probing Berry Curvature in Magnetic Topological Insulators through Resonant Infrared Magnetic Circular Dichroism. Phys Rev Lett 134:016601. https://doi.org/10.1103/PhysRevLett.134.016601
86. Oh T, Nagaosa N (2024). Unraveling the dynamics of magnetization in topological insulator-ferromagnet heterostructures via spin-orbit torque. npj Spintron 2:42. https://doi.org/10.1038/s44306-024-00045-0
87. Yang Y, Xie Z, Zhao Z, et al (2024). Harnessing synergy of spin and orbital currents in heavy metal/ferromagnet multilayers. Commun Phys 7:336. https://doi.org/10.1038/s42005-024-018 29-w
88. Seyler KL, Zhong D, Huang B, et al (2018). Valley Manipulation by Optically Tuning the Magnetic Proximity Effect in WSe_2/CrI_3 Heterostructures. Nano Lett 18:3823–3828. https://doi.org/10.1021/acs.nanolett.8b01105
89. Roemer R, Lee DHD, Smit S, et al (2024). Unraveling the electronic structure and magnetic transition evolution across monolayer, bilayer, and multilayer ferromagnetic Fe_3GeTe_2. npj 2D Mater Appl 8:63. https://doi.org/10.1038/s41699-024-00499-0
90. Puthirath Balan A, Kumar A, Reiser P, et al (2024). Identifying the Origin of Thermal Modulation of Exchange Bias in $MnPS_3/Fe_3GeTe_2$ van der Waals Heterostructures. Adv Mater 36:. https://doi.org/10.1002/adma.202403685
91. Quessab Y, Xu J-W, Ma CT, et al (2020). Tuning interfacial Dzyaloshinskii-Moriya interactions in thin amorphous ferrimagnetic alloys. Sci Rep 10:7447. https://doi.org/10.1038/s41598-020-64427-0

92. Och M, Martin M-B, Dlubak B, et al (2021). Synthesis of emerging 2D layered magnetic materials. Nanoscale 13:2157–2180. https://doi.org/10.1039/D0NR07867K
93. Liu S, Yuan X, Zou Y, et al (2017). Wafer-scale two-dimensional ferromagnetic Fe_3GeTe_2 thin films grown by molecular beam epitaxy. npj 2D Mater Appl 1:30. https://doi.org/10.1038/s41699-017-0033-3
94. Jovicevic-Klug M, Jovicevic-Klug P, McCord J, Podgornik B (2021). Investigation of microstructural attributes of steel surfaces through magneto-optical Kerr effect. J Mater Res Technol 11:1245–1259. https://doi.org/10.1016/j.jmrt.2021.01.106
95. Endoh T, Honjo H (2018). A Recent Progress of Spintronics Devices for Integrated Circuit Applications. J Low Power Electron Appl 8:44. https://doi.org/10.3390/jlpea8040044
96. Mo C-L, Lin H-C, Chen M-J (2025). Ultra-high remnant polarization in wake-up-free ferroelectric thin films realized by atomic layer annealing. Acta Mater 288:120855. https://doi.org/10.1016/j.actamat.2025.120855
97. Yuan S, Zhou Y, Yin L, et al (2025). Recent Progress in Chemical Vapor Deposition of 2D Magnetic Materials. Adv Phys Res 4:. https://doi.org/10.1002/apxr.202400169
98. Sampan-A-Pai S, Phoomatna R, Boonruesi W, et al (2023). Magnetization dynamics at finite temperature in CoFeB–MgO based MTJs. Sci Rep 13:2637. https://doi.org/10.1038/s41598-023-29597-7
99. Sun TH, Chen SC, Su WH, Chen JR (2013). Microstructures and magnetic properties of L10 FePt films deposited on NaCl-type films. Thin Solid Films 544:504–508. https://doi.org/10.1016/j.tsf.2013.01.038

Chapter 6
Multiferroic Thin Films for Device Applications

Vinay Pratap Verma, Upkar Kumar Verma, and Sandeep Kumar Pundir

6.1 Introduction

The requirement of miniaturization, higher functionality, and low-power consumption has incessantly driven the rapid advancement of modern electronics [1–3]. Since multiferroics are a representation of two or more ferroic order parameters in a single phase, this family of materials has been of peculiar interest among such possible candidates [1, 4]. The three fundamental ferroic orders are ferroelectricity, characterized by a spontaneous and switchable electric polarization [5, 6]; ferromagnetism, defined by spontaneous magnetic ordering; and ferro-elasticity, which involves spontaneous strain accompanied by domain switching under stress [7, 8]. These ferroic orders are schematically represented in Fig. 3.1 in which ferroelectric dipole alignment in $BaTio_3$, spin ordering in Fe, and ferro-elastic strain domains are elaborated [1].

Magnetoelectric (ME) coupling, is the potential coupling between ferroelectric and ferromagnetic properties, creates entirely new possibilities for device designing. For example, such coupling makes it possible to modulate electric polarization with a magnetic field or control magnetization with an electric field. The deployment of this double control mechanism could enable ultra-low-power and multifunctional devices, providing a remedy to the energy inefficiencies built into traditional charge-based structures [9].

The coexistence of multiple ferroic order simultaneously in a single material is remarkable in multiferroic materials which is not observed. Since, different electronic configurations are necessary to exhibit ferroelectricity and magnetism. Ferroelectricity existed in transition metal oxides with unoccupied d-orbitals, like Ti^{4+} in

V. P. Verma · S. K. Pundir (✉)
Department of Physics, University of Lucknow, Lucknow, Uttar Pradesh, India
e-mail: Pundir_s@lkouniv.ac.in

U. K. Verma
Department of Physics, B. S. N. V. P. G. College, Lucknow, Uttar Pradesh, India

S. Sharma (ed.), *Multifunctional Heterostructure Thin Films Handbook*,
https://doi.org/10.1007/978-3-032-18692-8_6

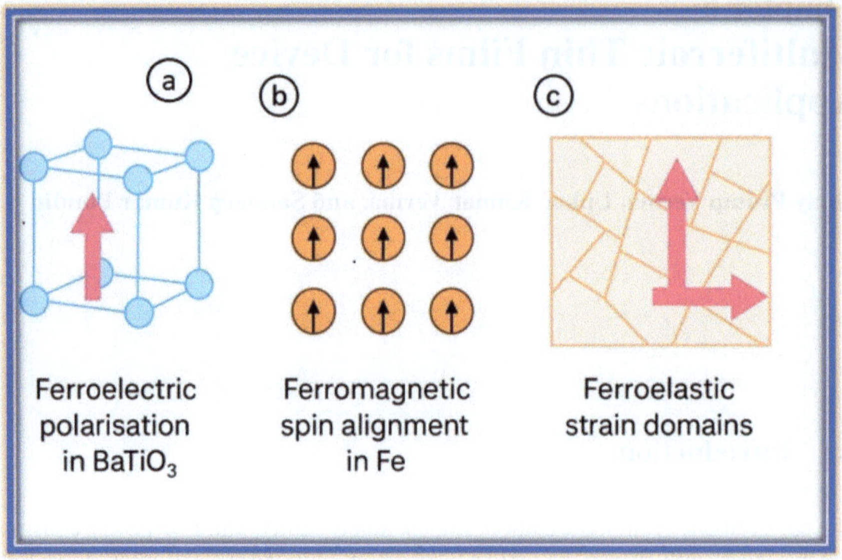

Fig. 3.1 Illustration of three primary ferroic orders: **a** ferroelectric polarization in BaTio3, **b** ferromagnetic spin alignment in Fe, **c** ferro-elastic strain domains

$BaTiO_3$, while magnetism is present due to half-filled d-orbitals to maintain localized magnetic moments, like Fe^{2+} or Mn^{3+} [9]. It is very crucial to coexistence of these odd properties in a single crystalline material. Thus, due to this extraordinary character few single-phase multiferroics are known. This exceptional property and toughness of it attract intention of researcher toward intrinsic single-phase materials, such as $BaFeO_3$and $YMnO_3$. Thus, development advanced methods for designing composite heterostructures with integration of ferroic functionalities establish a new era [10].

6.1.1 Multiferroic Thin Films Devices

Multiferroic thin films are emerging new exceptional material in this new era, such as bulk crystals, thin film exhibits unusual properties like flexibility in strain, composition, and interfaces. Using strain engineering, lattice mismatch between a substrate and a film can stabilize metastable phases having better ferroic properties that are not generally seen in bulk [11]. For example, $BaFeO_3$ thin films deposition with suitable strain engineering shows a tetragonal-like phase with a giant polarization, larger than its bulk. Thin film interfaces engineering also provide the way of controlling charge, orbital, and spin states, with artificial manipulation of weak or nonexistent coupling phenomena. Thin films are also naturally compatible in semiconductor fabrication and provide a straightforward route to incorporating multiferroics materials into electronic and spintronics devices [12].

6.1.2 Historical Background

Discovery of $BaFeO_3$ (BFO) in thin film form early was not considerable but the twenty-first century was become a turning point for BFO. $BaFeO_3$ is a compound that exhibit ferroelectricity near the Curie temperature at 1100 K and antiferromagnetism near the Neel temperature at 643 K [13]. BFO is the one of the few multiferroics that function at room temperature. Its adventures ferroic behavior, with the tunable epitaxial strain and defect engineering, has made it an ideal system for fundamental research and applied science to design new materials for thin film having better and advanced characteristics. $BaFeO_3$ thin films are also best suited for memory devices with tunnel junctions and spintronic structures. Thus, it is motivating the discovery of other perovskite-based and non-perovskite-based multiferroics having better properties [14].

Apart from $BaFeO_3$, there have been various multiferroic materials are known based on different mechanisms. Spin-induced ferroelectricity is found in orthorhombic rare-earth manganite, for example, in $TbMnO_3$ non-collinear magnetic order and electric polarization arises from spin orbit coupling. Hexagonal manganite like $YMnO_3$ shows geometric ferroelectricity due to structural distortion, and it serves as a prototype for improper ferroelectricity [15]. Though these materials exhibit multiferroicity at low temperatures, but they provide valuable insight on the microscopic mechanisms of ferroic coupling. Which help to researcher in the designing and development of room-temperature multiferroic materials [16].

6.1.3 Synthesis of Multiferroic Thin Films

Synthesis of multiferroic thin films is very crucial to maintain its properties during processing of it. It is done by sophisticated deposition methods at atomic level for accuracy and defect reduction [17]. Pulsed laser deposition (PLD) has been a very popular technique, well suited to complicated oxide materials, since it is compactible to stoichiometric transfer from substrate to target material. Molecular beam epitaxy (MBE) has highest control over growth of the thin film at the atomic level and useful in layer-by-layer build-up of heterostructures. But it is high expensive restricted its uses [18]. Atomic layer deposition (ALD) having scalability and conformality, that makes it a good choice for large-scale integration. While chemical solution deposition (CSD) is a low-cost method for the preparation of multiferroic films on large areas [19]. New emerging technologies enabled to control structural, electronic, and ferroic characteristics, that provide a wider space for designing multiferroic thin films.

6.1.4 Applications of Multiferroic Thin Films

Multiferroic thin films having very wide applications. In memory devices, ferroelectric random access memory (FeRAM) uses switchable polarization for non-volatile memory devices and magnetoelectric random access memory (ME-RAM) uses polarization with control of magnetization for higher speed and reduced energy operation [20]. In spintronics, multiferroic materials enables electrical manipulation of spin transport and provides a much-desired function for the realization of logics that are beyond-CMOS logic devices [21]. Their good detection characteristics with electric and magnetic disturbances makes it a good candidate for high-accuracy sensors and actuators devices in microelectromechanical systems (MEMS). In addition to, multiferroic composites have been used as energy harvesters, since it has ability to convert mechanical strain into useful electrical energy with coupled ferroic phenomena [22]. Even more in the future, multiferroics are emerging candidate for neuromorphic computing and devices, in which ferroic domain switching can emulate synaptic plasticity, providing new dimensions for brain-based computing.

6.1.5 Major Issues in Multiferroic Thin Films

Even with these impressive properties, there are still major issues faced in the field of multiferroic materials and its development. Which hinders the commercially exploitable of multiferroic thin films [23]. The critical concerns are leakage currents due to oxygen vacancies and other structural defects, fatigue, and retention loss of ferroelectric switching. The challenge of achieving uniformity and reproducibility over very large wafer areas is a crucial point. The inherent constraint is realizing room-temperature magnetoelectric coupling in single-phase materials. Composite heterostructures and artificial superlattices may reduce this [24]. At the same time, the observation of emergent phenomena like topological domain vortices, skyrmions, and exotic spin textures in thin films force to think the field is still very far from novel physical properties that can lead the establishment of new era of unseen applications [25].

6.1.6 Recent Ongoing Research in Multiferroic Thin Films

Recent progress in computational materials science with artificial intelligence (AI) has extended the scope of the field [26]. Machine learning (ML) algorithms are used to predict structure correlations and property that helps to discover promising novel compounds with optimize growth parameters. Similarly, such data-driven methodologies revolutionize the design and development of multiferroic thin film by hastening them in a never-before-seen scale. Most researches are focused on materials issues to minimize error, with adventures characteristics in supporting experimental work [27].

In short, the study of multiferroic thin films is the junction of basic condensed matter physics. Cutting-edge synthesis methods make it suitable to engineer future device [28]. Their capability to integrate several ferroic orders on a single platform make it good candidate for memory, spintronics, sensing, energy harvesting [29], and neuromorphic computing. Despite of these challenges in practice, improvements in thin film growth technology minimize these using proper tuning with defect engineering, and interface engineering. Coupled with predictive AI designed structures, provide a solid foundation for it so it becomes popular commercial devices. This chapter provides a detailed overview of multiferroic thin films, from their basic principles to their synthesis process [23], including structure and functional properties, and, at last, integration to devices. The discourse with considering how current limitations in practice are holding back their exploitation, and outline future research that must unlock their full potential [30].

6.2 Fundamentals of Multiferroics

Since in a single-phase multiple primary ferroic order are existed. So, composite multiferroic materials are becomes special with coexistence of multiple ferroic orders [30]. Ferroelectricity, ferromagnetism, and ferro-elasticity are the three main ferroic orders [31]. The coexistence of different orders and coupling among them make it capable for advanced functionalities that are not possible with traditional materials [31].

6.2.1 Ferroelectricity

Ferroelectric materials have spontaneous electric polarization, which can be reversed by an applied external electric field [32]. The non-centrosymmetric structure of ions in the crystal lattice induces a permanent dipole moment in the material. For example, in $BaTiO_3$, ion Ti^{4+} displacement within the oxygen octahedron causes a switchable dipole.

The main characteristics of ferroelectricity are the polarization hysteresis loop, like the magnetic hysteresis loop [33]. Curie temperature (T_c) is the temperature at which the material loses ferroelectricity property [34]. Thus, the domain structure in various regions of the crystal have polarization in different directions [35]. Ferroelectric thin films characteristics are mainly controlled by its epitaxial strain, film thickness, and interface conditions, that can significantly alter their polarization behavior.

6.2.2 Ferromagnetism

Ferromagnetism is generally observed due to the magnetic moments of atoms are aligned in a parallel direction by exchange interactions, that creates a net macroscopic magnetization in the absence of external field. Fe, Co, and Ni are model ferro materials, and perovskite-structured $La_{0.7}Sr_{0.3}MnO_3$-oxides also exhibit ferromagnetism nature. Magnetization hysteresis loop (M-H loop) is also a prominent characteristic of ferromagnetism. Magnetic domains are divided by domain walls and are essential to switching behavior [36]. Thus, above the Curie temperature (T_c), material behavior turns to paramagnetic [37]. Tuning ferromagnetism in thin films through dimensional confinement, interface exchange, strain, and magnetic properties make it suitable for spintronic applications.

6.2.3 Ferro-elasticity

When a mechanical stress is applied, ferro-elastic materials ordering spontaneously reoriented [38]. Structural domains such as distortions from the tetragonal-to-orthorhombic form are typically repositioned in ferro-elastic switching. Since polarization switching is accompanied by structural distortions, ferro-elasticity is often found together with ferroelectricity in the perovskite oxides.

Ferro-elasticity is significant in domain-wall movements, strain engineering, and mechanical coupling phenomena in multiferroic thin films despite less direct usage in devices [11].

6.2.4 Magnetoelectric (ME) Coupling

Magnetoelectric (ME) coupling, which is the coupling between magnetic and electric ordering, is the crucial characteristic of multiferroics. In single-phase multiferroics, this coupling can be direct. In composite multiferroics, it can be indirect. Figure 3.2 shows a schematic illustration of the main coupling mechanisms.

(a) **Single-phase coupling**: Ferroelectricity and antiferromagnetism coexist in some materials, such as $BiFeO_3$, and the interaction between the two order parameters results from Dzyaloshinskii–Moriya interactions, spin orbit coupling, or lattice distortions [39]. However, because of the conflicting electronic requirements ferromagnetism necessitates partially filled d-orbitals, whereas ferroelectricity typically favors empty d-orbitals strong coupling at room temperature is still uncommon. This mechanism shown in Fig. 3.2a.
(b) **Composite coupling**: Magnetoelectric coupling in artificial heterostructures thin films is accomplished by interface effects or strain mediation. For example, through lattice mismatch, a ferroelectric layer can strain a ferromagnetic

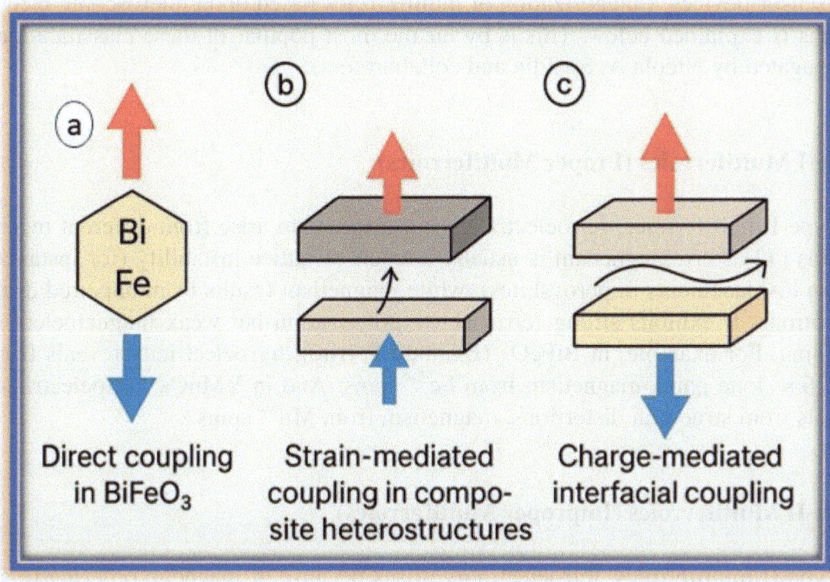

Fig. 3.2 Schematic of different magnetoelectric coupling mechanisms: **a** direct coupling in $BiFeO_3$, **b** strain-mediated coupling in composite heterostructures, **c** charge-mediated interfacial coupling

film, changing its magnetization (strain-mediated effect) [40]. As an alternative, exchange bias effects result from direct interfacial magnetic interactions, whereas charge modulation at interfaces can directly couple polarization and spin states (charge-mediated effect). Figure 3.2b, c shows examples of these. The strain-mediated coupling like piezoelectric-magneto-strictive effect is one of the main mechanisms. Using Interfacial carrier density modulation, or charge-mediated coupling [41], Direct interfacial magnetic interactions, or exchange bias coupling are also used for it. Epitaxial strain and nanoscale interfaces properly designed to maximize coupling efficiency.

6.2.5 Classification of Multiferroics

Multiferroics are compelling materials exhibiting varieties of ferroic order, i.e., ferroelectricity, ferromagnetism, ferro-elasticity, or ferrotoroidicity in a single phase. Owing to their coupling of electric, magnetic, and elastic properties, multiferroics are intensely investigated for new applications such as spintronics, storage devices, and sensor devices. Categorization of multiferroics based upon microscopic mechanisms is explained below. This is by far the most popular of these classifications (propagated by Nicola A. Spaldin and collaborators):

Type-I Multiferroics (Proper Multiferroics)

In Type-I multiferroics, ferroelectricity and magnetism arise from different mechanisms [42]. Ferromagnetism is usually a result of lattice instability (for instance, cation displacements in perovskites), while magnetism results from unpaired d- or f-electrons. It exhibits strong ferroelectric polarization but weak magnetoelectric coupling. For example, in $BiFeO_3$ (Bismuth Ferrite), ferroelectricity reveals from Bi^{3+}-6 s^2 lone pairs, magnetism from Fe^{3+} spins. And in $YMnO_3$, ferroelectricity reveals from structural distortions, magnetism from Mn^{3+} spins.

Type-II Multiferroics (Improper Multiferroics)

In Type-II multiferroics, ferroelectricity arises because of magnetic ordering (e.g., spiral spin structures) [43]. It exhibit strong magnetoelectric coupling but usually small polarization. For examples, in $TbMnO_3$, ferroelectricity reveals from spin-spiral ordering. In $Ni_3V_2O_8$, ferroelectric reveals polarization induced by frustrated magnetism.

This classification is essential for designing thin film devices, as Type-I materials are useful for high-polarization memory devices, whereas Type-II systems are attractive for sensors and spintronics. Table 3.1 summarizes the Types-I and Type-II multiferroics and their characteristics with application.

Table 3.1 Classification of multiferroics-based microscopic mechanism

Type	Mechanism	Strength of ME coupling	Example	Application
Type-I	Independent ferroelectric and magnetic origins	Weak	$BiFeO_3$, $YMnO_3$	Non-volatile memories
Type-II	Ferroelectricity induced by magnetic ordering	Strong	$TbMnO_3$, $DYMnO_3$	Sensors, spintronics

6.3 Thin Film Growth Techniques

Thin film formation is the keystone to extending multiferroic research, as structural, electronic, and ferroic properties are extremely sensitive to growth conditions [11]. Unlike bulk crystals, thin film geometries enable scientists to engineer epitaxial strain, control stoichiometry, design heterostructures, and leverage interface chemistry with exceptional accuracy. Nonetheless, such control relies ultimately on the deposition method adopted, since varying growth techniques produce films that range in crystallinity, defect density, surface topology, and scalability. In the past thirty years, numerous deposition schemes have been conceived and implemented, each with peculiar advantages and its share of difficulties. A few of the most popular ones are pulsed laser deposition (PLD), molecular beam epitaxy (MBE), atomic layer deposition (ALD), chemical solution deposition (CSD), and sputtering [44]. The comparisons of these growth techniques are summarized in Table 3.2, with prominent characteristics, advantages, and limitations of each growing multiferroic thin films technique.

6.3.1 *Pulsed Laser Deposition*

Pulsed laser deposition a traditional method used for oxide thin film formation. But it is fruitful and vast used for multiferroic compounds like $BiFeO_3$, $YMnO_3$ perovskite compounds in this era also [45]. In this methodology, a high-energy laser focused pulse falls on a ceramic target material, to remove material from its surface and creating a plasma plume having ions, atoms, and clusters of target material [46]. These plasma plume condense onto a substrate, and heated to grow a crystalline film under precisely controlled oxygen pressure. It can transfer target-to-film stoichiometry with incredible fidelity is potential strength of PLD [47]. This is important for the complex oxides, where proper ratio of cations can directly influence ferroelectric and

Table 3.2 Comparison of thin film growth techniques

Technique	Film quality	Stoichiometry control	Scalability	Cost	Typical application
PLD	Excellent (epitaxial)	High	Limited	Medium	Research and prototyping
MBE	Atomic precision	Excellent	Very low	Very high	Fundamental studies
ALD	Very good (uniform)	Moderate–high	High	High	Industrial integration
CSD	Moderate	Good	High	Low	MEMS, sensors
Sputtering	Good	Moderate	Very high	Medium	Large-scale fabrication

magnetic behavior. PLD is also capable for epitaxial growth of films with atomically smooth interfaces, is best suited for intrinsic coupling properties in multiferroics. This technique is not flawless, though. Its dependence on a focused laser pulse restricts to the deposition area. Thus, scalability at wafer-scale production is not easy. Also, the high-energy process generates microscopic droplets that may deposit to the substrate, introducing surface roughness [48]. With optimization process it is minimized, so this technique is still appropriate for laboratory-scale research, prototype equipment, but not for large-scale production [49].

6.3.2 Molecular Beam Epitaxy

Molecular beam epitaxy (MBE) is used in thin film deposition technology which provides good accuracy [44]. MBE process is done in ultra-high vacuum environments to get accuracy and defect-free thin film. In this process beams of elemental sources like metals, oxides, or organometallic compounds are thermally evaporated and focused on a heated substrate. Thin film growth is taking place in a layer-by-layer mode, followed by real-time techniques like reflection high-energy electron diffraction (RHEED). This provides controlling of film thickness at the monolayer level with facilitating the production of atomically sharp heterostructures, unprecedented cleanliness. The potential ability of multiferroics materials with interfaces engineering, where effects like spin orbit coupling, charge transfer, or orbital hybridization that are crucial for magnetoelectric coupling is well maintained during processing. Also, with artificial heterostructures between ferroelectric $BaTiO_3$ and ferromagnetic $SrRuO_3$ have been prepared with MBE to examine interfacial effects at the atomic scale [50]. The main disadvantages of MBE are slow growth rate and high proceeding cost. A film as thin as a few tens of nanometers in thickness takes hours to grow, while ultra-clean vacuum systems make the operation costly. MBE is useful in basic research and the discovery of new interfacial physics, but it is less convenient for bulk or industrial-scale production [51].

6.3.3 Atomic Layer Deposition

Atomic layer deposition sits in an entirely different niche in the world of film growth methods [52]. In contrast to PLD or MBE, the latter two of which are mainly suited for epitaxial growth on single-crystal substrates, ALD is a chemical process based on self-limiting surface chemistry. In this process, the substrate is alternately brought into contact with gaseous precursors, which only react with the surface, depositing one atomic layer at a time. Through repeated cycling of these reactions, thin films of higher thickness control and conformality over large substrates can be achieved. ALD is particularly attractive to embed multiferroic films in semiconductor devices because of its conformality and compatibility with commercial fabrication processes such as CMOS processing. For example, complex three-dimensional structures, which is significant in nanoscale memory and sensing applications, ALD is used. However, ALD has some limitations. Few precursors are appropriate for complex oxides such as $BaFeO_3$ or $TbMnO_3$, and low-temperature-grown films are typically non-crystalline and need subsequent annealing. Additionally, the growth rate is intrinsically low relative to sputtering or CSD [53]. Despite these issues, ALD is one of the most viable methods for scaling multiferroics from the lab to the industry, particularly as precursor chemistry becomes increasingly advanced.

6.3.4 Chemical Solution Deposition

Chemical solution deposition (CSD), or the sol- gel method, is another more readily available and less expensive method [54]. In this process, metal-organic precursors are dissolved in a solvent that results to a homogeneous solution and then spin-coated or dip-coated on a substrate. Heat treatment expels the organic material and crystallizes the oxide film. The potential benefits of CSD are it is used to large surfaces also, appropriate for basic level as well as at industrial level [55]. Solution chemistry modifications, easily control stoichiometry also add dopants to the films. This plethora of properties make it suitable technique to deposit multiferroic layers in capacitors, sensors, and MEMS devices [56]. But epitaxial growth is much more challenging using CSD than vacuum techniques. These films are usually having many tiny holes or voids, surface cracking, or nonuniformity of composition, especially in thicker films. An effort to bypass such issues, several coating and annealing sequences are usually required, adding complication and at cost of forfeiting uniformity. Consequently, although CSD will not likely displace methods like PLD or MBE for high-performance epitaxial films, it is still useful for those applications for which cost and scalability are more important than atomic precision.

6.3.5 Sputtering

Sputtering technique, like radio-frequency (RF) and direct current (DC) magnetron sputtering, is also known as a method that bridges the gap between laboratory-scale and large-scale film growth technique [57]. In sputtering, an inert gas ion plasma that is usually argon is focused on a target material. The high-energy ions knock out atoms from the target materials, that are deposited on the substrate to grow a thin film. Sputtering has great scalability, large-area coverage with uniformity, and establish good contact layer between film and substrate [58]. This technique is one of the most popular deposition techniques in semiconductor industries. In multiferroic thin films, sputtering is used for making composite and bilayer structures in which stoichiometry is less important than epitaxial single-phase films. Sputtering in multicomponent oxides is challenging because variation in the sputtering variables can result in non-stoichiometric films. Optimizing parameters like power, pressure, and substrate heating is play critical role to minimize this problem [59]. A well-regulated sputtering can grow thin films appropriate for commercial microelectronics. Which set up a vital step toward scaling up multiferroics research laboratories to industries [23].

In brief, these methods draw attention to deep research on multiferroic thin films growth technology. PLD and MBE are used for high-quality epitaxial films in fundamental research, but it uses is limited by their drawback like scalability and high expensive technology. ALD and sputtering are well correlated for industrial needs. But it is not well suited for defect-free crystallinity required to investigate intrinsic multiferroic phenomena. CSD is a low-cost, easily accessible alternative but epitaxial is facing issues with accuracy [60]. Thus, the choice of deposition method is depended on the availability of technology and needfulness. Such as, scientists exploring basic magnetoelectric interactions may uses MBE or PLD. In designing and development of commercial memory or sensing devices ALD or sputtering is used. Above discussion highlights the range of challenges in the field and importance in different growth technology from basic coupling mechanisms to the structures engineering of manufacturable device [61].

In recent years, scientist tried to hybridize these methods to resolve issues in multiferroic thin film growth technology. Such as, films grown by ALD can be annealed under optimized conditions through the application of knowledge derived from PLD or MBE research, results to merging scalability with crystallinity [62]. Similarly, sputtered films can be integrated into heterostructures that epitaxial templates are initially grown by using PLD [63]. Such type of synergistic approaches explores the interaction among various deposition techniques results to better outcomes. Thus, the future of the multiferroic thin film would not using one method, but uses optimized engineered mixes grown technology that provide better result. Thin film growth methodologies are at the center of multiferroic research and application [64]. The lab grown thin film with MBE provides good accuracy whereas at the scale of industry using ALD and sputtering. Each methodology advances itself uniquely day by day with help of new inventions [64]. Rigorous need for multifunctional, power efficient devices, requires the integration and optimization of these deposition methods to grow multiferroics thin film useful for real-world technologies. Figure 3.3 illustrates PLD, MBE, and sputtering with many aspects.

6.4 Structural and Physical Properties of Multiferroic Thin Films

The physical characteristics of multiferroic thin films is depends on their crystal structure, epitaxial strain, domain configurations, and defect chemistry of material [11]. The bulk multiferroics possess interesting properties like ferroelectricity, ferromagnetism, and ferro-elasticity. But the geometry shifts in thin film brings with

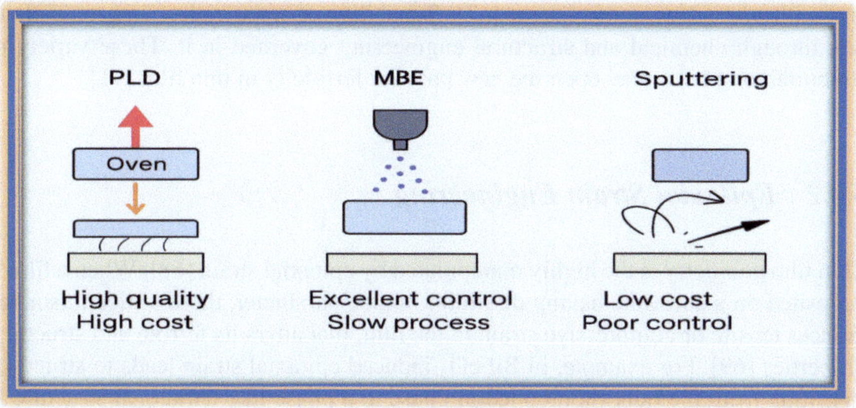

Fig. 3.3 Different deposition techniques used for multiferroic thin films formation

extra degrees of freedom. Ferroic orders are controlled by interface engineering, strain accommodation, and dimensional confinement, which enable the stabilization of new phases and functionalities that are not present in their bulks. Hence, deep understanding of these physical and structural explores main characteristics as well as the development of thin films with optimal functionalities in device applications [65].

6.4.1 *Perovskite Crystal Structure*

Perovskite crystal structure (PSC) having structures ABO_3 are best known multiferroic structures [66]. In the PSCs, off-center the B-site cations are responsible for ferroelectricity, like Ti^{4+} in $BaTiO_3$ or Fe^{3+} in $BiFeO_3$. Due to breaking of inversion symmetry results to a spontaneous electric polarization. Perovskites such as $BaFeO_3$ (BFO) has great interest now a days due to its combination of strong ferroelectricity near Curie temperature (T_c) at 1100 K and antiferromagnetic ordering near Néel temperature (T_n) at 643 K, both are well above room temperature. These remarkable features of BFO, make it an ideal material in the field of thin film research and development. Strain engineering and heterostructure modeling further extend its multifunctionality. Other than perovskites materials, other structural families also exhibit multiferroicity. For example, hexagonal manganite like $YMnO_3$ or $HoMnO_3$ exhibit improper ferroelectricity, in that polarization are generated by geometric distortions and off-center cation. The hexagonal materials are type-I multiferroics and shows different coupling between ferroic orders. Also, layered structures like Ruddlesden-Popper and Aurivillius phases has flexibility with perovskite-like units and with non-ferroelectric layers [67]. Tunability of polarization and magnetization through chemical and structural engineering governed in it. These variety of structural templates may open the new path for ferroicity in thin film [11].

6.4.2 Epitaxial Strain Engineering

Thin film geometry is the highly manipulated by epitaxial strain [68]. When a film is deposited on a substrate having dissimilar lattice parameter, these lattice mismatch induces tensile or compressive strain in the film, that alters its ferroic and structural properties [69]. For example, in $BiFeO_3$ induced epitaxial strain leads to structural transition from the bulk rhombohedral phase to a phase like tetragonal structure in thin films. The tetragonal-like phase is stabilized by using high compressive strain that possesses a giant polarization more than 150 μC/cm^2 which is not seen in bulk BFO. Figure 3.4 explores this strain-induced structural transition, with epitaxial engineering that stabilize new ferroic states. Furthermore, tensile strain can stabilize unconventional ferroelectric phases or polarize few structures, whereas compressive strain can enhance magnetic interactions in weakly ordered antiferromagnets. In addition to phase stabilization, strain also play a crucial role in magnetoelectric coupling and thus tunes the cross-control between electric and magnetic properties. These consequences make strain engineering one of the most useful approaches in the design and fabrication of multiferroic thin films [70].

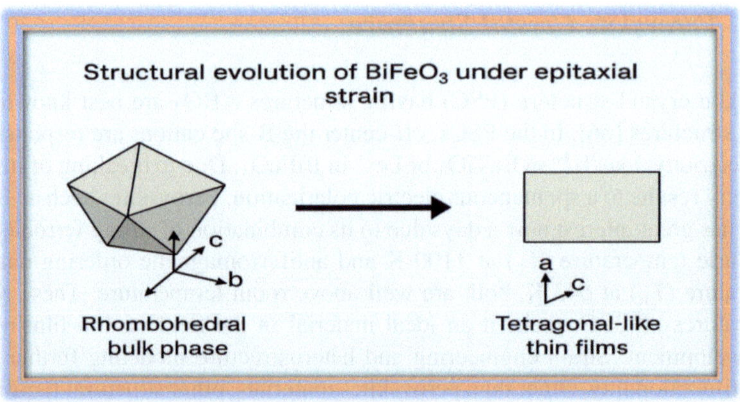

Fig. 3.4 Structural transition in $BaFeO_3$ under epitaxial strain

6.4.3 Domain Structures

Domain structures are another crucial dimension of thin film behavior [71]. In domains, polarization or magnetization adopts a uniform orientation, and this arrangement is highly influenced by epitaxial engineering, varying film thickness, and substrate symmetry. In ferroelectric thin films such as $BiFeO_3$, piezo-response force microscopy (PFM) has nanoscale stripe-like domain patterns, but the domain-wall density and orientation can be tuned by external strain or electric fields. Remarkably, the domain walls themselves are not merely passive boundaries but can host emergent functionalities. For example, the conductive domain walls in $BaFeO_3$ thin films, observed through local current mapping, shows enhanced electrical conductivity compared to the insulating domains. Figure 3.5 shows PFM images, representing the stripe-like ferroelectric domains and their conductive walls. This discovery has opened an entirely new avenue of "domain-wall nanoelectronics," where the walls rather than the bulk domains may serve as active channels for transport, potentially enabling ultraminiaturize devices. Similar phenomena are observed in magnetic domain structures, where the coupling between ferroelectric and magnetic domains further enriches the multifunctionality of thin films. Such coupled domain states allow electric fields to control magnetic textures and, conversely, magnetic fields to manipulate polarization, exemplifying the core promise of multiferroics for spintronic and magnetoelectric applications [72].

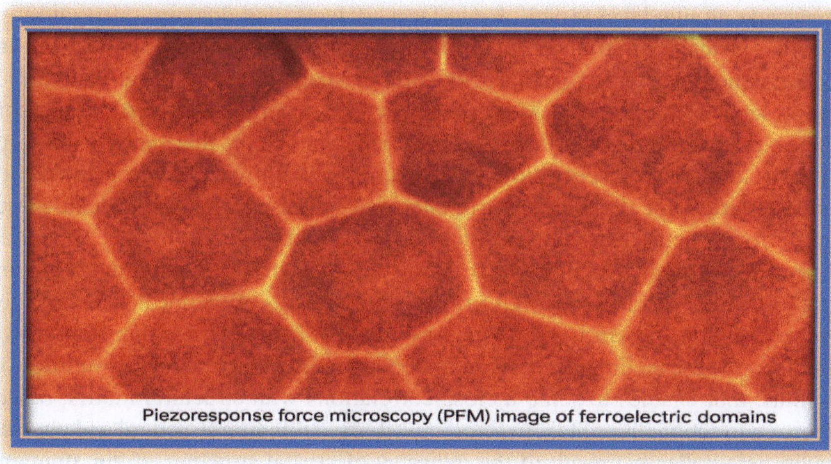

Fig. 3.5 Piezo-response force microscopy (PFM)—schematic image of ferroelectric domains in $BaFeO_3$ thin film, with conductive domain walls

6.4.4 Defects and Vacancies

While strain and domain configurations provide routes to enhance properties, defects, and vacancies play an equally decisive role, often with dual consequences [72]. These defects may degrade performance of system because of induced leakage currents, altered polarization, or introducing fatigue in repeated switching. Using proper defect engineering may tailor electronic and magnetic response under precise control condition. Oxygen vacancies in $BaFeO_3$ thin films may be tied with Fe^{3+}/Fe^{2+} redox sites that alters electronic structures and channels of conductivity is a such example. Oxygen partial pressures in growth conditions may be varied to balance and suppressing unwanted leakage current with advantage of vacancy sites results to higher functionality. In some of the cases, defect ordering itself introduces novel properties that are not seen in defect-free crystals. This suggests that point defects may serve as an additional design parameter rather than an obstacle [73].

6.4.5 Characterization Techniques

To understand the interplay of structure and ferroic behavior in multiferroic thin films, high-level and advanced characterization techniques are needed [11]. Standard X-ray diffraction (XRD) is wildly used to determine crystal structures, epitaxial orientation, and strain states. Transmission electron microscopy (TEM) is complements of XRD, that provide atomic level visual of interfaces, defects, and domains, also revealing exquisite existence of strain or dislocation. Atomic force microscopy (AFM) and PFM is a nanometer-scale visual provide direct visualization of domain-wall motion in the presence of applied field. Magnetic behavior is generally analyzed by superconducting quantum interference device (SQUID) magnetometry, that quantify magnetic hysteresis loops and susceptibility with exquisite sensitivity. Neutron diffraction provides is used for determination of magnetic ordering, that are specially used in antiferromagnetic materials in which X-rays is not more convenient technique. Modern spectroscopic techniques like X-ray photoelectron spectroscopy (XPS) and X-ray absorption spectroscopy (XAS) give precise details of chemical valence states, defect chemistry, and electronic structures which is beneficial for thin film growth technology. These techniques provide merely overall characteristics of multiferroic thin films, that enables to setup a close correlations with structural motifs, defect, and ferroic functionalities [11]. In short, multiferroic thin film structural and physical behavior is a collective result of the interplay of crystal structure, strain, domains, and defects [23]. The perovskite and hexagonal structure are the foundation of multiferroic families. $BaFeO_3$ is a model compound at room-temperature with strong ferroicity. Epitaxial strain engineering is a versatile tool to stabilize new phases that increase polarization, while domains and domain walls is a source of emergent functionality. Defects, that are usually negative, but under controlled conditions it can be beneficial and used for tuning parameters to optimize electric and magnetic response.

Use of appropriate characterization tools guide the rational design of thin films for memory, spintronic, sensing, and energy applications. The integration of these must ensure that multiferroic thin films are at the cutting edge of multifunctional materials research in condensed matter physics.

6.5 Applications of Multiferroic Thin Films Devices

In multiferroic thin films coexistence of ferroelectric and magnetic order are coupled by magnetoelectric (ME) interactions provide a great platform for new multifunctional device [74]. In the view of bulk multiferroics, thin film geometries are compatible with semiconductor processing with adjusting epitaxial strain and scalable to nanoscale dimensions, that are critical technological integration. Toward a whole host of device concepts, such as non-volatile memories and spintronics, energy harvesting, sensing, neuromorphic circuits, and flexi-electronics, as illustrated schematically in Fig. 3.6, multiferroic thin films are explored. Each one of such opportunities leverages the potential of being able to control ferroic orders in isolation, as well as across-coupling, to get around perennial limitations of classical device architectures [75].

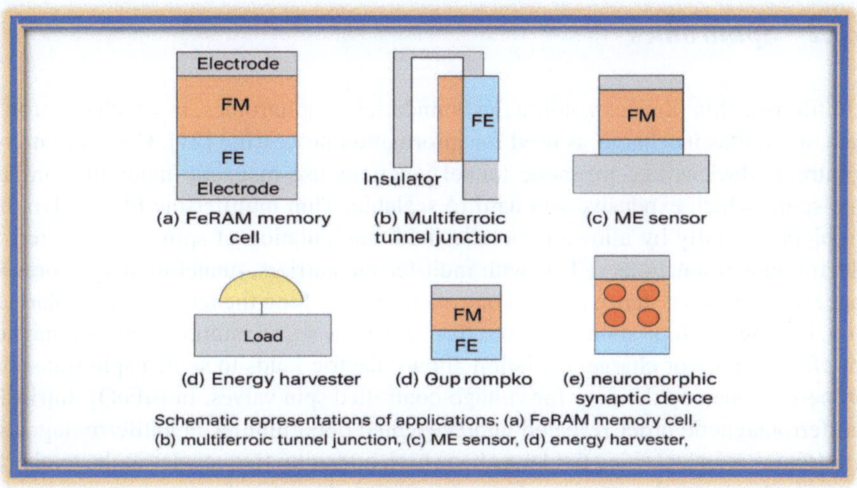

Fig. 3.6 Schematic representations of applications: **a** FeRAM memory cell, **b** multiferroic tunnel junction, **c** ME sensor, **d** energy harvester, and **e** neuromorphic synaptic device

6.5.1 Non-volatile Memories

One of the first and most visible applications of multiferroic thin films has been for non-volatile memories [76]. Bistable polarization states of ferroelectric materials are utilized by FeRAM Memory devices. Thin films of $BaFeO_3$ and Pb (ZrTi)O_3 have been used extensively for this application because they possess high-speed switching, low-energy loss, and resistance to radiation damage. But the commercial viability of FeRAM has been affected by problems like leakage currents and fatigue in ultrathin films, leading to degradation of endurance under repeated switching cycles. To address such limitations, researchers have shifted toward magnetoelectric random access memory (ME-RAM), in which electric fields directly control magnetization through ME coupling. In such an architecture, thin film heterostructures like $BaFeO_3$ in combination with ferromagnetic layers have shown switching of magnetization driven by voltage, providing access to ultra-low-power spintronics memories. An unusual example was published by [77] in which deterministic room-temperature switching of ferromagnetism was achieved with an electric field in a $BaFeO_3$ heterostructure. Such ME-RAM devices offer greater density, faster speeds, and significantly lower energy use than traditional magnetic RAM and are thus leading contenders for next-generation memory devices.

6.5.2 Spintronics

Multiferroic thin films expanding the boundaries of spintronics, in which electron's spin, as well as its charge, is used for information processing [21]. Conventionally, spintronic devices like magnetic tunnel junctions use magnetic fields to manipulate spins, which expensive and hard to scalable. Thin multiferroic films solve the problem elegantly by allowing electric-field manipulation of spin states. In ferroelectric tunnel junctions (FTJs) with multiferroic barriers, tunneling magnetoresistance as well as resistance can be controlled by switching the ferroelectric polarization, leading to a four-state resistance device. Likewise, on multiferroic/ferromagnet interfaces, strain or charge mediation allows electric fields to switch spin transport properties, opening the door for voltage-controlled spin valves. In $BiFeO_3$, intrinsic antiferromagnetic order revealed unprecedented opportunities in antiferromagnetic spintronics is an emerging field may have high operating frequencies in the terahertz range and enhanced robustness by external disturbance in ferromagnetic systems. Such exceptional abilities reveal that multiferroic thin film-based spintronics has the potential for future high-speed, low-power devices [78].

6.5.3 Sensors and Actuators

Multiferroic sensitivity is beneficial in sensors and actuators, due to its high sensitivity with electric, magnetic, and mechanical inputs [79]. Thin film-based magnetoelectric sensors can sense low magnetic fields by detecting voltage-induced magnetization changes, that reveals potential applications for biomedical imaging, defense, and navigation technologies. Thin film multiferroics are incorporated at the nanoscale for microelectromechanical and nanoelectromechanical systems (MEMS/NEMS), where electrical signals may be alters into mechanical motion through piezoelectric and magneto-strictive effects. Actuators are particularly useful in precision positioning systems, robotics, and in optics to develop energy-efficient devices. Also, ferroic effects in multiferroic thin films are used to design environmental sensors to sense stress, vibrations, or electromagnetic radiation with potential applications in IoT devices in smart environments. The coupling among different ferroic orders have a unique advantage over traditional single-phase materials. Thus, a single device can be able to sense and respond to multiple different external stimuli simultaneously [79].

6.5.4 Energy Harvesting

Multiferroic thin films are also used in energy harvesting [80]. Piezoelectric and magneto-strictive coupling in thin film composites can harvest mechanical vibrations. These devices are beneficial for self-sustaining power in low-energy electronics and wireless sensors, especially in remote area. $BiFeO_3$ and $CoFe_2O_4$ multilayer composites have magnetoelectric energy harvesting its efficiency. The potential of multiferroics into energy-efficient harvester its multifunctional nature. These systems can be miniaturized complex integrated silicon technology, and providing autonomous power solutions for the future generation devices like IoT devices, wearable sensors, and biomedical implants [81].

6.5.5 Neuromorphic and Quantum Devices

Multiferroic thin films is used in neuromorphic and quantum devices [23]. Neuromorphic computing has the potential to duplicate the human brain through the simulation of synaptic plasticity on artificial hardware. Multiferroic thin films can couple polarization and magnetization, constitute a natural platform for such systems [82]. Ferroelectric domain dynamics in thin films can mimic synaptic learning processes, that enable to design brain-inspired hardware. The simultaneous occurrence of ferroic orders and coupling of ferroic orders reveals the mimicry of synaptic weights and non-linear dynamics for cognitive computing. Aside from neuromorphic, multiferroic heterostructures are also used in spin-based quantum technologies in which electric-field-controlled spin states would be qubits that control the central element in quantum sensors. Strain and interface engineering in ultrathin multiferroics are projected to be crucial for achieving quantum-level control, but this is still an emerging research direction.

6.5.6 Wearable and Flexible

Another very promising direction is combining multiferroic thin films into wearable and flexible electronics. The development of flexible substrates has made it possible to deposit multiferroic films on bendable platforms without drastically disturbing their ferroic order. This provides new dimensions for wearable biomedical sensors, flexible memory devices, and bendable energy harvesters. For example, magnetoelectric (ME) sensors can be integrated to clothing or skin patches to measure physiological signals like heart rate or muscle activity, whereas self-powered harvesters can power these devices continuously by harvesting body motion or ambient vibration. The fundamental challenge, however, continues to be sustaining robust ferroic functionality with repeated mechanical bending and strain, since ferroic orders are inherently lattice-dispersion sensitive. Experimental attempts are thus directed toward new buffer layers, strain-relieving substrates, and nanostructured architectures to circumvent these obstacles [83].

These applications highlight the usefulness of multiferroic thin films [64]. By coupling multiple ferroic orders at nanoscale, reveals to new device functionalities that go beyond the limitations of traditional materials. Whether as ultra-low-power non-volatile memories, spintronic logic, self-powered sensors, neuromorphic circuits, or flexible wearables, multiferroic thin films furnish a roadmap for multifunctional devices at the technological frontier of electronics. While major challenges are still available in form of leakage, fatigue, interface matching that reduces overall performance and large-scale production. The advances in synthesis, characterization, and theoretical understanding will short out this obstacle in near future.

6.6 Challenges and Limitations

Although the multiferroic thin films have tremendous promise for next-generation spintronic and electronic devices. But their journey from laboratory to commercial availability has so far away [21]. Multiferroic thin film has a wide range of tools for tuning multifunctionality like strain engineering, interface tailoring, to integrate with silicon technology, but simultaneously also have a wide range of scientific and technological issues. These are arisen due to intrinsic material limitations, extrinsic defect, and incompatibility with standard semiconductor processing. To mitigate these problems is necessary for multiferroic thin films to revoke their full potential in memory storage devices, sensors, energy harvesters, and neuromorphic systems.

6.6.1 Leakage Current

Leakage current is one of the most crucial challenges in multiferroic thin films [84]. Ferroelectric thin films ideally have stable and reversible polarization, switching under electric field stress, should have low conduction losses. In practice high leakage currents are often observed, particularly in $BiFeO_3$-based thin films, where oxygen deficiency results to mixed Fe^{2+}/Fe^{3+} valence states. Also, Oxygen vacancies along with cation non-stoichiometry and interface defects, form parasitic conduction channel that degrade the performance of device. High leakage prevents the formation of clean polarization hysteresis loops, hence memory retention to be lost, and power consumption increases. To mitigate this defect engineering, stoichiometry optimization, inducing buffer layers, and substitution doping are employed. Rare-earth element doping has been employed to repress the formation of oxygen vacancies and stabilize the electronic structure, to lowers leakage current. This will have some success, but sustaining consistently low leakage current is difficult task as film thickness reduces to the nanometer scale during growth processing.

Strongly correlated with leakage is the issue of retention loss and fatigue, both of which have a highly significant impact on device reliability [85]. Fatigue corresponds to the deterioration of ferroelectric switching performance following multiple cycling, whereas retention loss appears as the time-dependent degradation of stored polarization. For thin films, the two are compounded by domain pinning at defects and by space charge trapping at interfaces. The real-world implications for FeRAM and ME-RAM devices are drastic, as they require long-term endurance and data stability. Fatigue-resistant electrode materials, including conductive oxides such as $SrRuO_3$ and $LaNiO_3$, have been found to contribute toward enhanced cycling stability by exhibiting superior chemical compatibility with multiferroic layers. Interface engineering to limit charge accumulation has also been investigated as a remedy. However, fatigue and retention loss are still unresolved issues for deployment at the scale of consumer electronics, especially under severe cycling and environmental stresses anticipated in consumer electronics.

6.6.2 Miniaturization

The device miniaturization creates another bottleneck as scaling to ultrathin thicknesses [86]. Previously integrated circuits have active layers of a few nanometers, but multiferroic properties enables to minimize the thickness of active layer. Polarization becomes unstable through depolarization fields and it will become more effective with decreasing thickness. Leakage currents, which are a difficulty in thicker films also, increase exponentially at the ultrathin scale, further constraining device performance. Finite-size effects can provide long-range ordering in magnetic thin films, reducing the varying ferroic properties that improves device function. Good ferroic behavior is observed below the thickness 10 nm, is a considerable challenge. Although new methods like utilization of ferroelectric-ferromagnetic superlattices or high-k dielectric buffers are used to stabilizing polarization. But reliable performance at industry scales has yet to be realized.

6.6.3 *Interface and Strain Issues*

Interface and strain tuning are also crucial challenge within it. Interfaces, which are most seen as opportunities for designing new functionality, also reveal structural and chemical instabilities. Interdiffusion at the film-substrate interface can generate non-stoichiometry, dislocation defects, and electrically or magnetically dead layers that creates polarization or magnetization close to electrodes. Strain relaxation creates dislocations that serve as nucleation sites for defects, degrading ferroic properties. Also, in epitaxial films achieving coherent strain on large scales, and local strain inhomogeneity are a big challenge that result to erratic behavior. Mitigation technique like buffer layer insertion, meticulous substrate choice, and epitaxial strain optimization have been used to counteract these issues. Substrates such as $DyScO_3$ and $NdGaO_3$ offer lattice matches that stabilize new phases of $BaFeO_3$ thin films, yet substrate choice also limits thermal budgets and cost, restricting their application in scalable industrial processes [87].

6.6.4 *Stable Room-Temperature*

Another highly limiting constraint is the availability of stable room-temperature multiferroicity. Although $BaFeO_3$ is the most extensively researched room-temperature multiferroic, with ferroelectric and antiferromagnetic orders coexisting above room temperature, the coupling strength between the ferroic orders of $BaFeO_3$ is quite weak. Most other single-phase multiferroics, like $TbMnO_3$, possess strong magnetoelectric coupling but only at low cryogenic temperatures and are therefore of no use for real applications [88]. Composite thin films where independent ferroelectric and ferromagnetic phases are coupled through strain or charge mediation provide greater coupling at room temperature, but they are accompanied by fabrication complexity and reliability challenges. For example, controlling interfaces in such composites to obtain reproducible coupling strength is a non-trivial endeavor. Artificial heterostructures and composites, therefore, present fascinating opportunities, while introducing along the way engineering challenges that are still open [89].

6.6.5 Material Reliability and Stability

Material reliability and stability under use conditions are also an issue. Thin films are especially sensitive to chemical and structural degradation due to long-term cycling, exposure to the environment, and thermal stress [90]. Diffusion of oxygen from environmental humidity can change stoichiometry degrade the device performance, and thermal fatigue can form micro-cracks and phase segregation decrease the life span. Electrical cycling over a long period result to cumulative damage and decreasing performance. Although protective encapsulation layers, robust electrode materials mechanism are used to mitigate these. Also, researcher is being researched to overcome these problems. But, the multiferroic thin films devices, cannot satisfactorily achieve the industrial norms of resilience and reliability necessary for consumer electronics.

6.6.6 Integration with CMOS Technology

The most crucial technological challenge is integrating multiferroic thin films with CMOS technology. To move from the laboratory into commercial products, multiferroics will have to be made compatible with silicon-based platforms, both at large-area scalability and processing temperature. High-cost, low-area-deposition, high-temperature deposition methods such as molecular beam epitaxy and pulsed laser deposition, though well suited for creating research-quality films, are not good for mass production. Methods like atomic layer deposition (ALD) and sputtering are more scalable and industrially applicable but tend to find it difficult to replicate the high crystalline quality and epitaxy required for enhanced multiferroic characteristics. In addition, silicon integration requires deposition at below the temperature 500 °C, while several multiferroics need increased growth temperatures to attain phase purity and ordering as desired [91]. Attaining large-area uniformity, defect-free interfaces, and low-temperature deposition is a challenge. These problems are significant impediment to commercialization.

The main obstacle in multiferroic thin films are intrinsic material limitations, extrinsic defect, environmental effects, and technological integration. High leakage currents, fatigue, scaling bottlenecks, and interface instabilities are critical scientific issues. The limited availability of room-temperature multiferroics with significant strength also limits applications. Also, chemical instability and unreliability limit industrial-scale deployment. The challenge of scaling up multiferroic thin films to CMOS-compatible, low-cost, large-area processing have the gulf between laboratory to practical device [92]. These challenges will need to be overcome by improvements in synthesis and defect engineering, also creative device architectures, multidisciplinary innovation, and close interaction between academia and industry are helpful. So that the outstanding multifunctionality of multiferroic thin films can be best utilized for next-generation electronic technologies.

6.7 Future Trends and Recent Progress

The last two decades many research activities are ongoing in multiferroic thin films, driven by advanced deposition techniques, computational material science, and nanoscale device technology [93]. The intrinsic difficulties of multiferroic materials, like leakage currents, room-temperature coupling limitation, and interface issues remain daunting. The key achievements will open their many uncovered thrilling new scientific frontiers. New research directions are schematically represented in Fig. 3.7. The transition from the traditional bulk systems to engineered heterostructures, two-dimensional (2D) materials, topological domain exotics, neuromorphic devices, and artificial intelligence-based material discovery platforms establish new era. Collectively, these developments are reshaping the horizon for multiferroic thin films, to enhance their application in memory, spintronics, energy, and quantum technology.

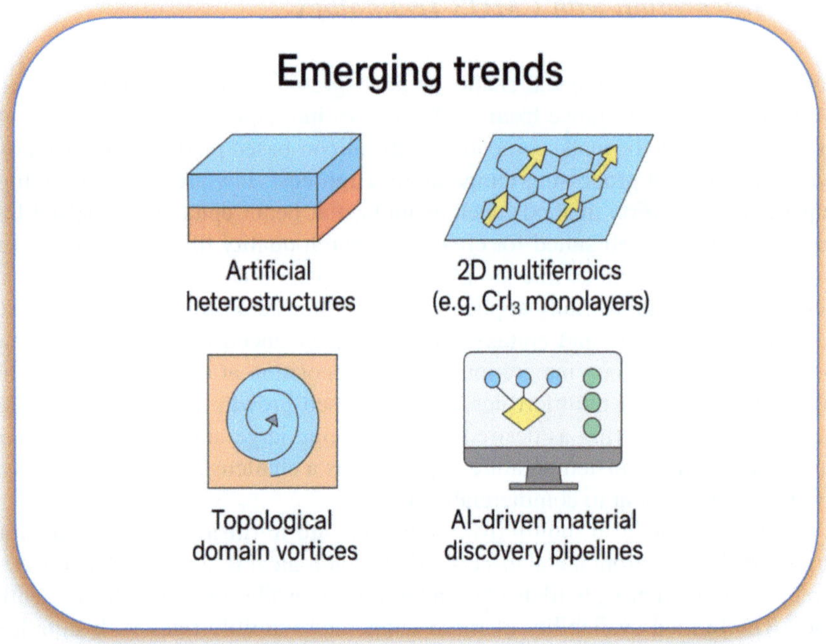

Fig. 3.7 Emerging trends in multiferroic thin film

6.7.1 Heterostructures and Superlattices

One of the most revolutionary developments is generation of artificial heterostructures and superlattices [94]. Alternating ferroelectric and ferromagnetic layers on the nanometer scale, scientists have opened new routes to increase magnetoelectric coupling using interfacial interactions that are not available in bulk materials. For example, $BaTiO_3/SrRuO_3$ superlattices have exhibit strain-stabilized polar magnetic phases that exist only because of interfacial effects. These artificial design approaches enable customized functionalities, as electric-field control the magnetization and magnetic modulation of polarization, that are not seen in conventional single-phase multiferroics. Interface engineering has now emerged as a key factor in thin film science, provide unprecedented material design and versatile functionality.

6.7.2 Strain Engineering

Strain engineering is equally potent method for stabilizing new phases in multiferroics [70]. In thin films, epitaxial strain leads to extreme alteration of structural and ferroic properties. For example, $BiFeO_3$ at high compressive strain stabilizes a tetragonal-like phase with giant polarization which is greater than 150 $\mu C/cm^2$, significantly higher than its bulk equivalent [95]. Strain gradients can control flexoelectricity and electric polarization in centrosymmetric systems opens the new design space for multiferroics. These strain-induced management provide avenues for optimizing functionalities, and enable the formation of new phases with correlated ferroic orders, under epitaxial control.

6.7.3 2D-Multiferroics

Two-dimensional multiferroic materials setup new dimensions in multiferroics [96]. Researcher are now focused on layered materials that have multiple ferroic behavior. 2D multiferroics like CrI_3, In_2Se_3, and $CuCrP_2S_6$ in certain cases exhibit ferroelectricity and magnetism that coexist even in atomically thin crystals. These properties have potential for ultrathin, flexible multiferroics that are appropriate for wearable electronics, flexible spintronic devices, and nanoscale sensors. In addition to, it quantum confinement effects that are inherent to 2D systems may provide a unique platform for new coupling phenomena that can be utilized in quantum information processing systems. Also, 2D material synthesis technology becomes compatible for Van der Waals heterostructures used as most electrifying frontiers for the future of multiferroics.

6.7.4 Exotic Domain

The recent discovery of exotic domain structures with topological nature plays a vital role in multiferroics thin film [25]. Interdisciplinary scientific analysis among materials science, physics, and chemistry has discovered complex domain topologies, such as vortex polarization states in ferroelectric thin films and skyrmion-like spin configurations in multiferroic magnets. Topological domains show emergent behaviors that are not observed in uniform ferroic domains. In $BaFeO_3$, increased domain-wall conductivity is seen that robust spin textures are immune to perturbation. The stabilization, patterning, and exploitation of these topological may unlock emergent application for next-generation memory devices in which information is stored and computed by virtue of domain-wall functionalities in instead of bulk polarization or magnetization. Now domain-wall nanoelectronics has emerged as the dominant field reveals novel device architectures with extraordinary scalability and energy-efficiency functionalities.

6.7.5 Structural and Domain Engineering

Tuning structural and domain engineering in multiferroic provide potential for neuromorphic and brain-inspired devices. The coupled and correlated ferroic states of these materials enable multi-level switching behavior make them ideal candidates for simulating synaptic plasticity in artificial neural networks. Research has reveals that ferroelectric domain switching in thin films can be used as artificial synapses with adaptive learning. The extension of memory capabilities beyond binary logic may be passible using these multiferroic-based devices and establish a hardware platform for neuromorphic and quantum computing platforms. The traditional CMOS-based computation encounters speed and energy bottlenecks but making brain-like computation can be achieved through new multiferroic material [97].

6.7.6 Using Artificial Intelligence and Machine Learning

Artificial intelligence and machine learning set up an era in each field. It is useful tool in the discovery and optimization of multiferroic thin films. Conventional experimental methodologies depend on trial and error, that consumes time and resource-extensive. Machine learning code can quickly predict structure-property correlations and optimize deposition conditions to achieve materials with ideal ferroic and coupling behavior. Data-driven technique is already used to design heterostructures with functionalities like to predict the thermodynamic stability of new 2D multiferroics. In addition to this, AI-assisted experimental workflows have real-time adaptive experimental design that can potentially transform the discovery rate in this area. By combining computational predictions with autonomous synthesis facilities have the potential to transform the gap between material conception and device realization.

Even with these promising developments are existed but the final objective is commercialization of multiferroic thin films. It can be achieved, reducing some key challenges. Scalable deposition processes like atomic layer deposition and sputtering must be optimized to support high-uniformity wafer-scale fabrication with low defect density. Low-temperature growth on silicon CMOS-compatible platforms is a requirement since present-day high-temperature deposition techniques are incompatible with commercial semiconductor processing. Reliability testing, to stringent industrial requirements, will be necessary to guarantee long-term operation under operating stress. Lastly, efficient interaction between academic researchers and industrial players will be crucial to close the gap between laboratory-scale demonstrations and commercial products.

Commercial potential of multiferroic thin films is especially promising due to the growing need for low-power, multifunctional, and miniaturized electronics [98]. Their potential uses range from non-volatile memory, spintronics, and energy harvesting to neuromorphic computing and flexible electronics, many of which are at the heart of the burgeoning Internet of Things (IoT) environment. With ongoing developments in heterostructure engineering, 2D materials, topological domain control, and AI-driven material discovery, multiferroic thin films are set to take a defining place in determining the future of electronic and quantum technology [99, 100].

6.8 Conclusion and Outlook

Multiferroic thin films belongs to family of functional thin films to be employed in the fabrication of advanced electronic, spintronic, and energy devices. The suitability of the thin films with ferroelectric, ferromagnetic, and ferro-elastic characteristics in conjunction with strong magnetoelectric (ME) coupling explores multifunctional applications that go beyond limitations of standard semiconductors.

This chapter has presented the essential concepts of multiferroicity, the dominant thin film preparation techniques, and the contributions of strain, their interfacial effects, and domains in realizing physical properties. Moreover, we have presented a few device applications, namely non-volatile memories systems, spintronics-based devices, sensors and actuators, and neuromorphic processing. Also, considerable challenges are discussed such as leakage currents, fatigue, limited multiferroic material at room temperature, and integration issues related to CMOS.

New innovations like artificial heterostructures, 2D multiferroics, topological domains, and AI-based material design are breaking these issues and provide new exciting possibilities. The integration of new advanced synthesis technique and computational design with prediction make it scalable, reproducible, and industrially viable. Also, ongoing innovation in material design, advanced synthesis, and computational technique, revolutionize the low-power electronics devices, Internet of Things devices, neuromorphic computers, and quantum technologies of the future.

References

1. Kim, T., Choi, C. H., Hur, J. S., Ha, D., Kuh, B. J., Kim, Y., … Jeong, J. K. (2023). Progress, challenges, and opportunities in oxide semiconductor devices: a key building block for applications ranging from display backplanes to 3D integrated semiconductor chips. *Advanced Materials, 35*(43), 2204663. https://doi.org/10.1002/adma.202204663
2. Atherton, W. (1984). Miniaturization of Electronics *From Compass to Computer: A History of Electrical and Electronics Engineering* (pp. 237–267): Springer. https://doi.org/10.1007/978-1-349-17365-5
3. Muhammad Saqib, Q., Mannan, A., Noman, M., Chougale, M. Y., Patil, C. S., Ko, Y., … Bae, J. (2024). Miniaturizing Power: Harnessing Micro-Supercapacitors for advanced microelectronics. *Chemical Engineering Journal, 490*, 151857. https://doi.org/10.1016/j.cej.2024.151857
4. Wang, K., Liu, J.-M., & Ren, Z. (2009). Multiferroicity: The Coupling Between Magnetic and Polarization Orders. *Advances in Physics - ADVAN PHYS, 58*, 321–448. https://doi.org/10.1080/00018730902920554
5. Ouyang, T., Cha, S., Sun, Y., Taniguchi, T., Watanabe, K., Gabor, N. M., & Lui, C. H. (2025). Electrically Switching Ferroelectric Order in 3R-MoS2 Layers. *Nano Letters, 25*(4), 1459–1465. https://doi.org/10.1021/acs.nanolett.4c05370
6. Kruse, M., Petralanda, U., Gjerding, M. N., Jacobsen, K. W., Thygesen, K. S., & Olsen, T. (2023). Two-dimensional ferroelectrics from high throughput computational screening. *npj Computational Materials, 9*(1), 45. https://doi.org/10.1038/s41524-023-00999-5
7. Achary, S. N., Jayakumar, O. D., & Tyagi, A. K. (2012). 4 - Multiferroic Materials. In S. Banerjee & A. K. Tyagi (Eds.), *Functional Materials* (pp. 155–191). London: Elsevier. https://doi.org/10.1016/B978-0-12-385142-0.00004-0
8. Yokoyama, T., Masuda, H., Suzuki, M., Ehara, K., Nogi, K., Fuji, M., … Toda, K. (2008). CHAPTER 1 - BASIC PROPERTIES AND MEASURING METHODS OF NANOPARTICLES. In M. Hosokawa, K. Nogi, M. Naito & T. Yokoyama (Eds.), *Nanoparticle Technology Handbook* (pp. 3–48). Amsterdam: Elsevier. https://doi.org/10.1016/b978-0-444-56336-1.50001-1
9. Nelson, C. W. (1963). Ferroelectricity and the Chemical Bond in Perovskite-Type Oxides. https://doi.org/10.21236/AD0406722
10. Kostishyn, V. G., Panina, L., Kozhitov, L. V., Timofeev, A., & Kovalev, A. N. (2015). Synthesis and Multiferroic Properties of M-type SrFe12O19 Hexaferrite Ceramics. *Journal of Alloys and Compounds, 645*. https://doi.org/10.1016/j.jallcom.2015.05.024
11. Fernandez, A., Acharya, M., Lee, H. G., Schimpf, J., Jiang, Y., Lou, D., … Martin, L. W. (2022). Thin-film ferroelectrics. *Advanced Materials, 34*(30), 2108841. https://doi.org/10.3390/mi16050522
12. Myny, K. (2018). The development of flexible integrated circuits based on thin-film transistors. *Nature electronics, 1*(1), 30–39. https://doi.org/10.1038/s41928-017-0008-6
13. Wu, J., Mao, S., Ye, Univ, Z.-G. S., Chem, D. (2010). Room-temperature ferromagnetic/ferroelectric BiFeO3 synthesized by a self-catalyzed fast reaction process. *Journal of Materials Chemistry, 20*. https://doi.org/10.1039/C0JM00729C
14. Dufour, P. (2023). *Structural and physical properties of antiferroic oxide thin films*. Université Paris-Saclay. https://theses.hal.science/tel-04099832
15. Li, M., Tan, H., & Duan, W. (2020). Hexagonal rare-earth manganites and ferrites: a review of improper ferroelectricity, magnetoelectric coupling, and unusual domain walls. *Physical Chemistry Chemical Physics, 22*(26), 14415–14432. https://doi.org/10.1039/d0cp02195d
16. Scott, J. F. (2013). Room-temperature multiferroic magnetoelectrics. *NPG Asia Materials, 5*(11), e72–e72. https://doi.org/10.1038/am.2013.58
17. Singh, M. K., Yang, Y., & Takoudis, C. G. (2009). Synthesis of multifunctional multiferroic materials from metalorganics. *Coordination Chemistry Reviews, 253*(23–24), 2920–2934. https://doi.org/10.1016/j.ccr.2009.09.003

18. Cho, A. Y., & Arthur, J. (1975). Molecular beam epitaxy. *Progress in solid state chemistry, 10*, 157–191. https://doi.org/10.1016/0079-6786(75)90005-9
19. Zhang, J., Li, Y., Cao, K., & Chen, R. (2022). Advances in atomic layer deposition. *Nanomanufacturing and Metrology, 5*(3), 191–208. https://doi.org/10.1007/s41871-022-00136-8
20. Takasu, H. (2000). The ferroelectric memory and its applications. *Journal of Electroceramics, 4*(2), 327–338. https://doi.org/10.1023/A:1009910525462
21. Gradauskaite, E., Meisenheimer, P., Müller, M., Heron, J., & Trassin, M. (2021). Multiferroic heterostructures for spintr onics. *Physical Sciences Reviews, 6*(2), 20190072. https://doi.org/10.1515/psr-2019-0072
22. Sasmal, A., Sen, S., Chelvane, J. A., & Arockiarajan, A. (2023). PVDF based flexible magnetoelectric composites for capacitive energy storage, hybrid mechanical energy harvesting and self-powered magnetic field detection. *Polymer, 281*, 126141. https://doi.org/10.1016/j.polymer.2023.126141
23. Ramesh, R., & Spaldin, N. A. (2007). Multiferroics: progress and prospects in thin films. *Nature materials, 6*(1), 21–29. https://doi.org/10.1038/nmat1805
24. Liang, X., Matyushov, A., Hayes, P., Schell, V., Dong, C., Chen, H., ... Martins, P. (2021). Roadmap on magnetoelectric materials and devices. *IEEE Transactions on Magnetics, 57*(8), 1–57. https://doi.org/10.1109/TMAG.2021.3086635
25. Tian, G., Yang, W., Gao, X., & Liu, J.-M. (2021). Emerging phenomena from exotic ferroelectric topological states. *APL Materials, 9*(2). https://doi.org/10.1063/5.0039139
26. Das, M., Perez, T. C., Shetty, D., Hiremath, P., Naik, N., & Bhat, R. (2024). An overview on the role of artificial intelligence in modern advancements of material science. *ES General, 5*, 1183. https://doi.org/10.30919/esg1183
27. Tiwari, P. C., Pal, R., Chaudhary, M. J., & Nath, R. (2023). Artificial intelligence revolutionizing drug development: Exploring opportunities and challenges. *Drug Development Research, 84*(8), 1652–1663. https://doi.org/10.1002/ddr.22115
28. Lu, Chengliang & Hu, Weijin & Tian, Yufeng & Wu, Tom. (2015). Multiferroic Oxide Thin Films and Heterostructures. Applied Physics Reviews. 2. 021304. https://doi.org/10.1063/1.4921545
29. Fox, C., Mao, Y., Zhang, X., Wang, Y., & Xiao, J. (2023). Stacking order engineering of two-dimensional materials and device applications. *Chemical Reviews, 124*(4), 1862–1898. https://doi.org/10.1021/acs.chemrev.3c00618
30. Montgomery, D. P. (2023). "This study is not without its limitations": Acknowledging limitations and recommending future research in applied linguistics research articles. *Journal of English for Academic Purposes, 65*, 101291. https://doi.org/10.1016/j.jeap.2023.101291
31. Schmid, H. (1994). Multi-ferroic magnetoelectrics. *Ferroelectrics, 162*(1), 317–338. https://doi.org/10.1080/00150199408245120
32. Martin, L. W., & Rappe, A. M. (2016). Thin-film ferroelectric materials and their applications. *Nature Reviews Materials, 2*(2), 1–14. https://doi.org/10.1038/natrevmats.2016.87
33. Jagla, E. A. (2005). Hysteresis loops of magnetic thin films with perpendicular anisotropy. *Physical Review B—Condensed Matter and Materials Physics, 72*(9), 094406. https://doi.org/10.1103/PhysRevB.72.094406
34. Fang, B.-J., Ding, C.-L., Liu, W., Li, L.-Q., & Tang, L. (2009). Preparation and electrical properties of high-Curie temperature ferroelectrics. *The European Physical Journal-Applied Physics, 45*(2), 20302.
35. Potnis, P. R., Tsou, N.-T., & Huber, J. E. (2011). A review of domain modelling and domain imaging techniques in ferroelectric crystals. *Materials, 4*(2), 417–447. https://doi.org/10.1051/epjap/2009004
36. Yamanouchi, M., Chiba, D., Matsukura, F., & Ohno, H. (2004). Current-induced domain-wall switching in a ferromagnetic semiconductor structure. *Nature, 428*(6982), 539–542. https://doi.org/10.3390/ma4020417
37. Zhernovoi, A., & D'yachenko, S. (2015). Measurement of the thermodynamic temperature using a paramagnetic thermometric material. *Technical Physics, 60*(4), 595–599. https://doi.org/10.1134/S1063784215040325

38. Abrahams, S. (1971). Ferroelasticity. *Materials Research Bulletin, 6*(10), 881–890. https://doi.org/10.1016/0025-5408(71)90066-3
39. Davydova, M., Zvezdin, K., Mukhin, A., & Zvezdin, A. (2020). Spin dynamics, antiferrodistortion and magnetoelectric interaction in multiferroics. The case of BiFeO3. *Physical Sciences Reviews, 5*(12). https://doi.org/10.1515/psr-2019-0070
40. Singh, S. (2023). Coupling of strain and magnetism in manganite-based complex oxide heterostructures. *arXiv preprint arXiv:2310.12573.*
41. Paudel, B., Vasiliev, I., Hammouri, M., Karpov, D., Chen, A., Lauter, V., & Fohtung, E. (2019). Strain vs. charge mediated magnetoelectric coupling across the magnetic oxide/ferroelectric interfaces. *RSC advances, 9*(23), 13033–13041. https://doi.org/10.1039/C9RA01503E
42. Khomskii, D. I. (2006). Multiferroics: Different ways to combine magnetism and ferroelectricity. *Journal of Magnetism and Magnetic Materials, 306*(1), 1–8. https://doi.org/10.1016/j.jmmm.2006.01.238
43. Tokura, Y., & Seki, S. (2010). Multiferroics with spiral spin orders. *Advanced Materials, 22*(14), 1554–1565. https://doi.org/10.1002/adma.200901961
44. Toma, F., Rahman, S., Hussain, K., & Ahmed, S. (2024). Thin film deposition techniques: A comprehensive review. *J. Mod. Nanotechnol, 4*(6), 1–16. https://doi.org/10.53964/jmn.2024006
45. Coy, E. (2016). Growth and characterization of new multiferroic materials. https://hdl.handle.net/2445/102414
46. York, T. H. (1996). *Pulsed laser deposition of electronic ceramics and analysis of the ablation plume*: University of London, University College London (United Kingdom). https://discovery.ucl.ac.uk/id/eprint/10100098
47. Schou, J. (2009). Physical aspects of the pulsed laser deposition technique: The stoichiometric transfer of material from target to film. *Applied Surface Science, 255*(10), 5191–5198. https://doi.org/10.1016/j.apsusc.2008.10.101
48. Mostafa, A. M. (2025). The Influence of various parameters on the ablation and deposition mechanisms in pulsed laser deposition. *Plasmonics*, 1–19. https://doi.org/10.1007/s11468-024-02706-6
49. Bramsiepe, C., Sievers, S., Seifert, T., Stefanidis, G., Vlachos, D. G., Schnitzer, H., ... Bruins, M. (2012). Low-cost small scale processing technologies for production applications in various environments—Mass produced factories. *Chemical Engineering and Processing: Process Intensification, 51*, 32–52. https://doi.org/10.1016/j.cep.2011.08.005
50. Velev, J., Jaswal, S., & Tsymbal, E. (2011). Multi-ferroic and magnetoelectric materials and interfaces. *Philosophical Transactions of the Royal Society A: Mathematical, Physical and Engineering Sciences, 369*(1948), 3069–3097. https://doi.org/10.1098/rsta.2010.0344
51. Brahlek, M., Gupta, A. S., Lapano, J., Roth, J., Zhang, H. T., Zhang, L., ... Engel-Herbert, R. (2018). Frontiers in the growth of complex oxide thin films: past, present, and future of hybrid MBE. *Advanced Functional Materials, 28*(9), 1702772. https://doi.org/10.1002/adfm.201702772
52. Johnson, R. W., Hultqvist, A., & Bent, S. F. (2014). A brief review of atomic layer deposition: from fundamentals to applications. *Materials today, 17*(5), 236–246. https://doi.org/10.1016/j.mattod.2014.04.026
53. Narayanan, V., Lommens, P., De Buysser, K., Vanpoucke, D. E., Huehne, R., Molina, L., ... Van Driessche, I. (2012). Aqueous CSD approach for the growth of novel, lattice-tuned $La_xCe_{1-x}O_\delta$ epitaxial layers. *Journal of Materials Chemistry, 22*(17), 8476–8483. https://doi.org/10.1039/C2JM15752G
54. Mackenzie, J. D., & Bescher, E. P. (2007). Chemical routes in the synthesis of nanomaterials using the sol–gel process. *Accounts of chemical research, 40*(9), 810–818. https://doi.org/10.1021/ar7000149
55. Tan, Q., & Li, H. (2024). Application of Computer Aided Design in Product Innovation and Development: Practical Examination on Taking the Industrial Design Process. *IEEE Access, 12*, 85622–85634. https://doi.org/10.1109/ACCESS.2024.3404963

56. Vyas, A. (2022). *On-chip electrochemical capacitors and piezoelectric energy harvesters for self-powering sensor nodes*: Chalmers Tekniska Hogskola (Sweden). https://doi.org/10.5555/AAI29229196
57. Ju Y, Ai L, Qi X, Li J, Song W. Review on Hydrophobic Thin Films Prepared Using Magnetron Sputtering Deposition. Materials (Basel). 2023 May 16;16(10):3764. https://doi.org/10.3390/ma16103764.
58. Wasa, K. (2012). *Handbook of sputter deposition technology: fundamentals and applications for functional thin films, nano-materials and MEMS*: William Andrew. https://doi.org/10.1016/C2010-0-67037-4
59. Spruit, R. G., Van Omme, J. T., Ghatkesar, M. K., & Garza, H. H. P. (2017). A review on development and optimization of microheaters for high-temperature in situ studies. *Journal of Microelectromechanical Systems, 26*(6), 1165–1182. https://doi.org/10.1109/JMEMS.2017.2757402
60. Koster, Gertjan & Rijnders, Guus & van Setten, Eelco & Slycke, Per & Rogalla, Horst. (2000). Epitaxial growth of oxides with pulsed laser interval deposition. Journal of Crystal Growth - J CRYST GROWTH. 211. 98–105. https://doi.org/10.1016/S0022-0248(99)00880-5.
61. Romig Jr, A., Dugger, M. T., & McWhorter, P. J. (2003). Materials issues in microelectromechanical devices: science, engineering, manufacturability and reliability. *Acta Materialia, 51*(19), 5837–5866. https://doi.org/10.1016/S1359-6454(99)00294-3
62. Meng, X. (2017). An overview of molecular layer deposition for organic and organic–inorganic hybrid materials: mechanisms, growth characteristics, and promising applications. *Journal of Materials Chemistry A, 5*(35), 18326–18378. https://doi.org/10.1039/C7TA04449F
63. Yu, J., Han, W., Suleiman, A. A., Han, S., Miao, N., & Ling, F. C. C. (2024). Recent Advances on Pulsed Laser Deposition of Large-Scale Thin Films. *Small Methods, 8*(7), 2301282. https://doi.org/10.1002/smtd.202301282
64. Roy, S., & Majumder, S. (2012). Recent advances in multiferroic thin films and composites. *Journal of Alloys and Compounds, 538*, 153–159. https://doi.org/10.1016/j.jallcom.2012.05.125
65. Acosta, E. (2021). Thin films/properties and applications *Thin Films*: IntechOpen. https://doi.org/10.5772/intechopen.95527
66. Khan, A. R., Bhardwaj, S., & Kumar, S. (2025). Perovskite ABO3 Multiferroics: Mechanisms, Advancements, and Challenges. *ECS Journal of Solid State Science and Technology, 14*(5), 053004. https://doi.org/10.1149/2162-8777/add296
67. Medvedeva, N. I., & Gubanov, V. A. (1996). Electronic structure and properties of aurivillius phases. *Journal of Structural Chemistry, 37*(3), 409–416. https://doi.org/10.1007/BF02578593
68. Liu, F., Wu, F., & Lagally, M. (1997). Effect of strain on structure and morphology of ultrathin Ge films on Si (001). *Chemical Reviews, 97*(4), 1045–1062. https://doi.org/10.1021/cr9600722
69. Hao, K.-R., Ma, X.-Y., Lyu, H.-Y., Zhu, Z.-G., Yan, Q.-B., & Su, G. (2021). The atlas of ferroicity in two-dimensional MGeX3 family: Room-temperature ferromagnetic half metals and unexpected ferroelectricity and ferroelasticity. *Nano Research, 14*(12), 4732–4739. https://doi.org/10.1007/s12274-021-3415-6
70. Li, T., Deng, S., Liu, H., & Chen, J. (2024). Insights into strain engineering: from ferroelectrics to related functional materials and beyond. *Chemical Reviews, 124*(11), 7045–7105. https://doi.org/10.1021/acs.chemrev.3c00767
71. Rao, R., Lavric, D., Nath, T., Eom, C., Wu, L., & Tsui, F. (1998). Three-dimensional strain states and crystallographic domain structures of epitaxial colossal magnetoresistive $La_{0.8}Ca_{0.2}MnO_3$ thin films. *Applied physics letters, 73*(22), 3294–3296. https://doi.org/10.1063/1.122749
72. Fert, A., Ramesh, R., Garcia, V., Casanova, F., & Bibes, M. (2024). Electrical control of magnetism by electric field and current-induced torques. *Reviews of Modern Physics, 96*(1), 015005. https://doi.org/10.1103/RevModPhys.96.015005

73. Mosquera-Lois, I., Kavanagh, S. R., Klarbring, J., Tolborg, K., & Walsh, A. (2023). Imperfections are not 0 K: free energy of point defects in crystals. *Chemical Society Reviews, 52*(17), 5812–5826. https://doi.org/10.1039/D3CS00432E
74. Kumar, M., Shankar, S., Kumar, A., Anshul, A., Jayasimhadri, M., & Thakur, O. (2020). Progress in multiferroic and magnetoelectric materials: applications, opportunities and challenges. *Journal of Materials Science: Materials in Electronics, 31*(22), 19487–19510. https://doi.org/10.1007/s10854-020-04574-2
75. Zweig, J. E., Kim, D. E., & Newhouse, T. R. (2017). Methods utilizing first-row transition metals in natural product total synthesis. *Chemical Reviews, 117*(18), 11680–11752. https://doi.org/10.1021/acs.chemrev.6b00833
76. Lotnyk, A., Behrens, M., & Rauschenbach, B. (2019). Phase change thin films for non-volatile memory applications. *Nanoscale Advances, 1*(10), 3836–3857. https://doi.org/10.1039/C9NA00366E
77. Heron, J., Bosse, J., He, Q., Gao, Y., Trassin, M., Ye, L., … Salahuddin, S. (2014). Deterministic switching of ferromagnetism at room temperature using an electric field. *Nature, 516*(7531), 370–373. https://doi.org/10.1038/nature14004
78. Flebus, B., Grundler, D., Rana, B., Otani, Y. C., Barsukov, I., Gubbiotti, G., … van der Sar, T. (2024). The 2024 magnonics roadmap. https://doi.org/10.1088/1361-648X/ad399c
79. Webb, J., Andersen, U., & Huck, A. 2a. Imaging electronic transport in materials using colour centers. *2024 Roadmap on Magnetic Microscopy Techniques and Their Applications in Materials Science.* https://doi.org/10.1088/2515-7639/ad31b5
80. Kumar, S., Sharma, V., Kumari, N., Kaur, G. A., Saha, A., Thakur, S., & Shandilya, M. (2024). Recent advances in perovskite materials: exploring multifaceted properties for energy harvesting applications. *Ionics, 30*(9), 5159–5188. https://doi.org/10.1007/s11581-024-05658-3
81. Abhinav, V., Basu, P., Verma, S. S., Verma, J., Das, A., Kumari, S., … Kumar, V. (2025). Advancements in Wearable and Implantable BioMEMS Devices: Transforming Healthcare Through Technology. *Micromachines, 16*(5), 522. https://www.mdpi.com/2072-666X/16/5/522#
82. Boyn, S., Grollier, J., Lecerf, G., Xu, B., Locatelli, N., Fusil, S., … Xavier, S. (2017). Learning through ferroelectric domain dynamics in solid-state synapses. *Nature communications, 8*(1), 14736. https://doi.org/10.1038/ncomms14736
83. Dasgupta, N. P., Sun, J., Liu, C., Brittman, S., Andrews, S. C., Lim, J., … Yang, P. (2014). 25th anniversary article: semiconductor nanowires–synthesis, characterization, and applications. *Advanced Materials, 26*(14), 2137–2184. https://doi.org/10.1002/adma.201305929
84. MacManus-Driscoll, J. L., Wu, R., & Li, W. (2023). Interface-related phenomena in epitaxial complex oxide ferroics across different thin film platforms: opportunities and challenges. *Materials Horizons, 10*(4), 1060–1086. https://doi.org/10.1039/D2MH01527G
85. Shea, H. R., Gasparyan, A., Chan, H. B., Arney, S., Frahm, R. E., López, D., … McConnell, R. P. (2004). Effects of electrical leakage currents on MEMS reliability and performance. *IEEE Transactions on Device and Materials Reliability, 4*(2), 198–207. https://doi.org/10.1109/TDMR.2004.826350
86. Wong, H., & Iwai, H. (2005). The road to miniaturization. *Physics World, 18*(9), 40. https://doi.org/10.1088/2058-7058/18/9/31
87. Micard, Q. G. R. (2021). Multiferroic BiFeO3 films and nanostructures for hybrid energy harvesters. https://doi.org/10.1039/D2SE00947A
88. He, Y., Iocozzia, J., & Lin, Z. (2018). Magnetoelectric effect in single-phase multiferroic materials *Nano/Micro-Structured Materials for Energy and Biomedical Applications: Latest Developments, Challenges and Future Directions* (pp. 49–75): Springer. https://doi.org/10.1007/978-981-10-7787-6_2
89. Van Der Giessen, E., Schultz, P. A., Bertin, N., Bulatov, V. V., Cai, W., Csányi, G., … Hütter, M. (2020). Roadmap on multiscale materials modeling. *Modelling and Simulation in Materials Science and Engineering, 28*(4), 043001. https://doi.org/10.1088/1361-651X/ab7150

90. Roesch, R., Faber, T., Von Hauff, E., Brown, T. M., Lira-Cantu, M., & Hoppe, H. (2015). Procedures and practices for evaluating thin-film solar cell stability. *Advanced Energy Materials, 5*(20), 1501407. https://doi.org/10.1002/aenm.201501407
91. Kim, D. H., Ning, S., & Ross, C. A. (2019). Self-assembled multiferroic perovskite–spinel nanocomposite thin films: epitaxial growth, templating and integration on silicon. *Journal of Materials Chemistry C, 7*(30), 9128–9148. https://doi.org/10.1039/C9TC02033K
92. Khan, S. M., Gumus, A., Nassar, J. M., & Hussain, M. M. (2018). CMOS enabled microfluidic systems for healthcare based applications. *Advanced Materials, 30*(16), 1705759. https://doi.org/10.1002/adma.201705759
93. Misiurev, D., & Holcman, V. (2024). Modeling of Magnetic Films: A Scientific Perspective. *Materials, 17*(6), 1436. https://doi.org/10.3390/ma17061436
94. Zhang, Z., Liu, P., Song, Y., Hou, Y., Xu, B., Liao, T., ... Sun, Z. (2022). Heterostructure engineering of 2D superlattice materials for electrocatalysis. *Advanced science, 9*(35), 2204297. https://doi.org/10.1002/advs.202204297
95. Islam, M. S. (2019). Development and characterization of spark plasma sintered bismuth ferrite (BiFeO3) multiferroic ceramics. http://lib.buet.ac.bd:8080/xmlui/handle/123456789/5403
96. Gao, Y., Gao, M., & Lu, Y. (2021). Two-dimensional multiferroics. *Nanoscale, 13*(46), 19324–19340. https://doi.org/10.1039/D1NR06598J
97. Aiello, C. D., Awschalom, D., & Bernien, H. Roadmap for unconventional computing with nanotechnology. https://doi.org/10.48550/arXiv.2301.06727
98. Lu, C., Hu, W., Tian, Y., & Wu, T. (2015). Multiferroic oxide thin films and heterostructures. *Applied physics reviews, 2*(2). https://doi.org/10.1063/1.4921545
99. Sahoo, R. R., & Choudhary, R. (2021). Various facets of ferroelectrics as functional materials: an informative review. *Ferroelectrics, 582*(1), 98–123. https://doi.org/10.1080/00150193.2021.1951039
100. Sivasubramani, S., Kishore, C., Acharyya, A., Georgiev, V., Shafik, R., & Prodromakis, T. (2025). Design Methodologies for Skyrmion-Based Circuits and Systems in AI-Driven Applications: Bi-Directional Integration [Feature]. *IEEE Circuits and Systems Magazine, 25*(3), 30–55. https://doi.org/10.1109/MCAS.2025.3587613

Dr. Sandeep Kumar Pundir is an Assistant Professor in the Department of Physics, Lucknow University, Lucknow, India, with over 10 years of teaching and research experience. He did his M.Sc. degree in physics from C.C.S. University Meerut and his Ph.D. degree in material science from CSIR-NPL, New Delhi, India. He has published numerous research papers and contributed to both national and international conferences and seminars. His research area of expertise includes computation condensed matter physics and experimental work, including synthesis and characterization of advanced functional materials. His work involved conducting density functional theory (DFT) calculations using the Full-Potential Linearized Augmented Plane Wave method, Optimization of Photovoltaic devices for energy conversion and has experience working with various advanced deposition techniques, including E-beam, sputtering, thermal evaporation, spin coating, etc., and characterization techniques such as SEM, HR-TEM, AFM, etc.

Chapter 7
Spintronic Devices Based on Heterostructure Thin Films

Nikita Rawal, Sangeeta Rawal, and Ranjana Pande

7.1 Introduction

The twenty-first century has brought significant developments for human beings worldwide regarding information and communication technology (ICT). In this era, the most vital items in any household are computers, smartphones, storage devices, MP3/MP4 players and the global network of the computer system (internet). All these devices are interconnected, since semiconductor integrated circuits (e.g. microprocessor and random access memory) and permanent magnetic data storage devices (hard disks and memory card) were the two different components that comprised these electronic devices (Fig. 7.1) [1]. Since the further reduction in size of a transistor implanted into the integrated circuits is in process, it is predicted that the transistor size will reach that of an atom. At this scale, writing and reading would be most difficult or impracticable due to the quantum size effect and thermal dissipation [2–4]. The disadvantage is that no transistor can be smaller than an atom, and this cannot be overcome by any means according to the present-day industry of electronic devices. Consequently, the expansion of the customary charge-based electronics equipment will cease within a short period of time [5]. With that day looming, various scientists in different fields across the globe are already developing a likely alternative technology known as spintronics (spin transport electronics). Spintronics is one of the technologies that has been proposed as possible solution to the fundamental issues noted in the current conventional electronic devices [6–8].

N. Rawal · R. Pande
Department of Life Science, Sharda School of Bio-Science & Technology, Sharda University, Greater Noida, Uttar Pradesh, India

S. Rawal (✉)
Department of Physics & Environmental Sciences, Sharda School of Engineering & Science, Sharda University, Greater Noida, Uttar Pradesh, India
e-mail: sangeeta.4@sharda.ac.in

S. Sharma (ed.), *Multifunctional Heterostructure Thin Films Handbook*,
https://doi.org/10.1007/978-3-032-18692-8_7

Fig. 7.1 Fundamental elements included in the contemporary devices of information and communication [1]

Spintronics is a field that focuses on using the spin of electrons, either by itself or in combination with their charge, to achieve the necessary functionalities for processing and storing digital information [9]. Conventional electronic equipment merely took advantage of the electric charge of electrons and disregarded their spin during operation. During its early history, spintronics was sometimes called spintronics (as it was commonly referred to at that time) along with magnetoelectronics, since both fields were interested in electronic phenomena due to the spin of electrons. But magnetoelectronics is mainly a term used to characterize devices based on magnetic effects and magnetic field interactions (switching magnetization), especially in metallic devices such as magnetic sensors and hard drives. However, spintronics is not limited to magnetic materials, but rather the involves the control of spin-dependent properties of semiconductors and nanostructures, and finds broader use in data storage and information processing. Therefore, magnetoelectronics represents a wider technical and scientific field in applying magnetic phenomena to electronic devices whereas spintronics represents a wider scientific and technological field, including the magnetic and quantum-mechanical nature of the spin of electrons.

Spintronics, on the other hands, makes use of both the electron's charge and its spin to transmit and process information, where the two spin orientations (spin $\uparrow$ and spin $\downarrow$) can represent binary states such as 0 and 1 [3, 10]. It combines the semiconductor characteristics that are sensitive to charge (integrated circuits) and the magnetic properties that exploit spin (storage hard drive) into a single chip, hence enabling all processes to be achieved through a single chip (Fig. 7.2). There are some virtues to this new technology as compared to the traditional electronic systems. In conventional electronics, the random access memory (RAM) used is volatile, meaning it loses all stored information when the power supply is switched off. In spintronics, the information may be left in that state because the spin of an electron is independent of energy (thus it is non-volatile) and does not vanish when power is turned off, all other things being equal. Moreover, the capacitors of analogy electronics are recharged (consume more power) continuously as they lose the charge [1]. Conversely, the energy needed to switch a spin state between 0 and 1 or vice versa is extremely low; hence spintronic devices use less power. Besides, they are also quicker and smaller than the traditional electronic systems because one spintronic chip is capable of performing several functions that would otherwise be done by a number of its counterparts. Considering that, we may conclude that the functionality, speed, storage and power consumption will be significantly improved [11].

While several materials such as magnetic semiconductors, graphene, organic conductors and topological insulators are some of the materials that have been studied in terms of spintronic aspects, we focus on heterostructure thin films. Based on these films, spin transport and interfacial interactions can be controlled precisely, and are required to produce spintronic devices that are efficient and miniaturized.

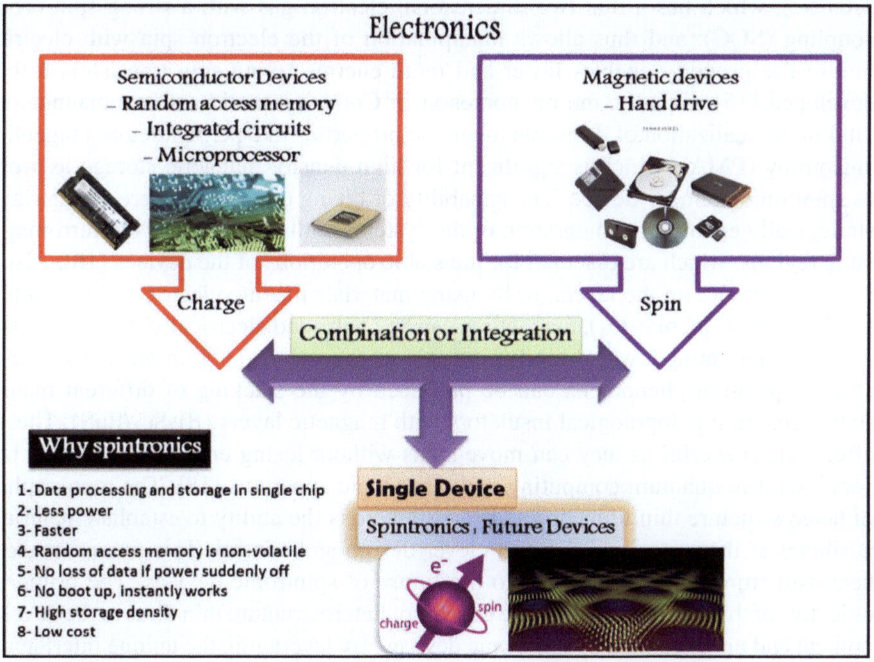

Fig. 7.2 Combination of semiconductor microelectronics and magnetic media, migration of electronics devices to spintronics devices [1]

Heterostructure thin films are important when it comes to increasing spintronic functionalities based on the use of unique interfacial phenomena and properties of the materials. These films, frequently composed of complex oxides and other compounds, make it possible to control spin currents and the magnetic characteristics, which are necessary for the development of sophisticated spintronic devices. The assembly of materials in heterostructures enables the opening up of new functionalities that include spin filtering, tunneling magnetoresistance and spin-orbit torque [12]. Heterostructure thin films are ones which have very thin layers of different materials on top of one another (potentially many different materials). These special structures help in improving the spintronic structures, which are technologies that utilize the spin of electrons (besides the charge) to store and process information. With a proper choice and alignment of materials, the scientists can control the movement and behaviour of the spins at the interfaces where the layers converge. In spintronics, a major difficulty has been to transferring spin-polarized electrons (electrons oriented in a certain direction of spin) out of a magnetic material and into a semiconductor with high efficiency. Normally, this is difficult due to an electrical property mismatch. The presence of an atomically thin insulating layer (such as MgO) between them can compensate for this insufficient performance by increasing the ease of tunneling transport of the electrons across the barrier [13, 14]. Such interfaces are also capable of developing special effects that have no occurrence in bulk materials. One can consider the interface between $LaAlO_3/SrTiO_3$ (Lanthanum Aluminate/Strontium Titanate), which has a thin two-dimensional electron gas with a strong spin-orbit coupling (SOC), and thus allows manipulation of the electron spin with electric fields. The property enables faster and more energy saving spin transistors to be developed [15]. Equally, the phenomenon in Co/Ni synthetic antiferromagnets is vital in the realization of desirable magnetic properties like perpendicular magnetic anisotropy (PMA), which is significant for high-density magnetic storage as well as spintronic memory device. The capability of laying down the different materials strategically enables the integration of the systems with compensated antiferromagnetic regions, which are essential for the stable operations of the devices [16]. Also, through pressure on the layers, or by using materials that are electric and magnetic (such as $BiFeO_3/CoFe_2O_4$), magnetism can be controlled electrically. This is significant for fabricating low-powered spintronic memory units [17]. In other instances, strange quantum phenomena can be produced by the stacking of different materials together, e.g. topological insulators with magnetic layers (Bi_2Se_3/EuS). These effects are powerful as they can move spins without losing energy so this may be very useful in quantum computing and other logic apparatus [18]. The topography of heterostructure thin films differs because there is the ability to establish material attributes at the nano level through clever design and control. This feature makes heterostructure thin films critical to the future of spintronic devices. The primary objective of this chapter is to explore the role of heterostructure thin films in the development and enhancement of spintronic devices. By leveraging the unique interfacial properties that emerge when combining different material layers such as ferromagnetic, semiconducting, insulating, multiferroic or topological phases, heterostructures offer advanced functionality not achievable with single-phase materials. This chapter aims to provide a comprehensive overview of how these engineered thin film systems contribute to improved spin injection, transport and manipulation, which are critical for next-generation memory, logic and quantum computing technologies.

7.2 Mechanism of Spintronics

Electron spin is an elementary quantum feature of electrons, which is an intrinsic kind of angular momentum. In traditional examples of spinning objects, rotation actually exists, whereas, in the case of electron spin, it is still a quantum-mechanical property that has two distinguishable orientations naturally known as spin-up and spin-down [13]. Spin has this binary quality that can be applied in the next generation of electronics, especially in the emerging spintronic technology, in which the spin of electrons is utilized in addition to their charge for information transfer and storage [19].

Spin polarization is, however, a macroscopic effect that exists when we have unequal numbers of spin-up and spin-down electrons. These spins are usually equal in non-magnetic substances thus giving rise to no net polarization. Nevertheless, in engineered heterostructures or a ferromagnetic material, a single orientation may prevail giving rise to a spin-polarized current [20]. It is this polarization of the spins that becomes important when using the concept of spintronic devices, e.g. magnetic tunnel junction devices, spin valve or spin transistor, where the level of spin polarization has a direct impact on device efficiency and sensitivity. Spin polarization is usually measured in terms of this difference over total number of spin-up and spin-down electrons to give a readily measurable design and optimization parameter of spin-based components [21].

The formula defining the degree of spin polarization along a certain direction quantitatively is:

$$P = (n \uparrow - n \downarrow) \div (n \uparrow + n \downarrow)$$

where n (↑) and n (↓) are the number of spin ↑ and spin ↓ electrons, respectively. $P = 1$ used to mean that a system is fully polarized (all spins in the same direction), and $P = 0$ means that there is no net degree of polarization at all (equal numbers of 'spin ↑' and 'spin ↓' electrons) [22].

When a material or structure becomes spin polarized then it becomes imperative to know what happens to these spin-aligned electrons within it. The mechanism for the propagation of spin-polarized electrons is known as spin transport.

Spin transport is a transfer of electrons of the same spin direction or a particular spin direction (i.e. up or down). In traditional electronics, the information flow is by means of electric charge. Yet, the amount of information is not (or rather, not necessarily) carried by the charge of electrons, but by their spin in spintronics. This flow of spin-polarized electrons is termed a spin current. Just as charge current can be used to power the devices and transfer signals, spin current could be utilized to store and process the information in new ways. The objective of spin transport is to ensure that spin-aligned electrons travel through the material while ensuring they do not lose their spin direction. This is relevant to the fabrication of effective spintronic (memory and logic) chips [23]. To transmit spin successfully spin transport requires the generation and inoculation of spin-polarized carriers into a non-magnetic material—a process referred to as spin injection. Spin injection is the initial step in the creation of spin-polarized currents in a device, and it allows the further movement and control of spins across interfaces. The level of spin polarization would quickly decay in the absence of effective spin injection to lower the performance and sensitivity of spintronic components. Thus, the detailed knowledge and optimization of spin injection processes is one of the key steps of the achievement of high-performance spin-based devices and circuits [24].

7.2.1 Spin Injection

Spin injection expresses the ability to introduce spin-polarized electrons into a non-magnetic material (typically a semiconductor). The target material (semiconductor, metal or superconductor) acquires a net spin population originating from a ferromagnet. This forms a very basic concept in spintronics, since it allows devices to utilize the electron spin in the processing, storing and transportation of data [24].

The spin injection involves transferring electrons with a preferred spin direction into a new substance. It normally begins with a ferromagnetic substance (such as iron or cobalt), which itself possesses a more substantial number of electrons with one orientation of spin (e.g. spin-up) than the alternative (spin-down). Such asymmetry develops a spin-polarized current, i.e. the electrons leaving have a dominant spin [25]. These electrons are spin polarized and subsequently transported along an interface, that is, they travel along the interface that borders the ferromagnet with another substance, usually a non-magnetic semiconductor. The quality of this interface is very important in determining how well this transfer works. When a surface is rough, or when energy levels (band alignment) are mismatched, or when a proper insulating barrier (such as MgO) is absent, then many of the electrons may lose their spin orientation or bounce back, and so spin injection becomes ineffective [26]. When the electrons manage to penetrate the target material, their direction of spin is expected to remain in that direction as long as possible. This generates an unequal distribution of spin orientations which is termed a non-equilibrium spin populations within the material. The presence of such spin imbalance is either detectable or can be used to affect other electronic properties, and so it is applicable in applications such as memory devices, sensors and more advanced computing [27].

7.2.2 Relaxation

Spintronics spin transport, with spin-polarized electrons injected into a material the next thing that is of interest is the way that those electrons move through the material, maintaining the orientation of their spin. Spin relaxation refers to the process in which electrons gradually stop acquiring the initial spin direction (either spin-up or spin-down) as they travel through material. [28, 29]. Spin relaxation mechanisms differ mainly in three broad modes:

(i) **Elliott-Yafet Mechanism**

Among the most prevalent spin relaxation sources, the Elliott-Yafet mechanism plays a leading role in metals and strongly scattering semiconductors. In this process, the spin is allowed to relax due to the collisions with either impurities or crystal lattice vibrations (also known as phonons). There is a minute probability of the spin of the electron flipping between up and down in every scattering process. Because these spin flips accumulate, over time, as the electrons undergo progressively more scattering events, this original spin alignment fades away. In this way, this mechanism requires a more rapid spin relaxation with the growing quantities of scattering [30].

(ii) **D'yakonov-Perel Mechanism**

This D'yakonov-Perel process is generally more common in non-centrosymmetric semiconductors, such as gallium arsenide (GaAs) semiconductors that are asymmetrical. The spin of these electrons is not immobile but starts to process (wobble, like a spinning top) due to the internal electric fields related to spin-orbit coupling. The relative orientation of this wobble is scrambled as it travels through the material, making the direction of the spin disordered. Amazingly, in this process, even higher scattering can reduce spin relaxation because the precession that the spins would have undergone is disrupted by constant collision with the lattice [31].

(iii) **Bir–Aronov–Pikus Mechanism**

The Bir–Aronov–Pikus mechanism plays a role in p-type semiconductors, where there is a high concentration of holes—which are simply vacant position of electrons. Here, spin relaxation occurs due to exchange interactions of electrons and holes. The interaction of electrons with these holes on a close-range basis causes them to change their spins and this phenomenon is referred to as relaxation. The higher the hole density in the material, that is, the higher the density of the holes, the more such interactions are possible, and the quicker the loss of spin information [32].

7.2.3 Diffusion

Diffusion explains how particles (like atoms, molecules or electrons) move from one place to another due to a difference in concentration or energy. This is a random, spontaneous process driven by thermal energy, and it plays a key role in many physical, chemical and biological systems [33]. In spintronics, this mechanism refers to the movement of spin-polarized electrons that have their spins pointing in the same direction in large part through a material. When spins are not equally distributed, a mismatch is created between the number of spins up and down electrons in a specific region that makes them scatter to the level up to which they are each in number. This is in the same way that the heat moves towards an area that is cold. When the spins are moving, they do not necessarily keep their initial orientation due to collisions with atoms, impurities, or the other electrons. Interactions between the spins may result in loss of alignment in them, a condition that is termed spin relaxation. Therefore, when the spin information diffuses into the material, it becomes lost gradually over the course of diffusion [34].

7.3 Fabrication of Heterostructure Thin Films

The spintronic technologies cannot progress without the fabrication of heterostructure thin films. These films are normally categorized into magnetic, non-magnetic, ferromagnetic or semiconductor. The spin orientation is controlled by magnetic layers, the spin transport is controlled by non-magnetic layers, spin polarization is controlled by the the ferromagnetic layers, and the integration with the conventional electronic systems is enabled by the semiconductor layers. The presence of these materials in one heterostructure enables the injection, transport and detection of spin to be done efficiently, which is indispensable for a high-performance spintronic device.

(i) **Magnetic**

Magnetic heterostructure thin films are a type of material whose fabrication involves depositing extremely thin layers of various materials with at least one of the layers being magnetic. The interfaces between the layers in these films are arranged in such a way that the new or enhanced magnetic properties they have are not present in the individual materials. They are materials that are sensitive to magnetic fields and can affect electron spin, a characteristic that is highly helpful in the sphere of spintronics [35] and some of the magnetic heterostructure are mention in Table 7.1. In magnetic heterostructures, the interface between layers creates the possibility to control spin and magnetic behaviour with more effectively. They are applied in devices such as magnetic sensors, memory storage devices (such as hard drives) and superior electronic components. With the combiation of magnetic and non-magnetic or even semiconductor layers, magnetic heterostructures can be prepared to effectively execute complex activities in an efficient, compact matrix [40].

Table 7.1 Types of magnetic heterostructures

Type of heterostructure	Overview	Example	References
Magnetic–non-magnetic	Combines magnetic and non-magnetic layers to control or modulate magnetic properties	2D magnets with graphene or metals	[33]
Magnetic–magnetic	Stack two or more different magnetic layers, enabling new effects like enhanced magnetism	CrI_3/Fe_3GeTe_2 (chromium triiodide/iron germanium telluride), $CuCr_2Te_4/Cr_2Te_3$ (copper chromium telluride/ chromium telluride)	[36]
Magnetic–topological Insulator	Pairs magnetic layers and topological insulators, leading to exotic quantum states	Cr-doped $(bi,Sb)_2Te_3$ (chromium-doped bismuth antimony telluride), $MnBi_2Te_4/bi_2Te_3$ (manganese bismuth telluride/bismuth telluride)	[37]
Multiferroic (magnetic–ferroelectric)	Integrates magnetic and ferroelectric control of magnetic and vice versa	$BiFeO_3$-based, ferromagnetic/ piezoelectric	[38]
Magnetic–semiconductor	Combine magnetic and semiconductor layers for spintronic and optoelectronic application	Magnetic TMDs (transition metal dichalcogenides) with semiconductors	[39]

(ii) **Non-magnetic**

The non-magnetic heterostructures are thin films synthesized using layers of materials that lack magnetic properties. These structures are designed either to be composited or to enhance properties such as electrical conductivity, optical properties or mechanical strength rather than magnetism. In fact, the interfaces between the various non-magnetic layers can be formed into new behaviours that do not exist in the separate materials, and these structure con be used in a large variety of electronic and optoelectronic devices [41], and some of the non-magnetic heterostructure are given in Table 7.2. Through layering non-magnetic materials with different energy levels or charge characteristics, scientists are able to regulate the actions of electrons within the system, culminating in improved efficiency and results in electronic gadgets [42, 45].

Table 7.2 Types of Non-magnetic heterostructures

Type of heterostructure	Overview	Example	References
Semiconductor heterostructures	Layers of different semiconductors (e.g. Si, Ge) to control electron flow and enhance devices	Quantum wells, transistors, lasers, solar cells	[42]
Insulator/ semiconductor	Combines insulating and semiconducting layers to manage charge and energy barriers	Sensors, memory devices, optoelectronic microcavities	[43]
2D material heterostructures	Stacks of atomically thin layers (e.g. graphene, hexagonal boron nitride, transition metal Dichalcogenides) for new electronic/optical traits	Flexible electronics, photodetectors, tunneling transistors, van der Waals heterostructures	[44]

(iii) **Ferromagnetic**

Ferromagnetic substances refer to those substances in which the atomic magnetic orientations run parallel to each other, resulting in highly intense and permanent magnetic properties even in the lack of an external magnetic field. When these materials are used in thin film heterostructures, meaning stacks of very thin layers, they can interact with other materials to create new or enhanced magnetic properties, which are important for advanced electronics and spintronic devices and some of them are given in Table 7.3 [46, 51].

(iv) **Semiconductor**

A semiconductor is a substance that has electrical conductivity between that of a conductor and an insulator, and the conductivity may be modulated by the addition of impurities or by external fields. The stacking of different semiconductor materials forms semiconductor heterostructures, which give precise control over electronic properties and optical properties, and some of the semiconductors heterostructures are given in Table 7.4. These heterostructures are foundational to modern electronics, enabling devices like transistors, lasers and solar cells [42, 49].

Table 7.3 Types of ferromagnetic heterostructures

Type of heterostructure	Overview	Example	References
Ferromagnetic/ ferromagnetic	Combines two or more different ferromagnetic layers to tailor properties like anisotropy and curie temperature	Vanadium Diiodide and vanadium Octaiodide heterostructure for tunable magnetic transitions and spintronic devices	[46]
Ferromagnetic/ non-ferromagnetic	Pairs a ferromagnetic layer with a non-magnetic or weakly magnetic material; used to control spin currents and study spin-orbit effects	Iron germanium telluride/ graphene for spin injection, detection, and spintronic device engineering	[47]
Ferromagnetic/ superconductor	Combines ferromagnetic and superconducting layers, enabling phenomena like spin-triplet superconductivity and quantum computation	Superconductor/ferromagnet (SC/FM) heterostructures for superconducting spintronics and π Josephson qubits	[48]
Ferromagnetic/ topological insulator	Stacks ferromagnets with topological insulators to create exotic states (e.g. skyrmions) for advanced spintronics	Iron germanium telluride/ hexagonal boron nitride for skyrmion formation and manipulation in 2D magnets	[49]
Ferromagnetic/ ferroelectric (multiferroic)	Combines ferromagnetic and ferroelectric layers, enabling electric field control of magnetism and multifunctional devices	Chromium germanium telluride/indium selenide and vanadium bismuth telluride/ indium selenide for electric control of magnetism and nonvolatile memory	[50]

7.4 Fabrication Techniques

All advanced electronic, photonic and sensing devices require precise control over thickness, composition, and interface quality of fabricated heterostructures. Molecular Beam Epitaxy (MBE) [56], Pulsed Laser Deposition (PLD) [57], Sputtering [58] and Atomic Layer Deposition (ALD) [59] are some of the common thin film deposition methods. The method, with their unique set of advantages and challenges for the growth of heterostructures are discussed below.

7.4.1 Molecular Beam Epitaxy (MBE)

MBE is a specialized method of growing extremely thin and high-quality films of materials commonly applied as part of an electronics system or nanotechnology

Table 7.4 Types of semiconductors heterostructures

Type	Overview	Example	References
Binary/compound Heterostructure	Layers of two or more different semiconductors, often with different band gaps or lattice constants	Silicon/germanium, gallium arsenide/ aluminium gallium arsenide	[42]
Type I, II, III heterojunctions	According to the band alignment at the interface: Type I (straddling), type II (staggered), type III (broken gap)	Functionalized 2D materials, e.g. scandium carbide difluoride/ scandium carbide dioxide	[52]
Quantum wells, dots, superlattices	Ultra-thin layers confining electrons/holes in one or more dimensions, leading to quantum effects	Quantum wells, quantum dots, superlattices	[53]
2D/Van der waals heterostructures	Stacks of atomically thin 2D materials with weak interlayer forces, enabling new properties	Graphene, molybdenum Disulfide/ tungsten Diselenide/ palladium Diselenide	[54]
Donor-acceptor (hybrid) structures	Combine organic and inorganic semiconductors for enhanced charge separation and transfer	Organic/inorganic hybrids	[55]

system. It takes place in a chamber that has an ultra-high vacuum, meaning that almost all the air and other undesirable particles are removed. This assists in maintaining the purity and cleanliness of the materials in the process [56]. The hot sources ionize the atoms or the molecules and they emerge in the form of beams within the chamber. These beams come in a straight line and land on a smooth, hot surface (this is called a substrate). When the atoms reach this surface, they get stuck, and the atoms behind them arrange themselves in an orderly manner, where they end up forming a crystal layer, that has the same structure as that of the bit of the surface. This phenomenon is known as epitaxial growth [60]. It is a procedural process, and every step is quite slow and highly regulated. It is exactly why the material can be made in layers, and the thickness and content of the component can be regulated with great accuracy. This aids the creation of complex structures created using several layers and the clean sharp interface between any two interfaces [61]. To ensure that everything is proceeding correctly, scientists are using instruments like Reflection High-Energy Electron Diffraction (RHEED) when the process of growth is being undertaken. This enables them to monitor the process of the formation of the layers as they are created, and that they are smooth and of quality grade [62].

7.4.2 *Pulsed Laser Deposition (PLD)*

Pulsed Laser Deposition (PLD) is a widely used technique to provide thin layering of a wide variety of materials including oxides, semiconductors, metals and polymers.

In PLD, a solid substance (known as the target) is subjected to a high-power pulsed laser beam in a vacuum chamber. The laser vaporizes or blasts off particles when it strikes the target, a process known as laser ablation. This action subsequently produces a hot cloud of particles, called a plasma plume, which jets out of the target [57]. The flying particles are then deposited in the chamber and land on another surface, which is referred to as the substrate. On reaching the substrate, they cool and condense to form a thin film layer by layer. The particles then organize themselves naturally, usually in line with the structure of the surface [63]. Atmospheric flexibility is one of the strong points of PLD; it can be performed under different gas pressures and gas types not only in vacuum. This allows the growth of complex materials to be easier and make it easier to adjust the properties of the thin film to be easier through control of parameters within the chamber [64].

7.4.3 *Sputtering*

Sputtering is a commonly used method for depositing thin films and making nanostructure through physically or chemically removing atoms of material from a solid target through high-energy particle impact.

(i) **Physical Sputtering:** Here, energetic ions (normally a gas plasma) are bombarded onto the surface of the target material. On collision with the surface, they transfer their energy, and they eject atoms or molecules for the target. These particles become ejected and end up fly across the chamber, dropping themselves onto the substrate as thin films [58].

(ii) **Chemical Sputtering:** In this process, the incoming particles used may be ions, electrons or photons that cause some chemical reactions to take place on the target surface. These reactions form volatile compounds (easily evaporated), and they exit the surface and fall to the substrate [65].

(iii) **Potential Sputtering:** This is the situation in which highly charged ions strike the surface. Fractions of these ions instead of relying solely on the property of speed (kinetic energy), release their electrical energy when they come in contact with materials. This energy assists in throwing the atoms from the surface even when the motion of the ion is low [66]. Overall, sputtering is a versatile and commonly accepted technique of thin film deposition in electronics, optics and materials science applications.

7.4.4 Atomic Layer Deposition (ALD)

Atomic Layer Deposition (ALD) is an atomically controlled, vapour phase deposition, conformal film growth process that functions by repeated sequential release of precursors which result in a highly uniform, conformal, pinhole free film.

The technique to generate ultra-thin films on an atom-by-atom basis involves the use of ALD. It is, in fact, achieved by periodically exposing the surface (substrate) to one or more gaseous chemical compounds (designated as precursors) of varying kinds. The reaction between any one of the chemicals is confined to the surface and is self-limiting in its nature in that the chemical interaction ceases when a single atomic layer is formed. This helps in the regulation of the thickness with high precision [59]. The system removes the gases and byproducts left after every exposure using an inert gas purge (e.g. nitrogen or argon) ensuring that unwanted reactions do not take place between steps in the air [67]. This repeated layer-to-layer process is done till it forms a thin film. Even complex or 3D surfaces have proved to be ideal surfaces to apply coating using ALD as it is able to apply coating in a homogeneous and uniform way even when the space is limited or the shape is complex [68].

7.5 Materials for Spintronic Heterostructures

The heterostructures based on spintronic materials are developed through the choice and integration of materials exhibiting different magnetic and electronic characteristics. These materials can be generally divided into three broad groups: magnetic metals and alloys (70), two-dimensional (2D) materials (73) and topological insulators and magnetic oxides (77). Magnetic metals and alloys are good spin injectors and detectors with good magnetic ordering. 2D materials, in their turn, provide unparalleled properties of spin transport and atomic-level tunable electronic behaviour. Current research is also directed towards the use of topological insulators and magnetic oxides due to their special spin–orbit interactions and interfacial properties that allow easy manipulation and storage of spin. Collectively, these materials lay the groundwork for the next-generation spintronic devices that will be more efficient, scalable and multifunctional.

7.5.1 Magnetic Metals and Alloys

Modern technology relies on magnetic metals and alloys as the basis of many modern devices, such as data storage, energy conversion and sensing. They are defined in terms of composition, structure and processing resulting in a very diverse range of materials with varying magnetic behaviour suited to a particular application [69]. These materials exist in different types and structures:

Ferromagnetic elements and alloys such as iron (Fe), cobalt (Co) and nickel (Ni) are the main ferromagnetic metals; they are usually alloyed with other substances to tailor their magnetic behaviour. The most common alloys, e.g. Fe-Co, Fe-Ni and Co-Ni, are generally tunable in terms of the magnetic moments and anisotropy through alloy composition [69]. Heusler alloys and high-entropy alloys (HEAs) provide adjustable ferromagnetic properties, high Curie temperatures and in some cases half-metallicity. They have potential for high-tech applications such as in spintronics and magnetocaloric applications [70]. In addition, amorphous and bulk metallic glasses, quickly cooled alloys, take on amorphous structures with special soft or hard magnetic properties and can be applied to transformer cores and magnetic sensors [71].

7.5.2 2D Materials

2D materials have become integral components in the design and development of spintronic heterostructures due to their exceptional electronic, magnetic and spin transport properties. Their atomically thin nature and high interface quality make them ideal platforms for efficient spin injection, transport and detection. Materials such as graphene provide long spin diffusion lengths, while TMDs like MoS_2 and WSe_2 exhibit strong spin-orbit coupling that enables effective spin manipulation [72]. Hexagonal boron nitride serves as an insulator spacer or tunnel barrier, minimizing spin scattering at interfaces, and magnetic 2D material such as CrI_3 and Fe_3GeTe_2 introduce controlled magnetism at the nanoscale [73]. When stacked into van der Waals heterostructures, these materials allow precise engineering of interfacial effects, resulting in tunable magnetic, electronic, and quantum properties. The unique combination of spin coherence, flexibility and interface control makes 2D materials a cornerstone, flexibility and interface control makes 2D materials a cornerstone in advancing next-generation spintronic devices with enhance performance and energy efficiency [74].

7.5.3 Topological Insulators (TIs) and Magnetic Oxides

Topological insulators and magnetic oxides represent some the most active fields in condensed matter research because of their exotic electronic properties and their promise to become the sources of recently discovered exotic quantum phenomena. The interplay of topological order and magnetism in these materials results in exotic effects with potential applications in spintronics, quantum computing and energy-efficient electronics.

Topological insulators (TIs) are unique materials that do not conduct electricity in their (bulk); however, the surfaces or edges of topological insulators act as excellent conductors of electricity. The band topology of this material, which is a quantum-physics property, maintains the stability of its surface states even in the face of impurities [75, 76]. By doping a topological insulator with magnetic atoms, bringing it into proximity with a magnetic material, or using an inherently magnetic material, introduces a breaking of time-reversal invariance [77].

7.6 Spintronic Devices Based on Heterostructures

Heterostructure spintronic devices utilize the combination of charge and spin of electrons to develop new features in data storage and logic. The major types of devices are Magnetic Tunnel Junctions (MTJs) [78], Spin Valves and Giant Magnetoresistance (GMR) [79] Devices and Spin Field-Effect Transistors (SFETs) [80]. MTJs are based on quantum tunnelling across layers of ferromagnet materials, which are separated by an insulating barrier, which is the heart of the MRAM technology. Spin valves and GMR devices use the changes in resistance because of magnetization alignment to sense and identify memory. Using electric fields, SFETs manipulate spin-polarized currents and provide a path to low-power and high-speed spin-based logic devices.

7.6.1 Magnetic Tunnel Junctions (MTJs)

Spintronic devices invariably include Magnetic Tunnel Junctions (MTJs) that con be used to conveniently control spin-polarized currents, but also detect their presence. Their unique structure and the quantum tunnelling phenomenon are the foundation of their adoption in memory, logic and sensing, and the ongoing development in materials and device engineering has also led continuously to higher performance [78].

Structure

Magnetic Tunnel Junctions are important building blocks in spintronics, especially for devices like MRAM (magnetic random-access memory). An MTJ is made up of two ferromagnetic layers (materials like iron or cobalt that have strong magnetism) with a very thin insulating layer in between. This insulating layer is often made of MgO or a 2D material like hexagonal boron nitride (h-BN) [78]. Although the middle layer is an insulator, quantum mechanics allows electrons to tunnel through it shown in Fig. 7.3. The amount of tunnelling (and consequently that of magnetic alignment of the two ferromagnetic layers) determines the electrical resistance of the device:

- When the magnetizations of the two are parallel (i.e. of the same direction), the resistance is small.
- When antiparallel (pointing opposite to one another), resistance is great.

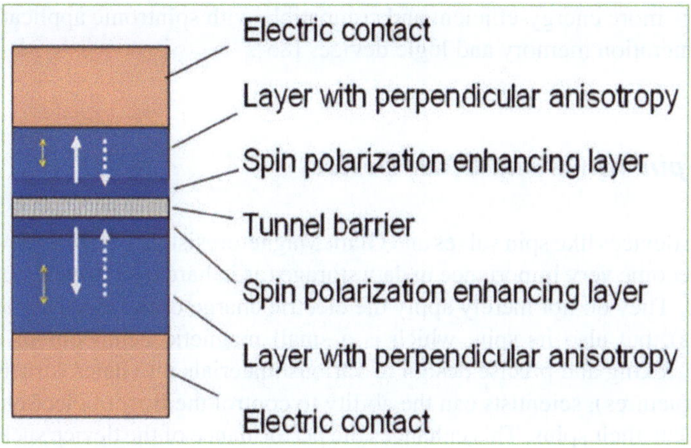

Fig. 7.3 Schematic perpendicular MTJ diagram [81]

This difference in resistance based on spin alignment allows MTJs to store data as binary 0s and 1s [82, 83]. MTJs use van der Waals heterostructures, which are made by stacking 2D materials with atomic-level precision. For example, structures like $Fe_3GaTe_2/WS_2/Fe_3GaTe_2$ or Fe_3GeTe_2/h-BN/Fe_3GeTe_2 are being explored. These offer very clean and sharp interfaces, which improve performance and open the door to new functionalities, such as better spin control, flexibility, and use in ultra-thin or transparent electronics [84].

Working Principle

The concept of operation of an MTJ is known as tunnelling magnetoresistance (TMR). It consists of two magnetic layers (electrodes) separated by a very thin insulating electrical layer (an MTJ). Although the barrier material is an insulator, quantum mechanics allows the electrons to tunnel through the middle layer because it is very thin.

The resistance of the MTJ is controlled by the relative orientation of the magnet directions magnetizations of each of the layers:

- When the magnetizations are parallel (in the same direction), one can easily transfer electrons through them by tunnelling; therefore, the resistance is also low.
- Fewer electrons can flow through when the magnetizations are antiparallel (in opposite directions) hence resistance is high.

This change in resistance is called TMR and is what allows the MTJ to store binary data like a switch between '0' and '1'. Some MTJs use special barrier materials or electric fields to filter and control the spins of electrons. This means only electrons with a certain spin direction are allowed to pass through. These techniques contribute to enhanced spin polarization, improve switching behaviour. This will make MTJs more energy-efficient and compatible with spintronic applications such as next-generation memory and logic devices [85].

7.6.2 *Spin Valves and GMR Devices*

Spintronic devices like spin valves and Giant Magnetoresistance (GMR) devices have recently become very importance in data storage (as in hard disk drives), sensors and even logic. They do not merely apply the electric charge of the electron (as in usual electronics), but also its spin, which is a small magnetic behaviour of electrons. Through stacking and precise design of various materials into heterostructures (thin layered structures), scientists can the ability to control the flow of electrons, as well as manipulate their spins. This enhances the performance of the device such that they are faster, smaller and consume less energy. Thus, the clever way of using materials and multilayered designs is what makes spintronic technology to be more effective and enable contemporary technology [79].

Structure

Spin valves and Giant Magnetoresistance (GMR) devices are prepared by stacking two ferromagnetic films (highly magnetic materials such as iron or cobalt) separated by a non-magnetic spacer (Fig. 7.5). This spacer may be metal or a semiconductor. The key principle is that the magnetization configuration of the layers determines the electrical resistance of the device:

- The resistance is low when the magnetizations are aligned (point towards the same direction).
- When they are antiparallel (pointing in opposite directions) resistance is high [87].

The outcome can be read and recorded as data, mainly in hard drives and sensors (Fig. 7.4). These devices can also be based on on van der Waals heterostructures formed by stacking low-dimensional 2D materials. They include 2D ferromagnet like Fe_3GeTe_2 or Co/Ni alloy, and 2D spacers like MoS_2, WS_2 or graphene. Designing with 2D materials enables the layers to possess exceptionally clean, atomically abrupt interfaces, which enhances performance. Such devices also can be flexible and bendable, which is useful in wearable electronics or next-generation sensors [88].

Working Principle

Giant Magnetoresistance (GMR) and spin valves rely on the working principle which deals with current transformation of the current in the device according to the direction of magnetization employed in the magnetic layers. The devices have two ferromagnetic layers that have a non-magnetic thin layer separating them. When the two magnetic layers are aligned in parallel there is less resistance to transfer of electrons (especially spin polarized electrons) through them and therefore the resistance is low. On the other hand, Whene the magnetizations are antiparallel, the scattering

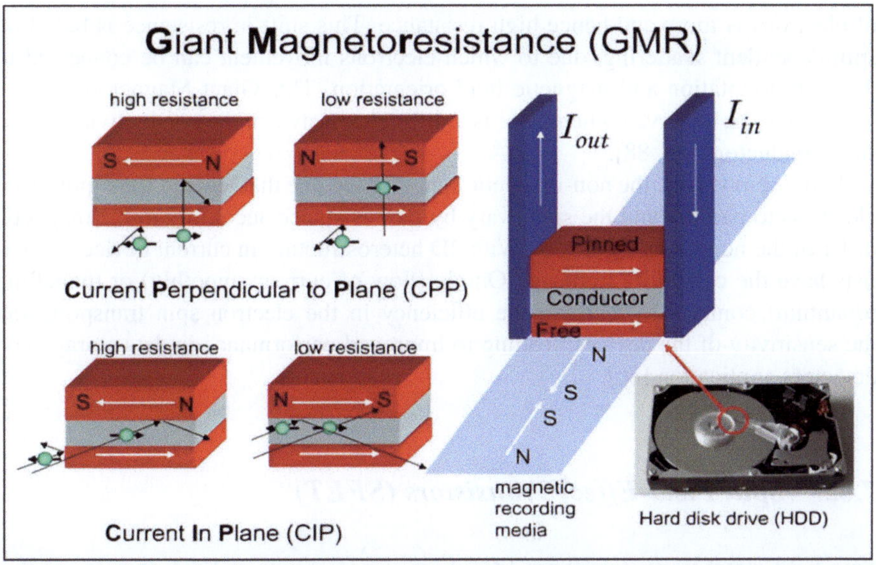

Fig. 7.4 Schematic diagram of Giant Magnetoresistance (GMR) effect with two-Giant Magnetoresistance (GMR) and two configurations, currents perpendicular to plane (CPP), and in-plane (CIP). The mechanical arrangement of the structure varies, places paralleling or antiparalleling the magnetic layers, and results in the low resistive or high resistance respectively. One of the popular utilizations of GMR is used in the hard disk drive (HDD) reading data [86]

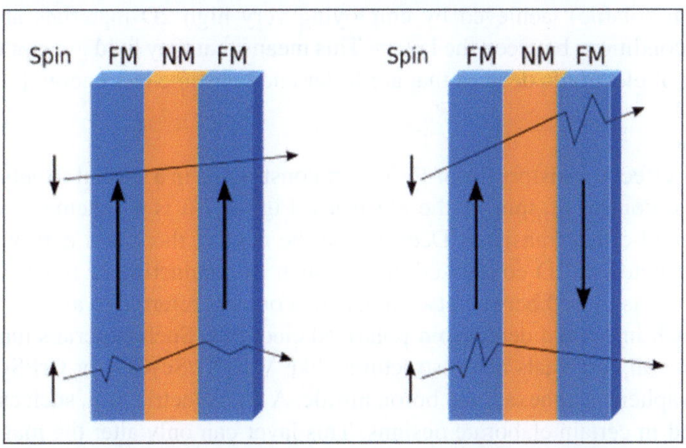

Fig. 7.5 Functional and schematic spin valve design. It contains 2 ferromagnetic layers with a non-magnetic spacer between them. The resistance of the device varies according to the orientation of the magnets; a small resistance can be achieved using parallel and large resistance with antiparallel. HDDs suggested magnet read head with the use of spin valves [86]

of electrons is more and hence high resistance. This shift in resistance is based on spin-dependent scattering, due to which electrons movement can be connected to the spin orientation and magnetic field orientation. This Giant Magnetoresistance (GMR) can be observed when there is a difference between the conductivity of two such conductors [86, 88].

In the spin valves, the non-magnetic spacer is the one that lets the spins-polarized electrons to pass by and the spins vary by the resistance according to the magnetic order of the neighbouring layers. With 2D heterostructure in current devices, scientists have the capability to design Ohmic (flow of current smoothly) or tunnelling (quantum) contacts to increase the efficiency in the electron spin transport, and the sensitivity of the devices, leading to improved performance in data storage and detectors application [86].

7.6.3 *Spin Field-Effect Transistors (SFET)*

Spin field-effect transistors (Spin-FETs) are a special type of electronic device that exploits not only the charge of electrons (as in normal transistors do), but also their spin, a small magnetic property in order to convey and manipulate information. These devices are constructed using heterostructures (materials that have differing properties due to being layered). Spin-FETs may operate at high frequencies with reduced power consumption and with data storage persisting even when the device is off (non-volatile), achieved by employing very high 2D materials and special interface conditions between the layers. This means that they hold great potential for the future of electronic devices that are faster and consume less energy [80].

Structure

Spin field-effect transistors (Spin-FETs) are constructed in a special structure, which aids in the control of spin of the electrons. Figure 7.6 is a schematic variant of the spin field-effect transistor. Deep inside the device, there is a narrow throughway constructed of 2D condensed matter, such as conducting or magnetic arrays. This channel is placed between two magnetic contacts referred to as the source and drain, which inject and detect spin-polarized electrons. Such materials may then be stacked in van der Waals heterostructures, like VSi_2N_4/Sc_2CO_2 or $CrPS_3/Sc_2CO_2$, or even graphene and hexagonal boron nitride. A ferroelectric film, such as Sc_2CO_2, is included in certain elaborate designs. This layer can only alter the magnetic and electrical property of the channel simply by applying an electric field and then the device can be switched off or on even without the magnetic field. This results in faster and energy-efficient Spin-FETs [80, 90, 91].

Working Principle

Spin-FETs working principle is premised on manipulation of energy direction of a certain spin of electrons. The injected spin-polarized electrons into the channel are first injected by the magnetic source. Next, electronic properties of the channel can

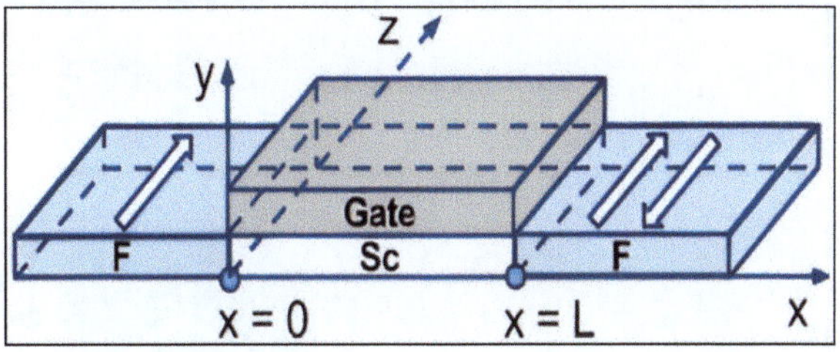

Fig. 7.6 Diagrammatic representation of spin field-effect transistor [89]

be altered by the application of a gate voltage or the electric field of a ferroelectric layer. This shift is able to change the channel state between those that do or do not permit some spins to directly impact with the amount of current flow like when the device is turned on or off [92]. In certain Spin-FETs, this control is achieved solely through electric fields without magnetic fields, and thus they are more energy-efficient and can hold their state even when there is no power present (non-volatile) [93]. Furthermore, some Spin-FETs are spin filters, allowing only electrons whose spin is in a certain direction will be able to pass through them, which make them even more valuable in spin-based electronics [94].

7.6.4 Spin-Orbit Torque (SOT) Devices

Recent types of spintronic technology include spin-orbit torque (SOT) devices, which allow the process of controlling the magnetization of a material to a manipulated with relatively high speed and using much less power than previous technologies and are of interest in the applications of memories and logic in electronics. They are made in stacks: in one material, there is strong coupling between spin and orbit (typically heavy metals), and the other is a magnet. The interface of these layers leads to the emergence of one-of-a-kind form of torque (force) when an electric current passes the device; it has the power to revert the direction of the magnetism on the layer of the magnet [95]. Figure 7.7 shows the spin-orbit torques (SOTs) have various applicable processes in the theoretical image. These encompass the spin Hall effect, Rashba-Edelstein effect, the orbital Hall effect, the heat gradient, the magnon (or spin wave) and the strain effect. They are involved in the creation and/or control of spin currents or torques of materials. Some of the materials that facilitate such effects are metals, metallic alloys, topological insulators, 2D materials and complex oxides. The magnetic layers used in SOT devices can be investigated as magnetic insulators, antiferromagnets, ferrimagnets, and each of them has benefits [96]. In SOT-MRAM, SOT-based devices are applied in both three terminal and two terminal magnetoresistive RAM architectures. Under these designs, the devices are small and energy conscious as the external magnetic fields are eliminated. In neuromorphic computing, SOTs are also being explored to be positioning themselves well, realizing brain-like circuits to process high values [98].

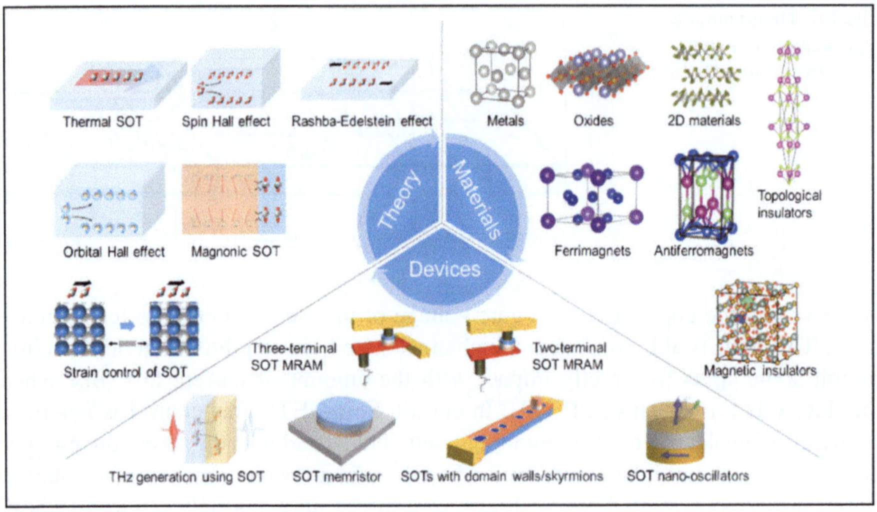

Fig. 7.7 Schematic illustration of the theoretical mechanisms (top-left), material platforms (top-right), and device applications (bottom) related to spin-orbit torque [96, 97]

Structure

The structure is formed in a spin-orbit torque (SOT) device through extremely thin layers of different materials that have been stacked together. Schematic of a SOT device to show the path of the write current in the SOT scheme, Fig. 7.8. One layer is non-magnetic yet possessing strong spin-orbit coupling; such a layer could be a heavy metal, topological insulator or two-dimensional materials such as WTe_2 (Tungsten Ditelluride). Immediately adjacent to it is a magnetic layer, e.g. Fe_3GeTe_2 (Iron Germanium Telluride). The interface between these two layers must be exceedingly clean and sharp, as in van der Waals heterostructures, so that spin signals can transit through them with ease [100, 101]. To enhance the functionality of the device further, additional interfacial layers such as HfOx, Ho or $MoTe_2$ may be inserted. These layers, or occasionally the application of voltage assists in controlling the interaction between the layers at their contact points which enhances the efficiency of the device [102].

Fig. 7.8 SOT device schematic showing the current path of the SOT Scheme [99]

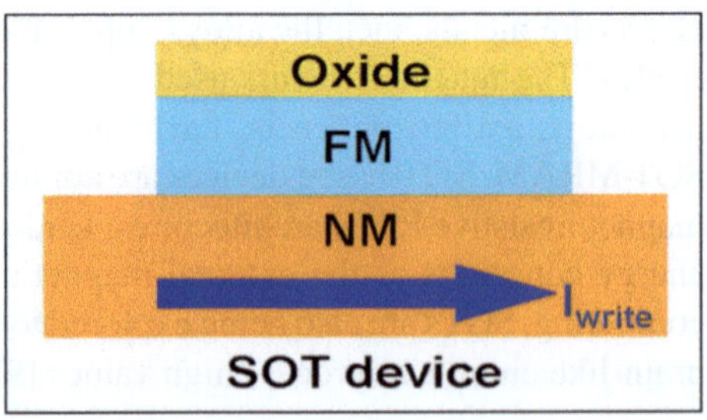

Working Principle

Spin-orbit torque (SOT) devices work by relying on the realationship between the spin of the electron and its motion. When an electric current flows through the non-magnetic layer (which has strong spin-orbit coupling), it not only carries charge but also creates a flow of spins this is called a spin current. This spin current moves into the neighbouring magnetic layer and exerts a torque on the magnetic moments (tiny magnetic directions of atoms), causing them to switch direction. This switching of magnetization happens without needing any external magnetic field [103].

In more advanced designs, the device structure is carefully engineered so that switching can happen reliably (called deterministic switching), even without any extra magnetic help. This makes SOT devices more practical and energy-efficient for real-world memory and logic applications [104].

7.7 Conclusion

Heterostructure thin film spintronic devices almost represent a radical change in the advancement of contemporary electronics. Spintronics, in contrast to traditional charge-based technologies, uses the combination of charge and spin of electrons to accomplish data storage and processing faster, more energy-efficiently and in non-volatile modes. The heterostructure thin films—layered thin films of well-designed architectures of materials (harvested off by ferromagnets, semiconductors, insulators, multiferroics, and topological insulators)—favour further control of spin injection, transport, and manipulation at the interface. Such structures enable the creation of new spintronic devices such as Magnetic Tunnel Junction (MTJ), spin valve, Spin-FET and spin-orbit torque (SOT) devices. Heterostructures bypass the drawbacks in common materials because of the effective exploitation of the phenomena of spin filtering, tunnelling magnetoresistance, spin-orbit coupling, in addition to perpendicular magnetic anisotropy at the interface. Moreover, such fabrication methods as MBE, PLD, sputtering and ALD provide precise layer control and high quality interface, which are crucial to the efficiency of the device. Due to the physical constraints related to scaling transistors, the semiconductor industry is in need of future memory, logic and quantum technologies: Spintronic heterostructures are a promising avenue. As the study of 2D materials, topological insulators, and advanced magnetic systems continues, such heterostructure devices have the chance to disrupt electronics by offering faster, higher storage density, improved energy efficiency and enhanced capabilities.

7.8 Future Outlook

The integration of spintronic heterostructure thin films with emerging computing paradigms such as neuromorphic and quantum technologies presents a promising frontier for future research. Spintronic devices have ideal properties to be used as artificial synapses and neurons in neuromorphic architectures due to their non-volatility, high-speed and low power of operation. To simulate synaptic plasticity, it is possible to use the tunable magnetic and electronic interactions of engineered heterostructures to permit adaptive and energy-efficient learning processes. At the same time, spin-based quantum devices utilize the quantum coherence and entanglement of electron spins and provide the future of scalable quantum information systems. With spintronic heterostructures coupled to superconductors, topological materials and 2D magnets, it might be possible to achieve strong spin qubit control and long coherence times. The future directions are believed to be optimization of interface quality, the improvement of spin coherence at room temperature, and the creation of hybrid systems that span the classics, neuromorphic and quantum worlds. This interdisciplinary integration may redefine the information processing technologies, which will lead to the development of intelligent and quantum-enhanced spintronic systems.

References

1. Yakout, S. M. (2020). Spintronics: future technology for new data storage and communication devices. *Journal of superconductivity and novel magnetism*, *33*(9), 2557–2580. https://doi.org/10.1007/s10948-020-05545-8
2. Shen, S. Q. (2008). Spintronics and spin current. *Physics*. http://hdl.handle.net/10722/59658
3. Bhatti, S., Sbiaa, R., Hirohata, A., Ohno, H., Fukami, S., & Piramanayagam, S. N. (2017) Spintronics based random access memory: a review. *Materials today*, *20*(9), 530–548. https://doi.org/10.1016/j.mattod.2017.07.007
4. Joshi, V. K. (2016). Spintronics: A contemporary review of emerging electronics devices. *Engineering science and technology, an international journal*, *19*(3), 1503–1513. https://doi.org/10.1016/j.jestch.2016.05.002
5. Ray, S. C. (2020). *Magnetism and spintronics in carbon and carbon nanostructured materials*. Elsevier. https://doi.org/10.1016/C2018-0-01884-8
6. Lu, J. W., Chen, E., Kabir, M., Stan, M. R., & Wolf, S. A. (2016). Spintronics technology: past, present and future. *International Materials Reviews*, *61*(7), 456–472. https://doi.org/10.1080/09506608.2016.1204097
7. Kalyani, V. L., & Agrawal, V. (2015). Spintronics—a vision for the future in electronics and computers. *J. Manag. Eng. Inf. Technol*, *2*, 30–36. http://www.jmeit.com/
8. El-Ghazaly, A., Gorchon, J., Wilson, R. B., Pattabi, A., & Bokor, J. (2020). Progress towards ultrafast spintronics applications. *Journal of Magnetism and Magnetic Materials*, *502*, 166478. https://doi.org/10.1016/j.jmmm.2020.166478
9. Dey, P., & Roy, J. N. (2021). An Overview of Spintronics. *Spintronics: Fundamentals and Applications*, 1–22. https://doi.org/10.1007/978-981-16-0069-2_1
10. Sarma, S. D. (2001). Spintronics: A new class of device based on electron spin, rather than on charge, may yield the next generation of microelectronics. *American Scientist*, *89*(6), 516–523. https://www.semanticscholar.org/author/S.-Sarma/2888542

11. Tian, Y., Bakaul, S. R., & Wu, T. (2012). Oxide nanowires for spintronics: materials and devices. *Nanoscale, 4*(5), 1529-1540. https://doi.org/10.1039/c2nr11767c
12. Datt, G. (2023). Spinel ferrite-based heterostructures for spintronics applications. In Ferrite Nanostructured Magnetic Materials (pp. 747–773). Woodhead Publishing. https://doi.org/10.1016/B978-0-12-823717-5.00034-6
13. Žutić, I., Fabian, J., & Sarma, S. D. (2004). Spintronics: Fundamentals and applications. *Reviews of modern physics, 76*(2), 323. https://doi.org/10.1103/RevModPhys.76.323
14. Yuasa, S., Nagahama, T., Fukushima, A., Suzuki, Y., & Ando, K. (2004). Giant room-temperature magnetoresistance in single-crystal Fe/MgO/Fe magnetic tunnel junctions. *Nature materials, 3*(12), 868–871. https://doi.org/10.1038/nmat1257
15. Hassan, M., Laureti, S., Rinaldi, C., Fagiani, F., Barucca, G., Casoli, F., … & Varvaro, G. (2022). Thin-film heterostructures based on Co/Ni synthetic antiferromagnets on polymer tapes: toward sustainable flexible spintronics. *ACS Applied Materials & Interfaces, 14*(45), 51496–51509. https://doi.org/10.1021/acsami.2c14000
16. Caviglia, A. D., Gabay, M., Gariglio, S., Reyren, N., Cancellieri, C., & Triscone, J. M. (2010). Tunable Rashba spin-orbit interaction at oxide interfaces. *Physical review letters, 104*(12), 126803. https://doi.org/10.1103/PhysRevLett.104.126803
17. Ramesh, R., & Spaldin, N. A. (2007). Multiferroics: progress and prospects in thin films. *Nature materials, 6*(1), 21–29. https://doi.org/10.1038/nmat1805
18. Insulator, M. T. (2013). Experimental Observation of the Quantum Anomalous Hall Effect in. *Science, 1234414*(167), 340. https://doi.org/10.1126/science.1234414
19. Xia, H., Hu, Y., Li, Z., Lan, H., & Zhang, J. (2025). Electron spin polarization in rechargeable batteries: theoretical foundation and practical applications. *Advanced Functional Materials, 35*(3), 2413491. https://doi.org/10.1002/adfm.202413491
20. Gurung, G., Elekhtiar, M., Luo, Q.-Q., Shao, D., & Tsymbal, E. Y. (2024). Nearly perfect spin polarization of noncollinear antiferromagnets. *Nature Communications*, 15(1). https://doi.org/10.1038/s41467-024-54526-1
21. Buryakov, A. M., Горбатова, A. B., Avdeev, P. Yu., Lebedeva, E., Brekhov, K., Овчинников, A. B., Gusev, N. S., Karashtin, E. A., Sapozhnikov, M. V., Мишина, E. Д., Tiercelin, N., & Preobrazhensky, V. (2023). Efficient Co/Pt THz spintronic emitter with tunable polarization. *Applied Physics Letters,* 123(8). https://doi.org/10.1063/5.0160497
22. Siegmann, H. C. (1983). Spin Polarized Electrons. *Europhysics News, 14*(2), 9–12. https://doi.org/10.1051/epn/19831402009
23. Ohnishi, K., Komori, S., Yang, G., Jeon, K.-R., Olthof, L. A. B. O., Montiel, X., Blamire, M. G., & Robinson, J. W. A. (2020). Spin-transport in superconductors. *Applied Physics Letters*, 116(13). https://doi.org/10.1063/1.5138905
24. Shao, D., Zhang, S., Li, M., Eom, C., & Tsymbal, E. Y. (2021). Spin-neutral currents for spintronics. *Nature Communications,* 12(1). https://doi.org/10.1038/s41467-021-26915-3
25. Wittmann, A., Schweicher, G., Broch, K., Novák, J., Lami, V., Cornil, D., McNellis, E. R., Zadvorna, O., Venkateshvaran, D., Takimiya, K., Geerts, Y., Cornil, J., Vaynzof, Y., Sinova, J., Watanabe, S., & Sirringhaus, H. (2020). Tuning Spin Current Injection at Ferromagnet-Nonmagnet Interfaces by Molecular Design. *Physical Review Letters*, 124(2). https://doi.org/10.1103/PhysRevLett.124.027204
26. Roundy, R. C., & Raĭkh, M. É. (2015). Spin injection from a ferromagnet into a semiconductor in the case of a rough interface. *Physical Review B*, 91(4). https://doi.org/10.1103/PhysRevB.91.045202
27. Liu, T., & Weiss, P. S. (2023). Spin Polarization in Transport Studies of Chirality-Induced Spin Selectivity. *ACS Nano*, 17(20), 19502–19507. https://doi.org/10.1021/acsnano.3c06133
28. Csősz, G., Dóra, B., & Simon, F. (2020). Entropy in Spin Relaxation, Spintronics, and Magnetic Resonance. *Physica Status Solidi* (b), 257(12). https://doi.org/10.1002/pssb.202000301
29. Boross, P., Dóra, B., Kiss, A., & Simon, F. (2013). A unified theory of spin-relaxation due to spin-orbit coupling in metals and semiconductors. *Scientific Reports*, 3(1). https://doi.org/10.1038/srep03233

30. Jureczko, P., Milivojević, M., & Kurpas, M. (2025). Strain-tuning of spin anisotropy in single-layer phosphorene: insights from Elliott-Yafet and Dyakonov-Perel spin relaxation. *New Journal of Physics. 27*(5), 053006.https://doi.org/10.48550/arXiv.2501.05911
31. Burkard, G., Ladd, T. D., Pan, A., Nichol, J. M., & Petta, J. R. (2023). Semiconductor spin qubits. Reviews of Modern Physics, 95(2). https://doi.org/10.1103/RevModPhys.95.025003
32. Hong, Jin, Z., Wang, L., & Guohong. (2011). Electron spin relaxation in intrinsic bulk InP semiconductor. *Journal of Applied Physics*, 109(2). https://doi.org/10.1063/1.3533965
33. Li, C., Wang, H., Yu, C. W., & Xie, D. (2021). Diffusion characteristics of the industrial submicron particle under Brownian motion and turbulent diffusion. Indoor and Built Environment, 31(1), 17–30. https://doi.org/10.1177/1420326X21991055
34. Tahir, H., Liu, K., Yang, Z.-Y., Baruah, K., Savoie, B. M., & Boudouris, B. W. (2025). Verdazyl radical polymers for advanced organic spintronics. *Nature Communications*, 16(1). https://doi.org/10.1038/s41467-025-56056-w
35. Jha, A., Kumar, N., Chaubey, S., & Sahni, M. (2016). Effects of substrate temperature, oxygen pressure and laser fluence on structural and magnetic properties of pulsed laser-deposited cobalt ferrite thin films. *Journal of Superconductivity and Novel Magnetism, 29*(3), 855–862. https://doi.org/10.1007/s10948-015-3345-6
36. Gibertini, M., Gibertini, M., Koperski, M., Morpurgo, A., & Novoselov, K. (2019). Magnetic 2D materials and heterostructures. *Nature Nanotechnology*, 14, 408–419. https://doi.org/10.1038/s41565-019-0438-6
37. Liu, J., & Hesjedal, T. (2021). Magnetic Topological Insulator Heterostructures: A Review. *Advanced Materials*, 35. https://doi.org/10.1002/adma.202102427
38. Hu, J., Chen, L., & Nan, C. (2016). Multiferroic Heterostructures Integrating Ferroelectric and Magnetic Materials. *Advanced Materials*, 28. https://doi.org/10.1002/adma.201502824
39. Onga, M., Sugita, Y., Ideue, T., Nakagawa, Y., Suzuki, R., Motome, Y., & Iwasa, Y. (2020). Antiferromagnet-semiconductor van der Waals heterostructures: interlayer interplay of exciton with magnetic ordering. *Nano letters. 20*(6), 4625–4630. https://doi.org/10.1021/acs.nanolett.0c01493
40. Awasthi, R. R., Mishra, V. K., Tripathi, P. K., Mishra, L., Verma, T., Nand, P., … & Vikram, K. (2025). Formation of pure and Sn-doped ZnO thin films via sol–gel spin coating technique: a study on structural, topographical, and sensing properties. *Journal of Materials Science: Materials in Electronics*, *36*(28), 1832. https://doi.org/10.1007/s10854-025-15786-9
41. Liu, Y., Zhang, S., He, J., Wang, Z., & Liu, Z. (2019). Recent Progress in the Fabrication, Properties, and Devices of Heterostructures Based on 2D Materials. *Nano-Micro Letters*, 11. https://doi.org/10.1007/s40820-019-0245-5
42. Rom, S., Ghosh, A., Halder, A., & Saha-Dasgupta, T. (2021). Machine learning classification of binary semiconductor heterostructures. *Physical Review Materials*, 5(4). https://doi.org/10.1103/PhysRevMaterials.5.043801
43. Bartolomeo, A. D., Giubileo, F., Grillo, A., Luongo, G., Iemmo, L., Urban, F., Lozzi, L., Capista, D., Nardone, M., & Passacantando, M. (2019). Bias Tunable Photocurrent in Metal-Insulator-Semiconductor Heterostructures with Photoresponse Enhanced by Carbon Nanotubes. *Nanomaterials,* 9(11), 1598–1598. https://doi.org/10.3390/nano9111598
44. Zhang, M., Liu, W., Gong, Y., Liu, Q., & Chen, Z. (2022). Graphene/Quantum Dot Heterostructure Photodetectors: From Material to Performance. *Advanced Optical Materials*, 10(24). https://doi.org/10.1002/adom.202201889
45. Paschoa, A., Gonzaléz, J. L., Checca-Huaman, N.-R., Nascimento, V. P., & Passamani, E. C. (2024). Colossal superconducting spin-valve effect in superconductor-non-magnetic metal heterostructure mediated by spin–orbit coupling. *Applied Physics Letters*, 124(22). https://doi.org/10.1063/5.0202337
46. Anjali, Gupta A., Tripathi B., Sahni M., Sharma K., Ranjan N. & Pandit S. (2024). Enhancing photocatalytic performance of Fe3O4 nanoparticles and Fe3O4@ ZnO nanocomposites. *Ionics, 30*(12), 8267–8279. https://doi.org/10.1007/s11581-024-05843-4
47. Eom, J., Lee, I., Kee, J., Cho, M., Seo, J., Suh, H., Choi, H., Sim, Y., Chen, S., Chang, H., Baek, S., Petrovic, C., Ryu, H., Jang, C., Kim, Y., Yang, C., Seong, M., Lee, J., Park, S., & Choi,

J. (2023). Voltage control of magnetism in Fe_3-xGeTe_2/In_2Se_3 van´der Waals ferromagnetic/ferroelectric heterostructures. *Nature Communications*, 14. https://doi.org/10.1038/s41467-023-41382-8
48. Wang, X., Li, C., Lü, W., Paudel, T., Leusink, D., Hoek, M., Poccia, N., Vailionis, A., Venkatesan, T., Coey, J., Coey, J., Tsymbal, E., A., & Hilgenkamp, H. (2015). Imaging and control of ferromagnetism in LaMnO3/SrTiO3 heterostructures. *Science (New York, N.Y.)*, *349*(6249), 716–719. https://doi.org/10.48550/arXiv.1409.0520
49. Chen, J., Wang, L., Zhang, M., Zhou, L., Zhang, R., Jin, L., Wang, X., Qin, H., Qiu, Y., Mei, J., Ye, F., Xi, B., He, H., Li, B., & Wang, G. (2019). Evidence for magnetic skyrmions at the interface of ferromagnet/topological-insulator heterostructures. *Nano letters. 19*(9), 6144–6151. https://doi.org/10.1021/acs.nanolett.9b02191
50. Gong, C., Kim, E., Wang, Y., Lee, G., & Zhang, X. (2019). Multiferroicity in atomic van der Waals heterostructures. *Nature Communications*, 10(1), 2657. https://doi.org/10.1038/s41467-019-10693-0
51. Tripathi, B., Pandit, S., Sharma, A., Chauhan, S., Mathuriya, A. S., Dikshit, P. K., ... & Singh, S. (2022). Modification of graphite sheet anode with iron (II, III) oxide-carbon dots for enhancing the performance of microbial fuel cell. *Catalysts*, *12*(9), 1040. https://doi.org/10.3390/catal12091040
52. Rom, S., Ghosh, A., Halder, A., & Dasgupta, T. (2021). Machine learning classification of binary semiconductor heterostructures. *Physical Review Materials*, 5, 043801. https://doi.org/10.1103/PhysRevMaterials.5.043801
53. Lee, Y., Hwang, Y., & Chung, Y. (2015). Achieving type I, II, and III heterojunctions using functionalized MXene. *ACS applied materials & interfaces*, 7 13, 7163–9. https://doi.org/10.1021/acsami.5b00063
54. Jmerik, V. (2022). Special Issue: Semiconductor Heterostructures (with Quantum Wells, Quantum Dots and Superlattices). *Nanomaterials*, 12.https://doi.org/10.3390/nano12101685
55. Li, G., Zhang, X., Wang, Y., Liu, X., Ren, F., He, J., He, D., & Zhao, H. (2024). A type-I van der Waals heterostructure formed by monolayer WS2 and trilayer $PdSe_2$ *Nanoscale. 16*(46), 21471–21481. https://doi.org/10.1039/D4NR02664K
56. Di, Y., Liu, J., Li, M., Zhang, S., You, M., & Lin, M. (2022). Donor-Acceptor Hybrid Heterostructures: An Emerging Class of Photoactive Materials with Inorganic and Organic Semiconductive Components. *Small*, e2201159. https://doi.org/10.1002/smll.202201159
57. Bauer, G., & Springholz, G. (1992). Molecular beam epitaxy—and applications. *Vacuum*, 43, 357–365. https://doi.org/10.1016/0042-207X(92)90038-X
58. Shepelin, N., Tehrani, Z., Ohannessian, N., Schneider, C., Pergolesi, D., & Lippert, T. (2023). A practical guide to pulsed laser deposition. *Chemical Society Reviews*, 52, 2294–2321. https://doi.org/10.1039/D2CS00938B
59. Aumayr, F., & Winter, H. (2003). Potential sputtering. *Philosophical Transactions of the Royal Society of London. Series A: Mathematical, Physical and Engineering Sciences*, 362, 102–77. https://doi.org/10.1098/rsta.2003.1300
60. Panish, M. (1980). Molecular Beam Epitaxy. *Science (New York, N.Y.)*, *208*(4446), 916–922. https://doi.org/10.1126/science.208.4446.916
61. Romanyuk, Y., Dengela, R., Stebounova, L., & Leoneb, S. (2007). Molecular beam epitaxy. https://doi.org/10.1016/B978-0-44-453153-7.00099-7
62. Joyce, B. (1985). Molecular beam epitaxy. *Reports on Progress in Physics*, 48, 1637–1697. https://doi.org/10.1016/B978-0-444-87074-2.50012-0
63. Ashfold, M., Claeyssens, F., Fuge, G., & Henley, S. (2004). Pulsed laser ablation and deposition of thin films. Chemical Society reviews, 33 1, 23–31. https://doi.org/10.1039/B207644F
64. Masood, K., Kumar, P., Malik, M., & Singh, J. (2021). A comprehensive tutorial on the pulsed laser deposition technique and developments in the fabrication of low dimensional systems and nanostructures. *Emergent Materials*, 1–18. https://doi.org/10.1007/s42247-020-00155-5
65. Garg, R., Gonuguntla, S., Sk, S., Iqbal, M., Dada, A., Pal, U., & Ahmadipour, M. (2024). Sputtering thin films: Materials, applications, challenges and future directions. *Advances in colloid and interface science*, 330, 103203. https://doi.org/10.1016/j.cis.2024.103203

66. Tu, Y., Chuang, T., & Winters, H. (1981). Chemical sputtering of fluorinated silicon. *Physical Review B*, 23, 823–835. https://doi.org/10.1103/PhysRevB.23.823
67. Johnson, R., Hultqvist, A., & Bent, S. (2014). A brief review of atomic layer deposition: from fundamentals to applications. *Materials Today*, 17, 236–246. https://doi.org/10.1016/j.mattod.2014.04.026
68. Pannock, D., & Banerjee, P. (2019). Atomic Layer Deposition. *Handbook of Synthetic Methodologies and Protocols of Nanomaterials*. https://doi.org/10.1142/11200-vol3
69. Oviroh, P., Akbarzadeh, R., Pan, D., Coetzee, R., & Jen, T. (2019). New development of atomic layer deposition: processes, methods and applications. *Science and Technology of Advanced Materials*, 20, 465–496. https://doi.org/10.1080/14686996.2019.1599694
70. Chauhan, S., Kumar, M., Chhoker, S., & Katyal, S. C. (2016). A comparative study on structural, vibrational, dielectric and magnetic properties of microcrystalline BiFeO3, nanocrystalline $BiFeO_3$ and core–shell structured BiFeO3@ SiO2 nanoparticles. *Journal of Alloys and Compounds, 666*, 454–467. https://doi.org/10.1016/j.jallcom.2016.01.116
71. Zhang, Y., Xu, P., Zhu, J., Yan, S., Zhang, J., & Li, L. (2023). The emergence of considerable room temperature magnetocaloric performances in the transition metal high-entropy alloys. *Materials Today Physics, 32*, 101031. https://doi.org/10.1016/j.mtphys.2023.101031
72. Egami, T. (1984). Magnetic amorphous alloys: physics and technological applications. *Reports on Progress in Physics*, 47. https://iopscience.iop.org/journal/0034-4885
73. Elbanna, A., Jiang, H., Fu, Q., Zhu, J., Liu, Y., Zhao, M., Liu, D., Lai, S., Chua, X., Pan, J., Shen, Z., Wu, L., Liu, Z., Qiu, C., & Teng, J. (2023). 2D Material Infrared Photonics and Plasmonics. *ACS nano*, 17(5) 4134–4179. https://doi.org/10.1021/acsnano.2c10705
74. Lin, Y., Torsi, R., Younas, R., Hinkle, C., Rigosi, A., Hill, H., Zhang, K., Huang, S., Shuck, C., Chen, C., Lin, Y., Maldonado-Lopez, D., Mendoza-Cortes, J., Ferrier, J., Kar, S., Nayir, N., Rajabpour, S., Van Duin, A., Liu, X., Jariwala, D., Jiang, J., Shi, J., Mortelmans, W., Jaramillo, R., Lopes, J., Engel-Herbert, R., Trofe, A., Ignatova, T., Lee, S., Mao, Z., Damian, L., Wang, Y., Steves, M., Knappenberger, K., Wang, Z., Law, S., Bepete, G., Zhou, D., Lin, J., Scheurer, M., Li, J., Wang, P., Yu, G., Wu, S., Akinwande, D., Redwing, J., Terrones, M., & Robinson, J. (2023). Recent Advances in 2D Material Theory, Synthesis, Properties, and Applications. *ACS nano, 17(11), 9694–9747*. https://doi.org/10.1021/acsnano.2c12759
75. Zhang, H. (2018). Introduction: 2D Materials Chemistry. *Chemical reviews*, 118 13, 6089–6090. https://doi.org/10.1021/acs.chemrev.8b00278
76. Tokura, Y., Yasuda, K., & Tsukazaki, A. (2019). Magnetic topological insulators. *Nature Reviews Physics*, 1, 126–143. https://doi.org/10.1038/s42254-018-0011-5
77. Wang, P., Ge, J., Li, J., Liu, Y., Xu, Y., & Wang, J. (2020). Intrinsic magnetic topological insulators. *The Innovation*, 2. https://doi.org/10.1016/j.xinn.2021.100098
78. Otrokov, M., Klimovskikh, I., Bentmann, H., Estyunin, D., Zeugner, A., Aliev, Z., Gass, S., Wolter, A., Koroleva, A., Shikin, A., Blanco-Rey, M., Hoffmann, M., Rusinov, I., Vyazovskaya, A., Eremeev, S., Koroteev, Y., Kuznetsov, V., Freyse, F., S'anchez-Barriga, J., Amiraslanov, I., Babanly, M., Mamedov, N., Abdullayev, N., Zverev, V., Alfonsov, A., Kataev, V., Buchner, B., Schwier, E., Kumar, S., Kimura, A., Petaccia, L., Santo, G., Vidal, R., Schatz, S., Kissner, K., Unzelmann, M., Min, C., Moser, S., Peixoto, T., Reinert, F., Ernst, A., Echenique, P., Isaeva, A., & Chulkov, E. (2018). Prediction and observation of an antiferromagnetic topological insulator. *Nature*, 576, 416–422. https://doi.org/10.1038/s41586-019-1840-9
79. Li, X., Zhu, M., Wang, Y., Zheng, F., Dong, J., Zhou, Y., You, L., & Zhang, J. (2022). Tremendous tunneling magnetoresistance effects based on van der Waals room-temperature ferromagnet Fe_3GaTe_2 with highly spin-polarized Fermi surfaces. *Applied Physics Letters, 122*(8). https://doi.org/10.1063/5.0136180
80. Stroski, P. N. (2017, August 2). *What is spintronics?* ElectricalLibrary.com.
81. Zhou, L., Huang, J., Tang, M., Qiu, C., Qin, F., Zhang, C., Li, Z., Wu, D., & Yuan, H. (2022). Gate-tunable spin valve effect in Fe_3GeTe_2-based van der Waals heterostructures. *InfoMat. 5*(3), e12371. https://doi.org/10.1002/inf2.12371

82. Yuan, Y., & Jiang, Y. (2024). Review on magnetic/nonmagnetic heterojunction interface effects on spintronic MTJ devices. *Semiconductor Science and Technology*, 39. https://doi.org/10.1088/1361-6641/ad22fd
83. Li, X., Lü, J., Zhang, J., You, L., Su, Y., & Tsymbal, E. (2019). Spin-dependent transport in van der Waals magnetic tunnel junctions with Fe_3GeTe_2 electrodes. *Nano letters. 19*(8), 5133–5139. https://doi.org/10.1021/acs.nanolett.9b01506
84. Jin, W., Zhang, G., Wu, H., Yang, L., Zhang, W., & Chang, H. (2023). Room-Temperature and Tunable Tunneling Magnetoresistance in Fe_3GaTe_2-Based 2D van der Waals Heterojunctions. *ACS applied materials & interfaces, 15*(30), 36519–36526. https://doi.org/10.1002/inf2.12504
85. Zhu, J. G. J., & Park, C. (2006). Magnetic tunnel junctions *Materials today, 9*(11), 36–45. https://doi.org/10.1016/S1369-7021(06)71693-5
86. Pan, H., Zhang, C., Shi, J., Hu, X., Wang, N., An, L., Duan, R., Deb, P., Liu, Z., & Gao, W. (2023). Room-Temperature Lateral Spin Valve in Graphene/Fe_3GaTe_2 van der Waals Heterostructures. *ACS Materials Letters. 5*(8), 2226–2232. https://doi.org/10.1021/acsmaterialslett.3c00510
87. Jia, Z., Zhao, M., Chen, Q., Tian, Y., Liu, L., Zhang, F., Zhang, D., Ji, Y., Camargo, B. C., Ye, K., Sun, R., Wang, Z., & Jiang, Y. (2025). Spintronic Devices upon 2D Magnetic Materials and Heterojunctions. ACS Nano. *19*(10), 9452–9483. https://doi.org/10.1021/acsnano.4c14168
88. Hassan, M., Laureti, S., Rinaldi, C., Fagiani, F., Barucca, G., Casoli, F., Mezzi, A., Bolli, E., Kaciulis, S., Fix, M., Ullrich, A., Albrecht, M., & Varvaro, G. (2022). Thin-Film Heterostructures Based on Co/Ni Synthetic Antiferromagnets on Polymer Tapes: Toward Sustainable Flexible Spintronics. *ACS applied materials & interfaces, 14*(45), 51496–51509. https://doi.org/10.1021/acsami.2c14000
89. Ciorga, M. (2024). Perspective on the spin field-effect transistor. *Journal of Physics D: Applied Physics*, 58. https://doi.org/10.1088/1361-6463/ad8003
90. Zhang, X., Liu, B., Huang, J., Cao, X., Zhang, Y., & Guo, Z. (2023). Non-volatile spin field effect transistor based on VSi_2N_4/Sc_2CO_2 multiferroic heterostructure. *Physical Review B* 109, (20) 205105. https://doi.org/10.1103/PhysRevB.109.205105
91. Abdelsalam, H., Sakr, M., Teleb, N., Abd-Elkader, O., Wang, Z., Liu, Y., & Zhang, Q. (2024). Highly efficient spin field-effect transistor based on nanographene and hBN heterostructures: spintronic and quantum transport properties. *Chinese Journal of Physics, 90*, 237–251 https://doi.org/10.1016/j.cjph.2024.05.012
92. Liu, B., Zhang, X., Hou, W., Feng, H., Dai, Z., & Guo, Z. (2025). High-Performance Non-volatile Spin FETs from 2D Metallic Ferromagnetic and Ferroelectric Multiferroic Heterostructure *Applied Physics Letters, 126*(24). https://doi.org/10.1063/5.0267449
93. Koo, H., Kim, S., Kim, H., Park, T., Choi, J., Kim, K., Go, G., Oh, J., Lee, D., Park, E., Hong, I., & Lee, K. (2020). Rashba Effect in Functional Spintronic Devices. *Advanced Materials*, 32. https://doi.org/10.1002/adma.202002117
94. Singh, S., & Romero, A. (2017). Giant tunable Rashba spin splitting in a two-dimensional BiSb monolayer and in BiSb/AlN heterostructures. *Physical Review B*, 95, 165444. https://doi.org/10.1103/PhysRevB.95.165444
95. Osintsev, D. (2014). Modeling spintronic effects in silicon (Doctoral dissertation, Technische Universität Wien). https://doi.org/10.34726/hss.2014.24842
96. Hu, S., Qiu, X., Pan, C., Zhu, W., Guo, Y.-D., Shao, D., Yang, Y., Zhang, D., & Jiang, Y. (2024). Frontiers in all electrical control of magnetization by spin orbit torque. Journal of Physics Condensed Matter, 36(25), 253001–253001. https://doi.org/10.1088/1361-648x/ad3270
97. Han, X., Wang, X., Wan, C., Yu, G., & Lv, X. (2021). Spin-orbit torques: Materials, physics, and devices. *Applied Physics Letters*, 118, 120502. https://doi.org/10.1063/5.0054656
98. Gupta, R., Bouard, C., Kammerbauer, F., Ledesma-Martin, J. O., Bose, A., Kononenko, I., Martin, S., Usé, P., Jakob, G., Drouard, M., & Kläui, M. (2025). Harnessing orbital Hall effect in spin-orbit torque MRAM. Nature Communications, 16(1). https://doi.org/10.1038/s41467-024-55437-x

99. Wu, S., Jin, T., Tan, F., Ang, C., Poh, H., Lim, G., & Lew, W. (2023). Enhancement of spin–orbit torque in Pt/Co/HfOx heterostructures with voltage-controlled oxygen ion migration. *Applied Physics Letters. 122*(12). https://doi.org/10.1063/5.0139443
100. Verma, G., Nisar, A., Dhull, S., & Kaushik, B. K. (2023). Neuromorphic Accelerator for Spiking Neural Network Using SOT-MRAM Crossbar Array. IEEE Transactions on Electron Devices, 70(11), 6012–6020. https://doi.org/10.1109/TED.2023.3317357
101. Shin, I., Cho, W., An, E., Park, S., Jeong, H., Jang, S., Baek, W., Park, S., Yang, D., Seo, J., Kim, G., Ali, M., Choi, S., Lee, H., Kim, J., Kim, S., & Lee, G. (2021). Spin–Orbit Torque Switching in an All-Van der Waals Heterostructure. *Advanced Materials*, 34. https://doi.org/10.1002/adma.202101730
102. Tang, W., Liu, H., Li, Z., Pan, A., & Zeng, Y. (2021). Spin-Orbit Torque in Van der Waals-Layered Materials and Heterostructures. *Advanced Science*, 8. https://doi.org/10.1002/advs.202100847
103. Ramaswamy, R., Lee, J. M., Cai, K., & Yang, H. (2018). Recent advances in spin-orbit torques: Moving towards device applications. *Applied Physics Reviews*, *5*(3). https://doi.org/10.1063/1.5041793
104. Ryu, J., Lee, S., Lee, K., & Park, B. (2020). Current-Induced Spin–Orbit Torques for Spintronic Applications. *Advanced Materials*, 32. https://doi.org/10.1002/adma.201907148
105. Li, W., Zeng, Y., Zhao, Z., Zhang, B., Xu, J., Huang, X., & Hou, Y. (2021). 2D Magnetic Heterostructures and Their Interface Modulated Magnetism. *ACS applied materials & interfaces*, *13*(43), 50591–50601. https://doi.org/10.1021/acsami.1c11132
106. Shao, Q., Li, P., Liu, L., Yang, H., Fukami, S., Razavi, A., … & Zhang, W. (2021). Roadmap of spin–orbit torques. *IEEE transactions on magnetics*, *57*(7), 1–39. https://doi.org/10.1109/TMAG.2021.3078583

Chapter 8
Investigation of Photovoltaic Performance and Stability of $Cs_2AgBiBr_{6\text{-}x}I_x$ (x = 0–6)-Based Perovskite Solar Cells Using DFT and SCAPS-1D

Ritu, Priyanka, Vinod Kumar, Ramesh Kumar, and Fakir Chand

8.1 Introduction

In recent years, technological developments and urbanization have increased the energy demand globally. Till now, the fossil fuels are the main sources of energy. As their availability is limited and their combustion aggravate environmental concerns like global warming, air pollution, etc. [1]. To combat the ensuring energy crisis, numerous alternative energy sources have been explored. Among these, the sun emerges as a primary and inexhaustible energy source, offering eco-friendly solutions for human energy needs.

Solar cells serve as the pivotal technology for converting solar energy into usable form. Historically, silicon (Si)-based solar cells have dominated the market owing to their remarkable power conversion efficiency (PCE) and stability. Nonetheless, the widespread adoption of Si-based solar cells is hindered by their costly production [2]. Perovskite solar cells (PSCs) represent a novel class of the photovoltaic technology that showcases PCE comparable to Si-solar cells but with reduced cost. However, the majority of investigated PSCs to date utilize organic perovskite materials in their absorber layers, which poses stability issues. The inherent volatility of the organic components in PSCs renders them highly susceptible to environmental factors such as moisture, oxygen, temperature, and light exposure [3]. Methylammonium lead iodide ($MAPbI_3$) is the most examined perovskite having high PCE but easily decomposes

Ritu · Priyanka · V. Kumar · F. Chand (✉)
Department of Physics, Kurukshetra University, Kurukshetra, India
e-mail: ritukhardia@kuk.ac.in; fchand@kuk.ac.in

R. Kumar
Department of Physics, Guru Jambheshwar University of Science and Technology, Hisar, India

S. Sharma (ed.), *Multifunctional Heterostructure Thin Films Handbook*,
https://doi.org/10.1007/978-3-032-18692-8_8

into PbI_2 and MAI, where MAI further reduces into either CH_3NH_2 and HI or CH_3I and NH_3. Along with stability issue, the presence of Pb makes it harmful for human as well as environment [4]. Although, a number of divalent atoms are proposed by researchers like Cu^{2+}, Ge^{2+}, Bi^{2+}, Mn^{2+} Sn^{2+} etc., whose outer cells and radius are identical to Pb. Out of these, Sn^{2+} proved preferred candidate due to its optimal band gap, which enhances the absorption in wider range of visible spectrum [5].

Another factor which favours Sn is its comparable radius (1.35 Å) with Pb (1.49 Å) non-toxic nature. Thus, replacement of Pb by Sn minimally disturbs perovskite structure [6]. However, despite excellent optoelectronic properties, Sn-based perovskites suffer with structural instability. Under ambident conditions, Sn^{2+} gets oxidized into Sn^{4+} very easily, which limits its usages in practical applications. Along this, the Sn-based PSCs also suffer from lower formation energy of vacancies which results into the degradation [7].

Slavney et al. [8] studied a Bi^{3+}-based double perovskite structures having Ag^+ as monovalent cation and found enhancement of carrier's life time, which increase its defect tolerance. Also, in various studies [9–11], it was found that the substitution Pb by Ag^+ and Bi^{3+} results high moisture stability, favourable band gap, small carrier effective mass, higher mobility, and extended carrier's life time for $Cs_2AgBiBr_6$ perovskite structure. Clure et al. [12] achieve suitable band gap and higher moisture stability in $Cs_2AgBiBX_6$ double perovskite as compared to organic $MAPbX_3$ perovskite solar cells. Furthermore, Gao et al. [13] found a high degree of crystallinity. Additionally, in $Cs_2AgBiBr_6$ double perovskite films, they demonstrated an ultra-smooth morphology, micro-sized grains, and exceptional thermal stability in these films, which render them suitable for photovoltaic applications. However, despite the enhanced stability of inorganic double perovskites against environmental factors, their lower efficiency remains a primary hurdle for commercialization [14, 15]. This reduced efficiency is attributed to their larger band gap and inferior light absorption capabilities compared to organic perovskite solar cells [16].

In past, concerted efforts are made by researchers to improve PCE of inorganic double PSCs. Gao et al. [13] studied the inverted ITO/Cu-NiO/$Cs_2AgBiBr_6$/C_{60}/BCP/Ag double PSCs yielding 2.23% PCE. Igbari et al. [17] analysed the optoelectronic properties and photovoltaic performance of $Cs_2AgBiBr_6$ films using deposition method and observed 2.51% PCE. Additionally, Raj et al. [18] investigated the FTO/TiO_2/$Cs_2AgBiBr_6$/Spiro-OMeTAD/Au structure, achieving a PCE of 6.68% through the optimization of parameters such as defect density, band gap, and resistance.

Tombe et al. [19] characterized a precursor solution prepared by mixing PbI_2 and MAX (Cl or Br) solutions. Their findings indicate that the absorbance and photoluminescence of I-Br-based PSCs are superior to those of pure I and I-Cl-based PSCs, resulting in improved optoelectronic properties.

Inspire ones by previous studies, on mixed halide inorganic perovskites here we studies various halide compositions such as $Cs_2AgBiBr_6$, $Cs_2AgBiBr_5I$, $Cs_2AgBiBr_4I_2$, $Cs_2AgBiBr_3I_3$, $Cs_2AgBiBr_2I_4$, $Cs_2AgBiBrI_5$, $Cs_2AgBiBr_6$, i.e. $Cs_2AgBiBr_{6-x}I_x$ (x = 0–6) using DFT and SCAPS-1D [20–22]. Optimization of absorber thickness demonstrates that perovskite compositions with equal proportions of Br and I serve as the most efficient absorber layers when coupled with SnO_2

as electron transport layer (ETL), Spiro-OMeTAD as hole transport layer (HTL), and FTO as front contact.

Similar to organic perovskite materials, the organic charge transport layers are susceptible to degradation when exposed to oxygen, light, heat, and moisture. Additionally, Spiro-OMeTAD, widely utilized as an HTL, requires a complex fabrication process and possesses organic characteristics, presenting significant challenges in the development of stable and cost-effective solar cells [19]. So, inorganic HTLs like NiO [23], CuSCN [24], CuI [25], $CuSbS_2$ [26], CZTSe [27], and $CuInSe_2$ [28] are explored. Due to inorganic nature, these materials exhibit superior chemical and thermal stabilities compared to the organic counterparts. Their wider band gap, higher carrier mobility, simpler fabrication techniques, and enhanced thermal stability make them promising alternatives to the traditional Spiro-OMeTAD. Simulation of HTLs is conducted via band gap alignment, aiding in understanding the efficient transport of holes and blocking the movement of electrons from the absorber to the HTL. Both factors contribute to reducing carrier recombination. Furthermore, the stability of the optimized structure is examined by analysing various factors such as defect density of absorber and interfaces, intensity, and temperature.

8.2 Methodology

8.2.1 *First Principle Calculations of $Cs_2Agbibr_{6\text{-}x}I_x$ Absorber Layer*

This study explores the structural, electronic, and optical characteristics of $Cs_2AgBiBr_{6\text{-}x}I_x$ using DFT implemented in the SIESTA code [29]. Generalized Gradient Approximations (GGA) parameterized by Perdew-Burke-Ernzerhof (PBE) are utilized to determine exchange and correlation energies [30]. For structural optimizations, the mesh cut-off energy of 300 R_y and double-ζ plus polarization basis sets are considered. To integrate over the first Brillouin zone, a Monkhorst-pack k-point grid of 04 × 04 × 04 is considered. Further, for structural optimizations, the Hellman-Feynman force on each atom kept to be less than 0.005 eV/Å and the convergence for energy is chosen 10^{-5} eV between two steps. The optical properties are computed by using complex dielectric function, $\varepsilon(\omega) = \varepsilon_1(\omega) + j\,\varepsilon_2(\omega)$. Here, $\varepsilon_2(\omega)$ is related to band structure and can be expressed as [31].

$$\varepsilon_2(\omega) = \frac{4\pi^2 e^2}{\Omega} \sum_{i,n} f_i(1 - f_n)|n|p|i|^2 \delta(E_n - E_i - \hbar\omega). \quad (1)$$

Here, Ω represents volume of supercell, f_i and f_n represent the fermi functions for initial $|i\rangle$ and final $\langle n|$ states having energies E_i and E_n respectively and p is momentum operator.

8.2.2 SCAPS-1D Numerical Simulations

The *n-i-p* structure and its corresponding band alignment are illustrated in Fig. 8.1a and b. This configuration incorporates an inorganic double perovskite layer as the absorber, situated between a HTL and an ETL. The PSC structure consists FTO as front contact, SnO_2 as ETL, $Cs_2AgBiBr_{6-x}I_x$ ($CABBr_{6-x}I_x$) as absorbing layer, Spiro-OMeTAD as HTL, and Au as back contact. Light incidents from the ETL side, where it is absorbed by the absorber layer to generate electron-hole pairs. Electrons are then transported from the absorber to front contact by the ETL, while holes move from the absorber to back contact by the HTL. This directional movement is facilitated by an appropriate band alignment depicted in Fig. 8.1b.

Simulations are conducted using one-dimensional simulator SCAPS-1D [20–22], which is based on fundamental semiconductor equations such as Poisson equation, drift-diffusion equation, and the continuity equation. This software enables tracking of the performance variation of solar devices in response to changes in various physical parameters including affinity, band gap thickness, doping density, and defect density. The performance in generals depicted in terms of four parameters: PCE, short-circuit current density (J_{sc}), fill factor (FF) and open-circuit voltage (V_{oc}).

The parameters for various materials required for simulation are extracted from various experimental and theoretical studies and the same are summarized in Tables 8.1 and 8.2 [32–41]. Here, t is optimized thickness, E_g is the band gap of material used, χ and ε_r are the affinity and relative permittivity, N_C and N_V are effective densities of states of conduction band and valence band respectively, μ_n and μ_p represent electron and hole mobilities. N_A and N_D respectively are acceptor and donor densities; V_{nt} and V_{pt} are electron and hole thermal velocities. Defect density of absorber used for initial optimization is 4×10^{14} cm^{-3} having Gaussian energy

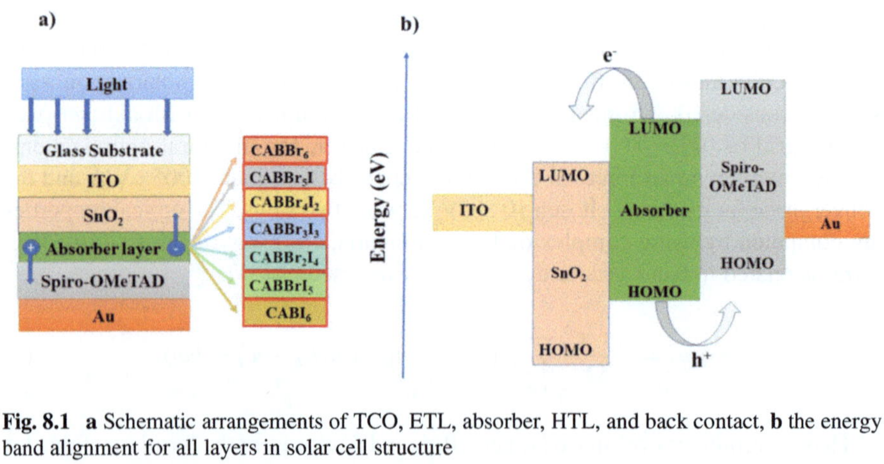

Fig. 8.1 **a** Schematic arrangements of TCO, ETL, absorber, HTL, and back contact, **b** the energy band alignment for all layers in solar cell structure

Table 8.1 Optimized parameters of TCO, ETL, and HTL used during simulation

Parameters	ITO	ETL(SnO_2)	HTL(Spiro-OMeTAD)
t (nm)	60	50	100
E_g (eV)	3.6	3.2	3.10
χ (eV)	4.5	4.2	2
ε_r	10	10	3
N_C (cm^{-3})	2.2×10^{16}	2.2×10^{18}	2.5×10^{20}
N_V (cm^{-3})	1.8×10^{18}	1.8×10^{19}	2.7×10^{18}
V_{nt} (cm s^{-1})	10^7	10^7	10^7
V_{pt} (cm s^{-1})	10^7	10^7	10^7
μ_n (cm^{2}V^{-1} s^{-1})	50	75	0.0001
μ_p (cm^{2}V^{-1} s^{-1})	75	100	0.0001
N_A (cm^{-3})	–	–	10^{19}
N_D (cm^{-3})	10^{17}	10^{19}	–

Table 8.2 Optimized parameters of various perovskite materials used during simulation

Parameters	$CABBr_6$	$CABBr_5I$	$CABBr_4I_2$	$CABBr_3I_3$	$CABBr_2I_4$	$CABBrI_5$	$CABI_6$
t (nm)	800	1100	900	1000	900	900	600
E_g (eV)*	1.27	1.14	1.23	1.29	1.28	1.21	1.2
χ (eV)	4.19	4.19	4.19	4.19	4.19	4.19	4.19
ε_r*	11.37	14.1	12.06	11.10	16.05	14.54	13.75
N_C (cm^{-3})*	1.35×10^{18}	1.51×10^{18}	1.51×10^{18}	2.02×10^{18}	2.77×10^{18}	4.39×10^{18}	2.16×10^{18}
N_V (cm^{-3})*	1.8×10^{18}	1.2×10^{18}	2.02×10^{18}	2.2×10^{18}	1.8×10^{18}	1.9×10^{18}	1.44×10^{18}
μ_n (cm^{2}V^{-1} s^{-1})	11.8	2	2	20	2	2	2
μ_p (cm^{2}V^{-1} s^{-1})	0.49	2	2	2	2	2	2
N_A (cm^{-3})	10^{13}	10^{13}	10^{13}	10^{13}	10^{13}	10^{13}	10^{13}
N_D (cm^{-3})	10^{13}	10^{13}	10^{13}	10^{13}	10^{13}	10^{13}	10^{13}

*Values determined by DFT calculations

distribution along with reference and characteristics energies 0.6 eV and 0.1 eV. The defect type chosen are neutral defects. During present simulation, the temperature is kept at 300 K with AM1.5G (1000 W/ m^2) illumination (Table 8.3).

Table 8.3 Optimized parameters of various HTLs materials used during simulation

Parameters	NiO [23]	CuSCN [24]	CuI [25]	$CuSbS_2$ [26]	CZTSe [27]	$CuInSe_2$ [28]
t (nm)	100	100	100	100	100	100
E_g (eV)	3	3.3	2.2	1.58	1.4	1.0
χ (eV)	2	1.7	2.7	3.8	4	3.9
ε_r	10.7	10	6.5	14.6	9	9
N_C (cm^{-3})	2.2×10^{18}	2.2×10^{21}	2.2×10^{19}	2.2×10^{18}	2.2×10^{18}	10^{18}
N_V (cm^{-3})	1.8×10^{19}	1.8×10^{21}	1.8×10^{20}	1.8×10^{19}	1.8×10^{19}	10^{19}
μ_n (cm^{2}V^{-1} s^{-1})	25	100	100	49	100	0.0002
μ_p (cm^{2}V^{-1} s^{-1})	12	25	44	49	12.5	0.0002
N_A (cm^{-3})	10^{19}	10^{19}	10^{19}	10^{19}	10^{19}	10^{19}

8.3 Results and Discussions

8.3.1 Analysis of DFT Results

Structural Properties

The crystal structure of $Cs_2AgBiBr_6$ compound belongs to cubic double perovskite derived system with space group Fm-3 m (#225) as shown in Fig. 8.2. The atomic positions of Cs, Ag, Bi and Br in unit cell of $Cs_2AgBiBr_6$ compound are as 8c (0.25, 0.25, 0.75), 4b (0.5, 0, 0), 4a (0, 0, 0), 24e (0.25, 0, 0), respectively.

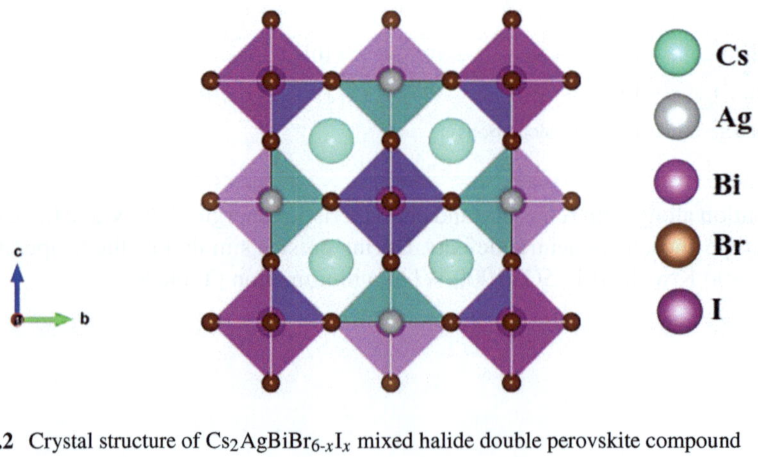

Fig. 8.2 Crystal structure of $Cs_2AgBiBr_{6\text{-}x}I_x$ mixed halide double perovskite compound

Table 8.4 Calculated values of lattice parameter (a), volume, and formation energy (E_f) for the mixed halide perovskite structures

Properties	$CABBr_6$	$CABBr_5I_1$	$CABBr_4I_2$	$CABBr_3I_3$	$CABBr_2I_4$	$CABBr_1I_5$	$CABI_6$
a (Å)	11.60	11.95	11.91	11.85	12.63	12.69	12.78
Volume (Å^3)	390.55	404.77	416.56	427.33	437.05	451.95	467.23
E_f (eV/atom)	– 1.20	– 1.15	– 1.11	– 1.07	– 1.03	– 0.99	– 0.95

Mixed halide double perovskite structures, $Cs_2AgBiBr_{6\text{-}x}I_x$ (x = 0–6), are obtained by replacing the Brions with Iions at different sites. To investigate the effects of mixed halide ions on various properties, three-dimensional crystal structures of $Cs_2AgBiBr_{6\text{-}x}I_x$ compounds are initially optimized. The optimized lattice parameters presented in Table 8.4 are in good agreement with previously reported experimental and theoretical studies [8, 42]. The data in Table 8.4 indicate that as the number of Iions increases, the volume of the unit cell structure also increases due to the larger atomic radii of the halide ions. Furthermore, the chemical stability of the mixed halide perovskite structures is confirmed through the analysis of formation energy (E_f), calculated using the following equation [43].

$$E_f = \frac{E_{\text{tot}}\left(\text{Cs}_2\text{AgBiBr}_{6-x}\text{I}_x\right) - \left(m_{\text{Cs}}E_{\text{Cs}} + n_{\text{Ag}}E_{\text{Ag}} + u_{\text{Bi}}E_{\text{Bi}} + v_{\text{Br}}E_{\text{Br}} + w_I E_I\right)}{N}, \tag{2}$$

where $E_{\text{tot}}(Cs_2AgBiBr_{6\text{-}x}I_x)$ represents the total energy of mix halide perovskite structure, E_{Cs}, E_{Ag}, E_{Bi}, $E_{\text{Br,}}$ and E_I represent the total energies of Cs, Ag, Bi, Br, and I atoms in stable bulk structures, n_{Cs}, n_{Ag}, $u_{\text{Bi,}}$ v_{Br}, w_{I}, and N represent the number of Cs, Ag, Bi, Br, I, and total atoms in their corresponding mixed halide perovskite structures. The values of E_f for all the above-mentioned structures are given in Table 8.4. The negative values of E_f establish the chemical stability of all these optimized structures.

Electronic Properties

The electronic properties carry important information about the material's bonding and light-absorbing behaviour and the same are in general are discussed using the band structure and density of states (DOS) plots. Figure 8.3 represents the band structure plots for various mixed halide perovskite structures, computed along the Γ-X-U-K-Γ-W-X high symmetry path. It is observed from Fig. 8.3 that the valence and the conduction bands are neither crossing or overlapping at the fermi level, indicating the existence of finite band gap (E_g) in all the mixed halide perovskite structures.

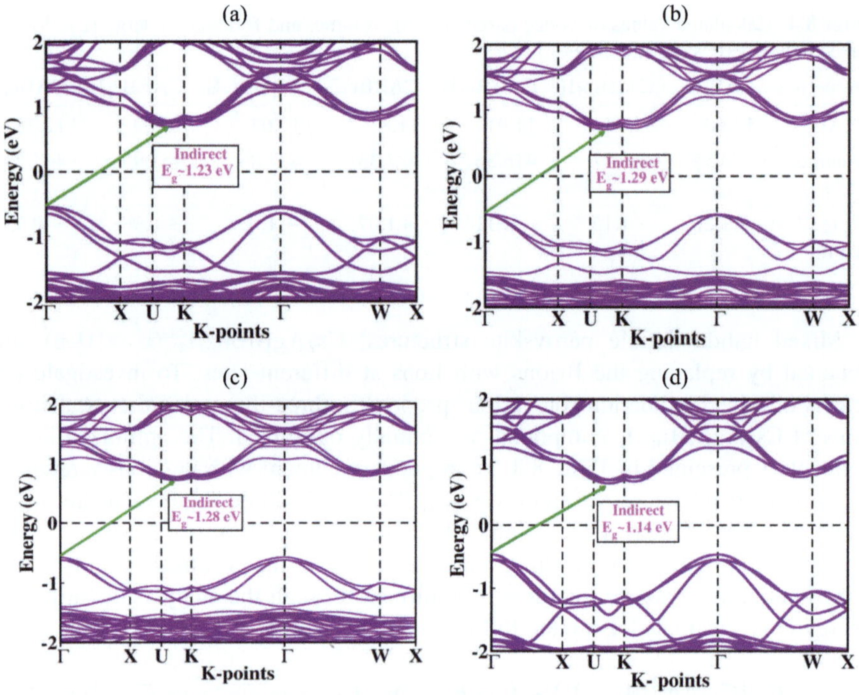

Fig. 8.3 Band structure plots for **a** $CABBr_6$, **b** $CABBr_5I_1$, **c** $CABBr_4I_2$, **d** $CABBr_3I_3$, mixed halide perovskite structures

The data of E_g of all the seven structures are given in Table 8.2, and the same have a close agreement with earlier reported studies [35]. Based on these bandgap values, it can be inferred that all the seven structures exhibiting semiconductor behaviour, with E_g ranging from 1.14 eV to 1.29 eV. Additionally, the nature of the bandgap can be determined by the relative positions of the valence band maxima (VBM) and conduction band minima (CBM) in the Brillouin zone. If both VBM and CBM align along the same K-point, the bandgap is termed as direct; conversely, if VBM and CBM lie on different K-points, the bandgap is indirect one. From Fig. 8.3, it is evident that for all seven structures, the VBM is located at the Γ-point while the CBM lies between the U-K high symmetry points, indicating an indirect bandgap.

To further analyse the contribution of different orbitals to band structures, density of states (DOS) plots is given in Fig. 8.4. Figure 8.4a shows that for the case of $CABBr_6$, in the energy range from −2 eV to 0 eV, below the Fermi level, the band formation by the Br-4p orbitals whereas, in the energy range of 0 e–2.5 eV, above the Fermi level, Bi-6p and Br-4p orbitals contribute the most to the formation of bands. As the number of iodide ions increases, there is a decrease in contribution of Br-4p orbitals below Fermi level, accompanied by an increase in the contribution of I-5p orbitals. Conversely, above the Fermi level, a similar trend is observed where the contribution of Br-4p orbitals reduced while the contribution of I-5p orbitals increases with an increase in the number of iodide ions. However in the case of $CABI_6$, where all the Br atoms are substituted by I atoms, the VBM is solely attributed to the I-5p orbitals, while the CBM is formed from a combination of Bi-6p and I-5p orbitals.

Optical Properties

In addition to electronic properties, optical properties aid in comprehending a material's response to electromagnetic radiations. Optical properties are closely related to the electronic band structure. Here, various optical parameters including dielectric function, absorption coefficient, refractive index, etc., are elaborated upon in detail. The optical properties are computed in terms of the dielectric function, $\varepsilon(\omega)$, represented by $\varepsilon(\omega) = \varepsilon_1(\omega) + j\varepsilon_2(\omega)$, where $\varepsilon_1(\omega)$ and $\varepsilon_2(\omega)$ represent the real and imaginary parts $\varepsilon(\omega)$, respectively. Here, $\varepsilon_1(\omega)$ gives the amount of energy stored within material, whereas $\varepsilon_2(\omega)$ is associated with the energy losses occurring within a material. Using $\varepsilon_1(\omega)$ and $\varepsilon_2(\omega)$, other optical parameters as absorption coefficient, reflectance, refractive index, etc. can calculated.

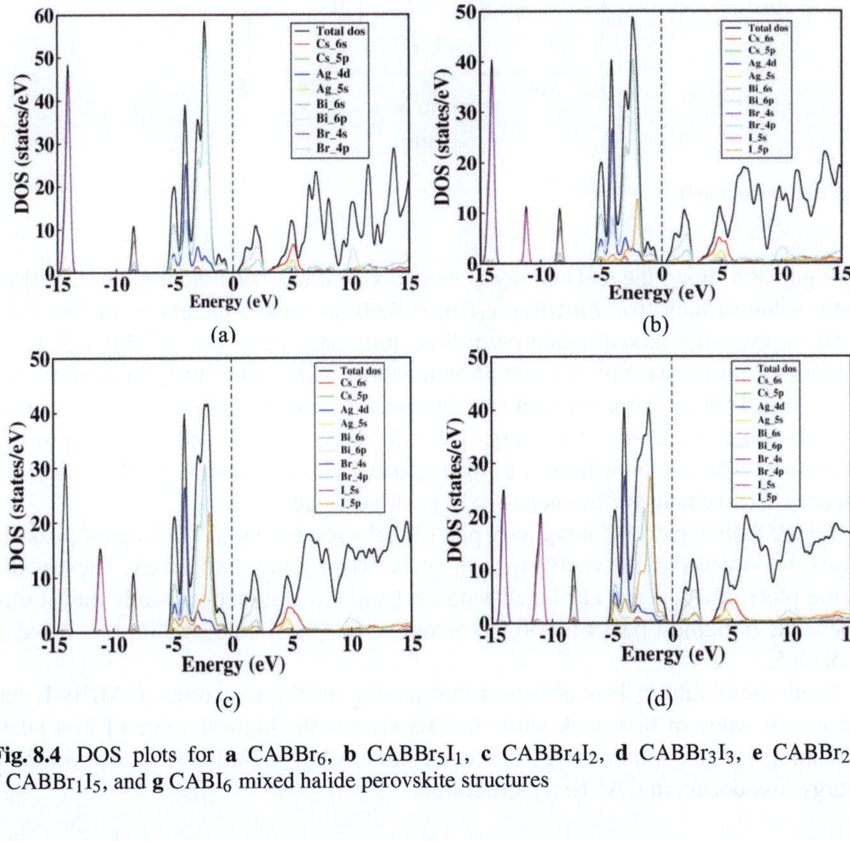

Fig. 8.4 DOS plots for **a** $CABBr_6$, **b** $CABBr_5I_1$, **c** $CABBr_4I_2$, **d** $CABBr_3I_3$, **e** $CABBr_2I_4$, **f** $CABBr_1I_5$, and **g** $CABI_6$ mixed halide perovskite structures

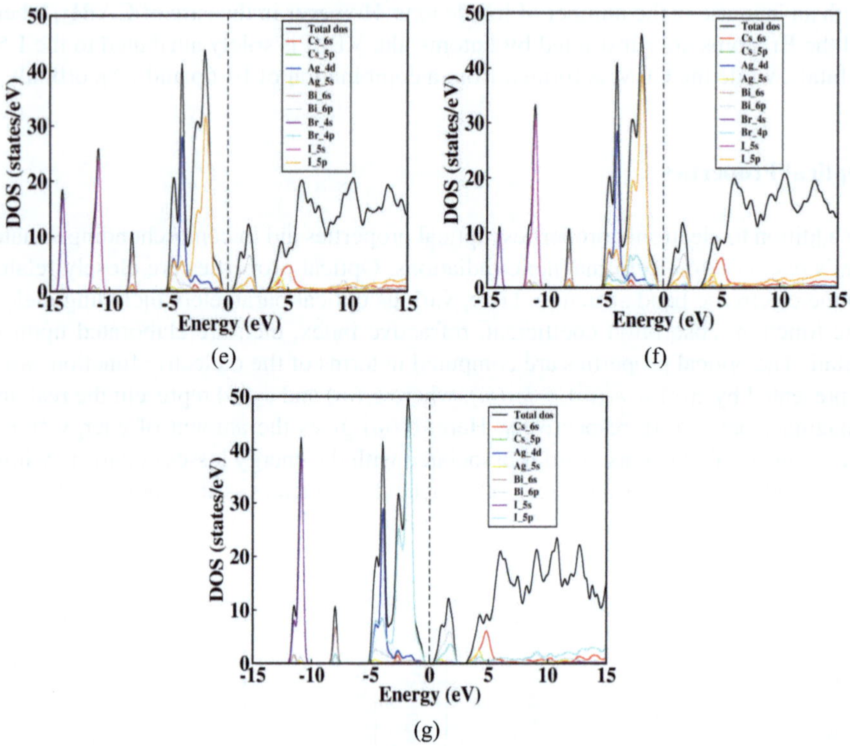

Fig. 8.4 (continued)

Figure 8.5 shows the real and imaginary parts of dielectric function. For $CABBr_6$, static value of dielectric function, $\varepsilon_1(0)$, is 4.60, as visible clearly from Fig. 8.5a. $\varepsilon_1(0)$ for the other mixed halide perovskite structures are given in Table 8.5. $\varepsilon_1(\omega)$ increases with increase of incident photon energy and reaches to a maximum value of 10.55 at 1.96 eV, implying that energy storage capacity of the material increases. $\varepsilon_1(\omega)$ becomes negative in the energy range 3.54 eV–3.90 eV and becomes positive afterwards. This region is known as the region of plasma frequency (PF), where the curve of $\varepsilon_1(\omega)$ changes from negative to positive range.

Figure 8.5b depicts the imaginary part of dielectric function. The highest peaks of $\varepsilon_2(\omega)$ are observed in the visible region, i.e. between 1.5 eV and 3.0 eV. These peaks in the plots of $\varepsilon_2(\omega)$ are attributed valence bands to conduction bands transitions. The value of highest peaks for all the mixed halide perovskite structures is listed in Table 8.5.

From these values, it is observed that among all the structures, $CABBr_3I_3$ has the lowest value of first peak while $CABBr_2I_4$ has the highest value of first peak, indicating that maximum energy loss occurs in $CABBr_2I_4$ structure while minimum energy loss occurs in $CABBr_3I_3$ structure.

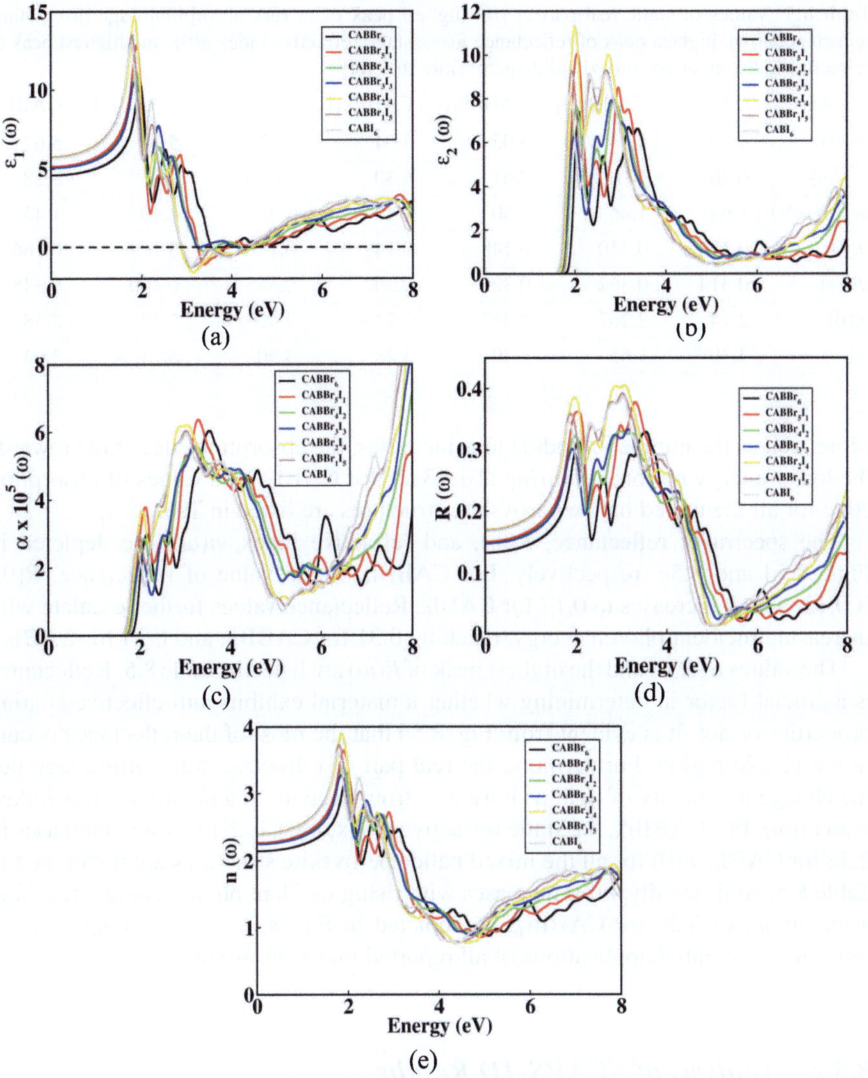

Fig. 8.5 Plots **a** real, **b** imaginary parts of dielectric function, **c** absorption coefficient, **d** reflectance, and **e** refractive index for various mixed halide perovskite solar cells

Figure 8.5c shows the absorption coefficient, $\alpha(\omega)$, spectra for various mixed halide perovskite structures. The peaks in the absorption spectra occur due to transitions from valence to conduction bands. These transitions are most likely to occur in those two bands for which the energy difference E_c–E_v, is approx constant over wide range of k-values. This leads to the formation of the parallel bands. Transitions from such bands lead higher values of $\alpha(\omega)$ [44, 45]. An absorption edge for $CABBr_6$ is observed at 1.64 eV, become absorption increases with rise in incident photon energy.

Table 8.5 Values of static real part ε_1 (0), highest peak of ε_2 (ω), absorption edge $\alpha(\omega)$, static reflectance $R(0)$, highest peak of reflectance $R(\omega)$, static refractive index $n(0)$, and highest peak of refractive index $n(\omega)$, for mixed halide perovskite structures

Properties	$CABBr_6$	$CABBr_5I_1$	$CABBr_4I_2$	$CABBr_3I_3$	$CABBr_2I_4$	$CABBr_1I_5$	$CABI_6$
ε_1 (0)	4.60	5.14	5.05	5.04	5.77	5.74	5.67
ε_2 (ω)	6.70	10.06	7.61	6.30	11.29	9.26	8.28
α (ω) (eV)	1.64	1.46	1.50	1.49	1.48	1.40	1.43
$R(0)$	0.132	0.150	0.148	0.147	0.170	0.169	0.166
$R(\omega)$	0.314	0.362	0.322	0.302	0.385	0.360	0.345
$n(0)$	2.14	2.267	2.247	2.246	2.40	2.39	2.38
$n(\omega)$	3.30	3.65	3.40	3.28	3.90	3.73	3.63

Moreover, as the number of iodide ions increases, the absorption edge shifts towards the lower energy region, occurring at 1.43 eV for $CABI_6$. The values of absorption edge for all the mixed halide perovskite structures are listed in Table 8.4.

The spectra of reflectance, $R(\omega)$, and refractive index, $n(\omega)$, are depicted in Figs. 8.5d and 8.5e, respectively. For $CABBr_6$, static value of reflectance, $R(0)$, is 0.13, which increases to 0.17 for $CABI_6$. Reflectance values further escalate with increasing incident photon energy, reaching 0.31 for $CABBr_6$ and 0.34 for $CABI_6$.

The values of $R(0)$ and the highest peak of $R(\omega)$ are listed in Table 8.5. Reflectance is a crucial factor in determining whether a material exhibits antireflective coating properties or not. It is evident from Fig. 8.5d that the most of the reflectance occurs in the visible region. Furthermore, the real part of refractive index, $n(\omega)$, signifies the change in velocity of light as it transits from vacuum to a medium and is linked with $\varepsilon_1(\omega)$. For $CABBr_6$, the static refractive index, $n(0)$, is 2.14, which increases to 2.38 for $CABI_6$. $n(0)$ for all the mixed halide perovskite structures are mentioned in Table 8.5. Additionally, $n(\omega)$ increases with rising incident photon energy, reaching a maximum of 3.30 for $CABBr_6$, as depicted in Fig. 8.5e. These values of $n(\omega)$ indicate its potential applications of all reported materials as solar panels.

8.3.2 Analysis of SCAPS-1D Results

Absorber Thickness

The absorbing layer plays a critical role in determining the performance of a solar cell as it has direct influence on diffusion length of carriers. In this layer, both charge carriers generate and transport. To analyse the effect of the absorber layer on device performance, its thickness varied from 200 nm to 1500 nm. As shown in Fig. 8.6, the increase of absorber thickness results reduction in V_{oc} and FF due to increased resistance within the absorber material, while J_{sc} increases owing to greater absorption. The performance parameter PCE initially shows enhancement with increasing absorber thickness, attributed to higher absorption resulting in a greater number of photogenerated carriers available for conduction. However, further increase in thickness leads to a decrease in PCE due to higher recombination rates. In a thicker absorber, carriers do not have sufficient diffusion length required to transport them from absorber to external contacts [46]. From Fig. 8.6, it is evident that out of all seven structures, $CABBr_3I_3$ offers better performance at 1000 nm absorber thickness as given in Table 8.6.

The performance enhancement in case of $CABBr_3I_3$ is due to its favourable band gap, which closely aligns with ideal band gap value ~1.38 eV, for perovskite materials.

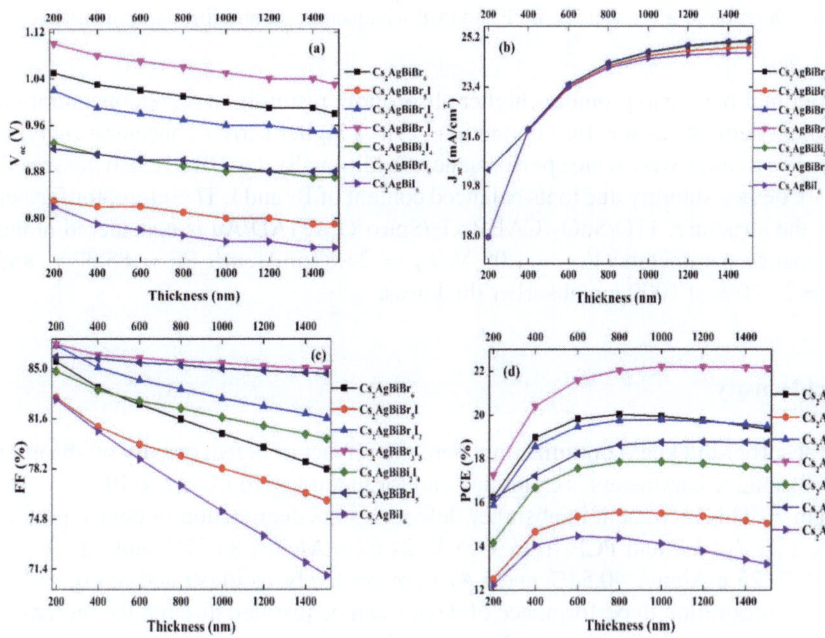

Fig. 8.6 Variation of **a** V_{oc}, **b** J_{sc}, **c** FF, and **d** PCE with increase of absorber thickness

Table 8.6 Effect of different absorber layers on performance parameters

Parameters	$CABBr_6$	$CABBr_5I$	$CABBr_4I_2$	$CABBr_3I_3$	$CABBr_2I_4$	$CABBrI_5$	$CABI_6$
t (nm)	800	1100	900	1000	900	900	600
E_f (eV/atom)	− 1.20	− 1.15	− 1.11	− 1.07	− 1.03	− 0.99	− 0.95
V_{oc}(V)	1.01	0.88	0.88	1.05	0.96	0.81	0.78
J_{sc}(mA/cm^2)	24.26	24.90	24.87	24.67	24.67	24.16	23.37
FF (%)	81.57	84.86	81.13	85.37	82.98	79.06	78.85
PCE (%)	20.03	18.76	17.78	22.20	19.78	15.56	14.47

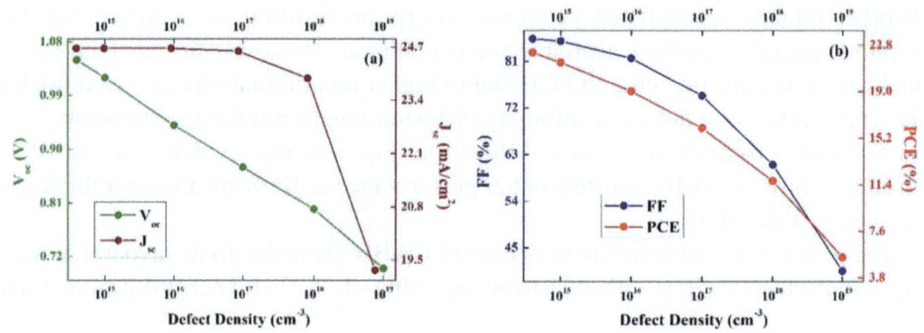

Fig. 8.7 Variation of **a** V_{oc} and J_{sc}, **b** FF and PCE with increase of absorber defect density

This optimal band gap promotes higher absorption, resulting in more concentration of photogenerated carriers for conduction [19]. A higher carrier concentration typically leads to improved device performance. Additionally, $CABBr_3I_3$ also possessed superior device stability due to its balanced content of Br and I. Therefore, for further study, the structure: ITO/SnO_2/$CABBr_3I_3$/Spiro-OMeTAD/Au is considered along performance parameters: $V_{oc} = 1.05$ V, $J_{sc} = 24.67$ mA/cm^2, FF = 85.37%, and PCE = 22.20% at 1000 nm absorber thickness.

Defect Density

After absorber thickness optimization, here the impact of defect density of absorber on performance parameters is examined and for that it varied from 4×10^{14} cm^{-3} to 10^{19} cm^{-3}. This increment in absorber defects results degradation of device performance V_{oc}, J_{sc}, FF and PCE from 1.05 V, 24.67 mA/cm^2, 85.37% and 22.20% to 0.70 V, 19.25 mA/cm^2, 40.51% and 5.47%, respectively, as illustrated in Fig. 8.7.

The deterioration in performance of device can be justified through the increased number of trapping centres with N_t, which in turn reduces the carriers' lifetime and their diffusion length. These factors collectively increase the recombination rate [47]. Hence, it is imperative to minimize the absorber defect density. However, it is important to note that, in practice, defects cannot be reduced to an arbitrarily low level. Therefore, to achieve optimal performance, the defect density is kept at its minimum value 4×10^{14} cm^{-3}.

HTL Selection

The role of the hole transport layer is crucial in determining the performance and stability of solar cells. In the present proposed structure, all layers are of inorganic nature but the HTL has organic nature. The organic nature of Spiro-OMeTAD renders it susceptible to reactivity towards incident radiation and moisture. To address this

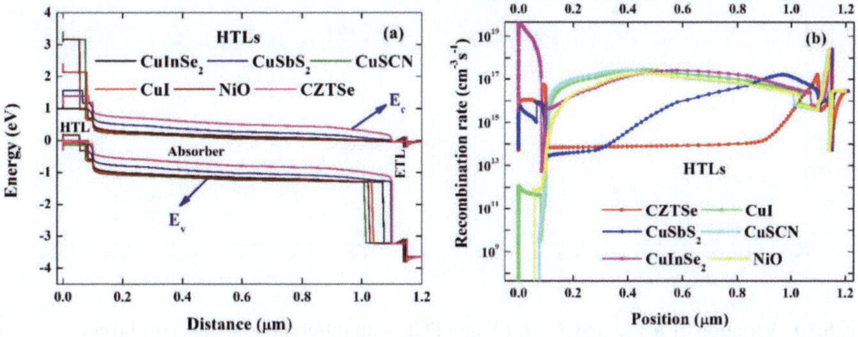

Fig. 8.8 **a** Energy band alignment and **b** variation of recombination rates for various inorganic HTL materials

vital stability issue, we studied various inorganic materials which could serve as HTLs based on their band alignment, as represented in Fig. 8.8a.

The band shifts observed with different HTLs indicate that the presence of CZTSe induces a greater shift in the HTL band, while causing a lesser shift in the ETL band. This band alignment facilitates the easy transportation of holes toward the back contact and blocks the movement of electrons, as illustrated in Fig. 8.9a and b. This separation results in carriers for conduction and effectively reduces the recombination rate, as demonstrated in Fig. 8.8b.

This study clearly reveals that the CZTSe offers higher performance as compared to other counter parts. The variation of performance parameters with different HTLs are shown in Fig. 8.10 as well as in Table 8.7.

The performance of all HTLs materials is also summarized in form of *J*–*V* and QE curves as depicted in Fig. 8.11.

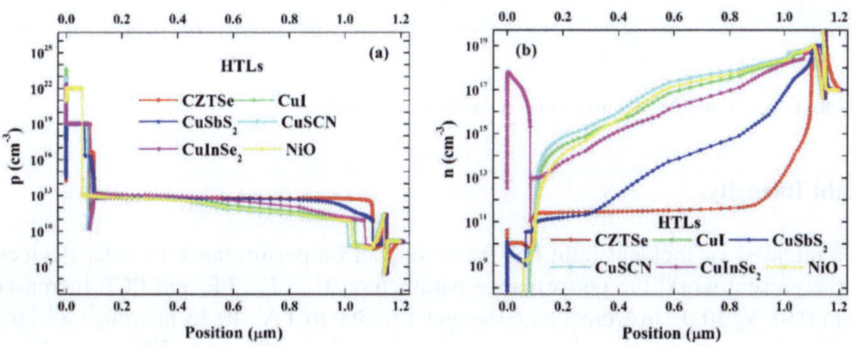

Fig. 8.9 Variation of **a** holes and **b** electrons concentrations with different HTL materials

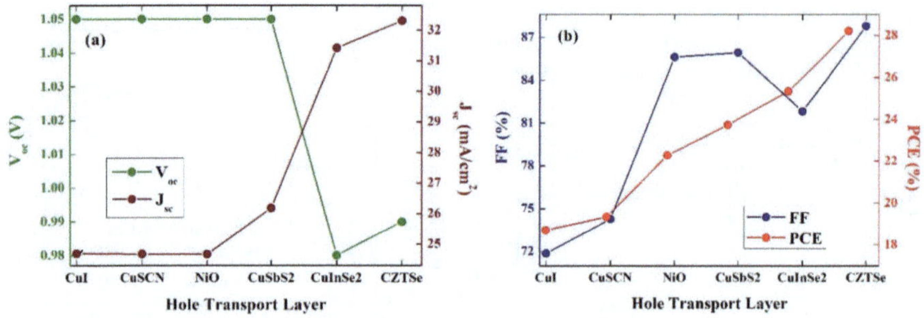

Fig. 8.10 Variation of **a** V_{oc} and J_{sc}, **b** FF and PCE with different hole transport layers

Table 8.7 Effect of HTLs on performance parameters of the proposed device

Parameters	CuSCN	CuI	$CuSbS_2$	CZTSe	$CuInSe_2$	NiO
V_{oc}(V)	1.05	1.05	1.05	0.99	0.98	1.05
J_{sc}(mA/cm^2)	24.68	24.69	26.18	32.31	31.42	24.67
FF (%)	74.28	71.86	85.94	87.79	81.82	85.63
PCE (%)	19.33	18.70	23.73	28.21	25.32	22.27

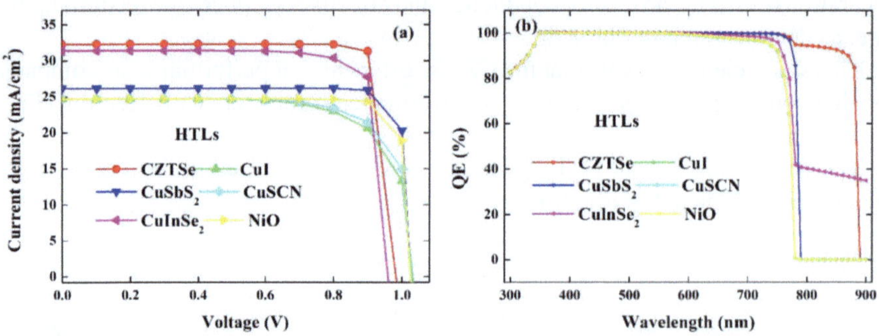

Fig. 8.11 **a** J–V and **b** QE curves for various HTL materials

Light Intensity

The intensity of incident light has major impact on performance of solar devices. In the present work, the performance parameters, V_{oc}, J_{sc}, FF, and PCE increased from 0.98 V, 20.02 mA/cm^2, 87.61% and 17.23% to 1 V, 40.36 mA/cm^2, 87.76% and 35.46%, respectively, when intensity is varied from 300 nm to 500 nm as shown in Fig. 8.12. This enhanced performance is justified in term of higher generation rate at higher intensity [48]. So, the intensity is kept at 500 nm for better performance of device.

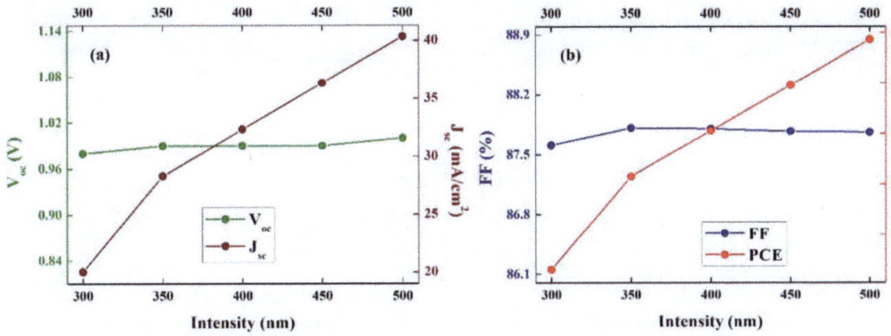

Fig. 8.12 Variation of **a** V_{oc} and J_{sc}, **b** FF and PCE with increase of intensity

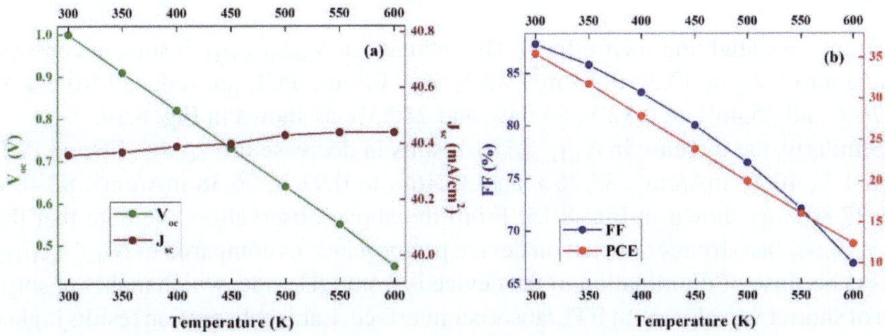

Fig. 8.13 Variation of **a** V_{oc} and J_{sc}, **b** FF and PCE with increase of temperature

Temperature

Temperature is a critical parameter that affects both the stability and performance of solar devices. In this study, the temperature is varied from 300 K to 600 K to analyse its impact on device performance. This variation results in decline in device performance parameters: V_{oc}, J_{sc}, FF and PCE from 1 V, 40.36 mA/cm^2, 87.76% and 35.46% to 0.45 V, 40.44 mA/cm^2, 66.92% and 12.36%, respectively, as shown in Fig. 8.13a and b. This reduced device performance is due to poor interconnectivity between various layers of solar structure and higher scattering rate, which increase recombination rate [49]. So, the device exhibits better performance at 300 K.

Interface Defect Density

Interfaces are inseparable part of solar structures due to the presence of a number of layers. The interface defect density has major impact on carrier concertation during transportation from one layer to other. Both absorber/HTL ($N_{\text{absorber-HTL}}$) and ETL/absorber ($N_{\text{ETL-absorber}}$) interface defects densities are varied from 10^{10} cm^{-2} to

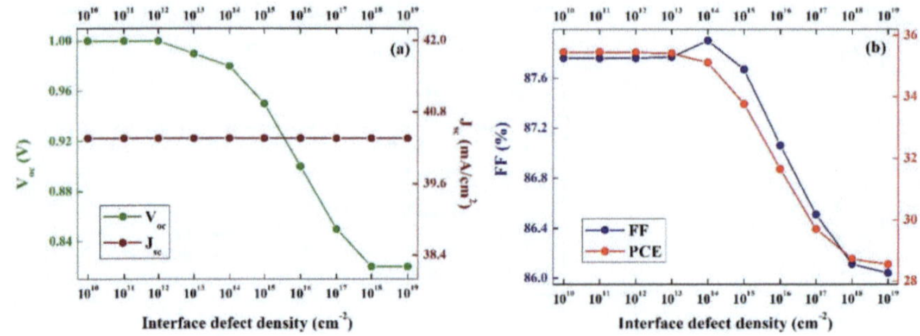

Fig. 8.14 Variation of **a** V_{oc} and J_{sc}, **b** FF and PCE with increase of absorber/HTL interface defect density

10^{19} cm^{-2} for studying their effects. The increase in $N_{\text{absorber-HTL}}$ results in constant behaviour of J_{sc} at 40.36 mA/cm^2, while V_{oc}, FF and PCE are reduced from 1 V, 87.76% and 35.46% to 0.82 V, 86.04% and 28.55% as shown in Fig. 8.14.

Similarly, the increase in $N_{\text{ETL-absorber}}$ results in decrease in V_{oc}, J_{sc}, FF and PCE from 1 V, 40.36 mA/cm^2, 87.76% and 35.46% to 0.92 V, 36.38 mA/cm^2, 82.48% and 27.80% as shown in Fig. 8.15. From the above observation, we note that the $N_{\text{ETL-absorber}}$ has stronger impact on device performance as compared to $N_{\text{absorber-HTL}}$. This is because of illumination of the device is from ETL side, which makes absorption of shorter wavelength at ETL/absorber interface. Larger absorption results higher carrier concentration and recombination rate [50]. For obtaining better performance, the value of interface defects should be kept at their minimum value, 10^{10} cm^{-2}.

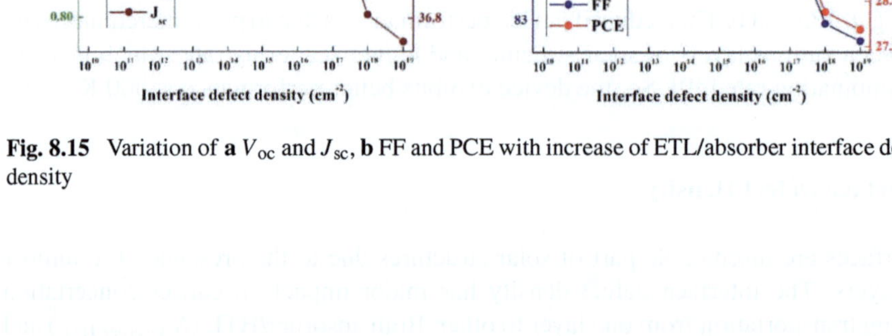

Fig. 8.15 Variation of **a** V_{oc} and J_{sc}, **b** FF and PCE with increase of ETL/absorber interface defect density

8.4 Conclusions

This study focuses on improving the efficiency and stability of double inorganic perovskite solar cells using a combination of DFT and SCAPS-1D simulations. The calculated bandgap values closely match with those reported in earlier studies. Analysis of the DOS plots reveals that as the number of iodide ions increases, the contribution of Br-4p orbitals to the valence and conduction bands decreases, while the contribution of I-5p orbitals increases. Examination of the optical parameters indicates that all seven mixed halide perovskite structures exhibit light absorption within the visible energy region. The values of $n(\omega)$ for these structures range between 2.1 and 2.4, which qualify them as promising candidates for applications in solar panels. The optimization of absorber thickness using SCAPS-1D reveals that out of $Cs_2AgBiBr_6$, $Cs_2AgBiBr_5I$, $Cs_2AgBiBr_4I_2$, $Cs_2AgBiBr_3I_3$, $Cs_2AgBiBr_2I_4$, $Cs_2AgBiBrI_5$, Cs_2AgBiI_6 inorganic perovskites, the $Cs_2AgBiBr_3I_3$ structure possess highest PCE, i.e. 22.20% (V_{oc} = 1.05 V, J_{sc} = 24.67 mA/cm^2 and FF = 85.37%). This highest performance can be justified in terms of equal content of Br and I, which enhances the absorbance of incident light.

Furthermore, the inclusion of both Br and I in $Cs_2AgBiBr_3I_3$ contributes to enhanced stability, as supported by defects density simulations. The presence of organic transport layers can lead to degradation of the solar cell structure. Therefore, various inorganic materials have been examined as potential substitutes for the existing Spiro-OMeTAD. This simulation involves band matching, where proper band alignment of CZTSe facilitates the extraction of holes and blocks electron movement, so reducing the re-combinations of carriers. Additionally, the impact of intensity, temperature, and interface defects density has been analysed. Performance increases with increasing intensity due to enhanced absorbance, whereas performance decreases with increasing temperature and interface defects density due to elevated recombination rates. This optimized structure: ITO/SnO_2/$Cs_2AgBiBr_3I_3$/CZTSe/Au offers 35.46% PCE (V_{oc} = 1.00 V, J_{sc} = 40.36 mA/cm^2 and FF = 87.76%). These findings may lay the groundwork for designing more efficient, eco-friendly, and stable inorganic perovskite solar cells.

Acknowledgements All the authors would like to thank Prof. Marc Burgelman (University of Gent, Belgium) for providing the SCAPS-1D software.

Data Availability Statement The data are available from the corresponding author upon reasonable request.

References

1. Ritu, Priyanka, V. Kumar, R. Kumar, and F. Chand, A theoretical comparison of $MAPbI_3$, $FAPbI_3$ and $(FAPbI_3)_{1-x}MAPb(Br_{3-y}Cl_y)_x$ based solar cells. J. Opt., 1-6, (2023) https://doi.org/10.1007/s12596-023-01474-6.

2. A. Husainat, W. Ali, P. Cofie, J. Attia, and J. Fuller, Simulation and analysis of methylammonium lead iodide ($CH_3NH_3PbI_3$) perovskite solar cell with Au contact using SCAPS 1D simulator. Am. j. opt. Photonics, 7(2), 33, (2019), https://doi.org/10.11648/j.ajop.20190702.12.
3. Ritu, Gagandeep, R. Kumar, and F. Chand, Optimization of mixed cation organic–inorganic lead halide perovskite solar cell performance. Int. J. Mod. Phys. B, 2450361, (2023), https://doi.org/10.1142/S0217979224503612.
4. K. Chakraborty, M.G. Choudhury, and S. Paul, Numerical study of Cs_2TiX_6 (X= Br^-, I^-, F^- and Cl^-) based perovskite solar cell using SCAPS-1D device simulation. Sol. Energy, 194, 886, (2019), https://doi.org/10.1016/j.solener.2019.11.005.
5. N. Gamal, S.H. Sedky, A. Shaker, and M. Fedawy, Design of lead-free perovskite solar cell using $Zn_{1-x}Mg_xO$ as ETL: SCAPS device simulation. Optik, 242, 167306, (2021), https://doi.org/10.1016/j.ijleo.2021.167306.
6. G. Pindolia, S.M. Shinde, and P.K. Jha, Optimization of an inorganic lead free $RbGeI_3$ based perovskite solar cell by SCAPS-1D simulation. Sol. Energy, 236, 802, (2022), https://doi.org/10.1016/j.solener.2022.03.053.
7. S. Yasin, T. Al Zoubi, and M. Moustafa, Design and simulation of high efficiency lead-free heterostructure perovskite solar cell using SCAPS-1D. Optik, 229, 166258, (2021), https://doi.org/10.1016/j.ijleo.2021.166258.
8. A.H. Slavney, T. Hu, A.M. Lindenberg, and H.I. Karunadasa, A bismuth-halide double perovskite with long carrier recombination lifetime for photovoltaic applications. J. Am. Chem. Soc., 138(7), 2138, (2016), https://doi.org/10.1021/jacs.5b13294.
9. H. Lei, D. Hardy, and F. Gao, Lead-free double perovskite $Cs_2AgBiBr_6$: fundamentals, applications, and perspectives. Adv. Funct. Mater., 31(49), 2105898, (2021), https://doi.org/10.1002/adfm.202105898.
10. S.C. Yadav, A. Srivastava, V. Manjunath, A. Kanwade, R.S. Devan, and P.M. Shirage, Properties, performance and multidimensional applications of stable lead-free $Cs_2AgBiBr_6$ double perovskite. Mater. Today Phys., 26, 100731, (2022), https://doi.org/10.1016/j.mtphys.2022.100731.
11. C.N. Savory, A. Walsh, D.O. Scanlon, Can Pb-free halide double perovskites support high-efficiency solar cells?. ACS Energy Lett., 1(5), 949, (2016), https://doi.org/10.1021/acsenergylett.6b00471.
12. E.T. McClure, M.R. Ball, W. Windl, and P.M. Woodward, Cs_2AgBiX_6 (X= Br, Cl): new visible light absorbing, lead-free halide perovskite semiconductors. J. Mater. Chem., 28(5), 1348, (2016), https://doi.org/10.1021/acs.chemmater.5b04231.
13. W. Gao, C. Ran, J. Xi, B. Jiao, W. Zhang, M. Wu, and Z. Wu, High-quality $Cs_2AgBiBr_6$ double perovskite film for lead-free inverted planar heterojunction solar cells with 2.2% efficiency. ChemPhysChem, 19(14), 1696, (2018), https://doi.org/10.1002/cphc.201800346.
14. M.K. Chini, S.G. Srinivasan, N.K. Tailor, D. Salahub, and S. Satapathi, Lead-free, stable mixed halide double perovskites $Cs_2AgBiBr_6$ and $Cs_2AgBiBr_{6-x}Cl_x$–A detailed theoretical and experimental study. Chem. Phys., 529, 110547, (2020), https://doi.org/10.1016/j.chemphys.2019.110547.
15. D. Wu, Y. Tao, Y. Huang, B. Huo, X. Zhao, J. Yang, and X. Tang, High visible-light photocatalytic performance of stable lead-free $Cs_2AgBiBr_6$ double perovskite nanocrystals. J. Catal., 397, 27, (2021), https://doi.org/10.1016/j.jcat.2021.03.007.
16. B. Wang, N. Li, L. Yang, C. Dall'Agnese, A.K. Jena, T. Miyasaka, and X.F. Wang, Organic dye/$Cs_2AgBiBr_6$ double perovskite heterojunction solar cells. J. Am. Chem. Soc., 143(36), 14877, (2021), https://doi.org/10.1021/jacs.1c07200.
17. F. Igbari, R. Wang, Z.K. Wang, X.J. Ma, Q. Wang, K.L. Wang, and Y. Yang, Composition stoichiometry of $Cs_2AgBiBr_6$ films for highly efficient lead-free perovskite solar cells. Nano Lett., 19(3), 2066, (2019), https://doi.org/10.1021/acs.nanolett.9b00238.
18. A. Raj, M. Kumar, A. Kumar, A. Laref, and A. Anshul, Investigating the potential of lead-free double perovskite $Cs_2AgBiBr_6$ material for solar cell applications: A theoretical study. Int. J. Energy Res., 46(10), 13801, (2022), https://doi.org/10.1002/er.8099.

19. S. Tombe, G. Adam, H. Heilbrunner, D.H. Apaydin, C. Ulbricht, N.S. Sariciftci, and M.C. Scharber, Optical and electronic properties of mixed halide (X= I, Cl, Br) methylammonium lead perovskite solar cells. J. Mater. Chem. C, 5(7), 1714, (2017), https://doi.org/10.1039/c6tc04830g.
20. M. Burgelman, P. Nollet, S. Degrave, "Modelling polycrystalline semiconductor solar cells", Thin solid films, 361, 527, (2000), https://doi.org/10.1016/S0040-6090(99)00825-1.
21. M. Burgelman, K. Decock, S. Khelifi, A. Abass, "Advanced electrical simulation of thin film solar cells", Thin Solid Films, 535, 296, (2013), https://doi.org/10.1016/j.tsf.2012.10.032.
22. F. Hazeghi, S.M.B. Ghorashi, "Simulation of perovskite solar cells by using CuSCN as an inorganic hole-transport material," Mater. Res. Express., 6(9), 095527, (2019), https://doi.org/10.1088/2053-1591/ab2f1b.
23. A.N. Abena, A.T. Ngoupo, F.A. Abega, and J.M.B. Ndjaka, Numerical investigation of solar cells based on hybrid organic cation perovskite with inorganic HTL via SCAPS-1D. Chinese J. Phys., 76, 94, (2022), https://doi.org/10.1016/j.cjph.2021.12.024.
24. M. K. Hossain, M.H.K. Rubel, G.I. Toki, I. Alam, M.F. Rahman, and H. Bencherif, Effect of various electron and hole transport layers on the performance of $CsPbI_3$-based perovskite solar cells: A numerical investigation in DFT, SCAPS-1D, and wxAMPS frameworks. ACS omega, 7(47), 43210, (2022), https://doi.org/10.1021/acsomega.2c05912.
25. S. Mushtaq, S. Tahir, A. Ashfaq, R.S. Bonilla, M. Haneef, R. Saeed, and N. Amin, Performance optimization of lead-free $MASnBr_3$ based perovskite solar cells by SCAPS-1D device simulation. Sol. Energy, 249, 401, (2023), https://doi.org/10.1016/j.solener.2022.11.050.
26. A. Hima, and N. Lakhdar, Enhancement of efficiency and stability of $CH_3NH_3GeI_3$ solar cells with $CuSbS_2$. Opt. Mater., 99, 109607, (2020), https://doi.org/10.1016/j.optmat.2019.109607.
27. Y.H. Khattak, F. Baig, H. Toura, S. Beg, and B.M. Soucase, CZTSe kesterite as an alternative hole transport layer for $MASnI_3$ perovskite solar cells. J. Electron. Mater., 48(9), 5723-5733, (2019), https://doi.org/10.1007/s11664-019-07374-5.
28. V. Pandey, A.K. Gupta, and M. Shriwastav, Numerical simulation and optimization of lead-based perovskite solar cell with inorganic HTL using SCAPS-1D. J. Opt., 1-9, (2023), https://doi.org/10.1007/s12596-023-01372-x.
29. J. M. Soler, E. Artacho, J. D. Gale, A. Garcia, J. Janquera, P. Ordejon, and D. Sanchez Portal, The siesta method for ab initio order-n materials simulation, J. Phys. Condens. Matter 14, 2745 (2002), https://doi.org/10.1088/0953-8984/14/11/302.
30. J. P. Perdew, K. Burke, and M. Ernzerhof, Generalized gradient approximation made simple, Phys. Rev. Lett. 77, 3865 (1996), https://doi.org/10.1103/PhysRevLett.77.3865.
31. Priyanka, R. Kumar, and F. Chand, Tuning of electronic and optical properties of PtS_2 monolayer using stacking engineering, Phys. Stat. Sol. B 260, 2300132 (2023), https://doi.org/10.1002/pssb.202300132.
32. M.K. Hossain, A.A. Arnab, R.C. Das, K.M. Hossain, M.H.K. Rubel, M.F. Rahman, and R. Pandey, Combined DFT, SCAPS-1D, and wxAMPS frameworks for design optimization of efficient Cs_2BiAgI_6-based perovskite solar cells with different charge transport layers. RSC Adv., 12(54), 35002, (2022), https://doi.org/10.1039/D2RA06734J.
33. A. Mohandes, M. Moradi, and H. Nadgaran, Numerical simulation of inorganic $Cs_2AgBiBr_6$ as a lead-free perovskite using device simulation SCAPS-1D. Opt. Quantum Electron., 53, 1-22, (2021), https://doi.org/10.1007/s11082-021-02959-z.
34. M. Al-Hattab, Y. Chrafih, E. Oublal, M. Sahal, O. Bajjou, and K. Rahmani, Ab Initio Investigation for Solar Technology on the Optical and Electronic Properties of Double Perovskites Cs_2AgBiX_6 (X= Cl, Br, I). ECS J. Solid State Sci. Technol., 12(9), 094004, (2023), DOI https://doi.org/10.1149/2162-8777/acf7ed.
35. H. Wu, A. Erbing, M.B. Johansson, J. Wang, C. Kamal, M. Odelius, and E.M. Johansson, Mixed-Halide Double Perovskite Cs_2AgBiX_6 (X= Br, I) with Tunable Optical Properties via Anion Exchange. ChemSusChem, 14(20), 4507, (2021), https://doi.org/10.1002/cssc.202101146.
36. D. Kumari, and S.K. Pandey, Comprehensive study and performance analysis of an eco-friendly double perovskite $Cs_2AgBiBr_6$ on Si tandem solar cell. J. Opt. Soc. Am. B, 39(3), 756, (2022), https://doi.org/10.1364/JOSAB.443938.

37. N. Shrivastav, J. Madan, M.K. Mohammed, A.K. Al-Mousoi, M.K. Hossain, M. Amami, and R. Pandey, Optimizing the performance of $Cs_2AgBiBr_6$ based solar cell through modification of electron and hole transport layers. Mater. Today Commun., 36, 106761, (2023), https://doi.org/10.1016/j.mtcomm.2023.106761.
38. I. Alam, R. Mollick, and M.A. Ashraf, Numerical simulation of $Cs_2AgBiBr_6$-based perovskite solar cell with ZnO nanorod and P_3HT as the charge transport layers. Physica B Condens. Matter, 618, 413187, (2021), https://doi.org/10.1016/j.physb.2021.413187.
39. M.T. Islam, M.R. Jani, S.M. Al Amin, M.S.U. Sami, K.M. Shorowordi, M.I. Hossain, and S. Ahmed, Numerical simulation studies of a fully inorganic $Cs_2AgBiBr_6$ perovskite solar device. Opt. Mater., 105, 109957, (2020), https://doi.org/10.1016/j.optmat.2020.109957.
40. Ritu, Gagandeep, R. Kumar, and F. Chand, Performance optimization of $FASnI_3$ based perovskite solar cell. Mat. Today Proc., (2023), https://doi.org/10.1016/j.matpr.2023.05.516.
41. S. Karthick, S. Velumani, and J. Bouclé, Experimental and SCAPS simulated formamidinium perovskite solar cells: A comparison of device performance. Sol. Energy, 205, 349, (2020), https://doi.org/10.1016/j.solener.2020.05.041.
42. M. A. Hadi, M. N. Islamb, and J. Podder, Indirect to direct band gap transition through order to disorder transformation of $Cs_2AgBiBr_6$ via creating antisite defects for optoelectronic and photovoltaic applications, RSC Adv., 12, 15461 (2022), https://doi.org/10.1039/D1RA06308A.
43. A. F. Wani, B. Rani, S. Dhiman, U. B. Sharopov, and K. Kaur, SiH monolayer: A promising two-dimensional thermoelectric material, Int.. J. Energy Res. 46 (8), 10885 (2022), https://doi.org/10.1002/er.7889.
44. Priyanka, S Chowdhury, Ritu, V Kumar, R Kumar, and F Chand, Detailed investigations on stability and optoelectronic characteristics of the 1T-PdS_2 monolayer, Phys. Scr. (2024), https://doi.org/10.1088/1402-4896/ad1a0f.
45. Priyanka, R Kumar, and F Chand, Tuning of optical performance of CrS_2 monolayer using strain engineering, Mat. Today Proc. (2023), https://doi.org/10.1016/j.matpr.2023.05.205.
46. R. Jaiswal, R. Ranjan, N. Srivastava, A.K. Sharma, M. Yoshimura, L. Chang, and R.N. Tiwari, Numerical study of eco-friendly Sn-based Perovskite solar cell with 25.48% efficiency using SCAPS-1D. J. Mater. Sci. Mater. Electron., 34(8), 753, (2023), https://doi.org/10.1007/s10854-023-10171-w.
47. Y. El Arfaoui, M. Khenfouch, and N. Habiballah, Efficient all lead-free perovskite solar cell simulation of $FASnI_3$/$FAGeCl_3$ with 30% efficiency: SCAPS-1D investigation. Results Opt., 13, 100554, (2023), https://doi.org/10.1016/j.rio.2023.100554.
48. Ritu, Gagandeep, R. Kumar, and F. Chand, Numerical simulation of a mixed-halide perovskite solar cell using doping gradient. J. Comput. Electron., 22(5), 1532, (2023), https://doi.org/10.1007/s10825-023-02085-x.
49. H. Dixit, N.K. Bansal, S. Porwal, D. Kumar, and T. Singh, Performance assessment of earth-abundant Kesterite and lead-free chalcogenide-based perovskite solar cells using SCAPS-1D. Optik, 295, 171474, (2023), https://doi.org/10.1016/j.ijleo.2023.171474.
50. M.M. Haque, S. Mahjabin, S. Khan, M.I. Hossain, G. Muhammad, M. Shahiduzzaman, M. Akhtaruzzaman, Study on the interface defects of eco-friendly perovskite solar cells. Sol. Energy, 247, 96, (2022), https://doi.org/10.1016/j.solener.2022.10.024.

Chapter 9
Thin Film Sensors: Applications in Environmental and Biological Detection

Zaira Jocelyn Hernández Simón, José Alberto Luna López, Gabriel Omar Mendoza Conde, and Eduardo Flores

9.1 Introduction

The earliest documented use of thin films dates back approximately 5000 years, when the Egyptians utilized gold foil for decorative purposes. However, it was not until 1852, when Grove discovered the deposition of thin films using the sputtering phenomenon, revealing that they could be obtained in a vacuum environment for the first time, that further research into this type of material began [1].

While there is a consensus regarding the thickness of "thin films," there is not a universally accepted limit for this property. The range of this dimension may extend up to a few hundred nanometers for some researchers, while for others, it may encompass up to a couple of microns. Unquestionably, this class of materials has been revolutionary in scientific and technological applications, thereby making substantial contributions to the fields of nanoscience and nanotechnology [2].

The focal point of a considerable number of research projects of significance across diverse disciplines is on thin films, given their distinctive physical and chemical properties, frequently exhibiting substantial disparities from those of the bulk material. This phenomenon can be attributed to three primary factors that result in alterations to the properties of these materials. Initially, the dimensional constraints

Z. J. H. Simón (✉) · J. A. L. López
Centro de Investigaciones en Dispositivos Semiconductores (CIDS-ICUAP), Benemérita Universidad Autónoma de Puebla (BUAP), Puebla, México
e-mail: zaira.hernandez@alumno.buap.mx

G. O. M. Conde
Coordinación de Electrónica, Instituto Nacional de Astrofísica, Óptica y Electrónica (INAOE), Puebla, México

E. Flores
Departamento de Física Aplicada, Centro de Investigación y de Estudios Avanzados (Cinvestav), Mérida, México

S. Sharma (ed.), *Multifunctional Heterostructure Thin Films Handbook*,
https://doi.org/10.1007/978-3-032-18692-8_9

imposed on these materials result in size-dependent effects, including quantum effects, optical interference, and nonlinear mass transport [3]. The second factor is the effects caused by the interfaces between the films, which, given their size, are decisive in their properties [4]. The third factor consists of structural defects and impurities present in thin films that substantially modify the material's behavior [5].

A clear example of the modifications that a material experience when is converted into a thin film can be seen, for example, in metallic thin films. The manifestation of the so-called metal–semiconductor transition, accompanied by a transition in conductivity toward that of a semiconductor, can be attributed to effects such as the "quantum size effect." This effect has multiple applications, one of which is, for example, their use in the formation of field-effect transistors (TFTs) [6].

The numerous applications of thin films have led to their indispensable status in a variety of fields, including energy, electronics, aerospace, and medicine, to mention a few [7]. While each of these areas necessitates the progress of distinct types of sensors for optimal performance, it is evident that the most recent applications are centered on the development of sensors that enhance our quality of life. These sensors encompass a wide range of applications, including environmental sensors that monitor natural resources essential for human existence and biological sensors that facilitate the precise and personal assessment of health status. In order to elucidate the operating principles and technological advances of environmental and biological sensors based on thin films, a detailed description of these sensors and their transduction principles will be provided. Subsequently, the environmental applications of these sensors will be examined, with a particular focus on the detection of air quality (an exploration of the types of gas sensors available) and water quality (a review of the pollutants that these sensors can detect). Finally, the theme of biological applications in sensors based on thin films will be examined, with emphasis placed on specific types of biosensors, including glucose sensors, immunosensors, piezoelectrical, and optical biosensors.

9.2 Sensors

A sensor can be defined as a particular device, designed to detect variations or alterations in its surrounding environment. This variation or alteration is then communicated to other processors as a form of signals or information. The classification of sensors is a subject that has been approached in a variety of manners by different specialists and researchers. In the classification process, the sensors are initially divided into two categories. The first category is referred to as "active sensors," while the second category is referred to as "passive sensors." The functionality of active sensors is conditional upon the presence of an additional excitation signal. Conversely, passive sensors function independently of external power sources, generating an output response. Passive sensors also referred to as self-generated sensors, that can produce their own electric signal without the need for external power sources.

Examples of such sensors include thermal sensors, electric field sensors, and metal detectors [8].

Another system to classify sensors is through their transduction principles, which are mentioned below:

9.2.1 Optical Detection

In the field of optical transduction, alterations in how molecules interact with photons due to fluctuations in their electronic configuration are identified by measuring light absorption, fluorescence, luminescence, or light scattering.

Colorimetry is a method of detection that involves observing changes in the color of a material due to optical interactions when the concentration or presence of a material exhibits variation. Despite the simplicity of this detection method, it generally lacks quantification, providing only qualitative data [9].

On the other hand, fluorescence contains a substantial amount of information about the molecule. In this process, light absorption causes an electronic transition. Molecules possess an excess of vibrational states, and it is probable that a molecule will transition to an excited vibrational state. Subsequent relaxation into the vibrational ground state may occur prior to the emission of a photon. For differences between the vibrational states (initial and final), the emitted photon energy will exhibit a slight deviation from the energy of the originally absorbed photon. In the context of sensing, fluorescence quenching stands as the most frequently exploited phenomenon, wherein the prevention of fluorescence is achieved by offering alternative (non-radiative) relaxation pathways for excited molecules [10, 11]. A notable disadvantage associated with this method is the necessity for advanced instrumentation to detect fluorescence. Its advantages include the potential to enhance selectivity and achieve lower detection limits compared to colorimetry.

Finally, chemiluminescence emerges when a chemical reaction is initiated by the introduction of a reagent that facilitates the excitation of the species of interest. The molecules are electronically excited by the chemical reaction, reaching an excited state. In accordance with the principles that govern other established luminescence processes, the emission of photons occurs during the relaxation of the molecules that have been excited.

In a recent period, substantial advancements have been observed in the area of optoelectronic device fabrication. These progresses have resulted in the emergence of high-quality and cost-effective precision light detectors and emitters, along with advanced fiber optic technologies [10]. However, significant challenges in optical transduction are associated to chromatic interference, this phenomenon has the capacity to impede signal transmission or modify baseline measurement parameters. The incorporation of receptor molecules or nanoparticles into a thin film has been demonstrated to be an effective method for minimizing sample interferences [12].

9.2.2 Mechanical Detection

Chemical reactions invariably cause modifications to physical dimensions because molecules (atoms or ions) occupy a certain volume. This mechanism is employed by certain catalysts and enzymes to achieve selectivity by focusing on specific substrates. It is therefore evident that mechanical sensors have the capacity to detect the spatial extent occupied by an analyte species (strain in a thin film, porous or surface layer) to ascertain its presence. Additionally, substantial alterations can be identified through modifications in the resonance frequency of materials. The conversion of mechanical signals to an optical or electrical is the final stage of signal transduction [10, 13].

9.2.3 Electrochemical Detection

In the context of electrochemical measurements, the analyte medium is incorporated into the electrical circuit as one of the electrodes, because no current flow is necessary for the measurement. In addition to this electrode, there is the working electrode, which generates potential based on the concentration of species [14]. To accurately quantify the working potential, it is imperative to establish a stable reference potential, the necessity of a reference electrode constitutes the most significant disadvantage associated with electrochemical sensors [15]. The optimal reference electrode must demonstrate stability, allowing for the maintenance of consistent potential over extended periods. In addition, the material under consideration should possess the attribute of reusability, in conjunction with compatibility with the chemical species present in its environment. This combination of properties is instrumental in the prevention of adverse reactions. Currently, there are multiple reference electrodes that have proven to be reasonably robust and are employed extensively, with the silver/silver chloride electrode being a notable example [16]. Electrochemical sensors have found extensive applications, particularly in the domains of pH measurement and redox-related parameters [17–19].

9.2.4 Electrical Detection

Despite the simplicity and robustness of electric transduction, it necessitates contact between the transducer and the analyzed substance. In comparison to electrochemical detectors, they offer the benefit of a simple configuration that eliminates the requirement for a reference electrode.

The difference between electrical and electrochemical sensors is that an electrochemical sensor's analyte solution is an integral part of the electrical circuit. In contrast, a sensor that is exclusively electrical in nature will interact with the analyte;

nevertheless, their properties are not the object of measurement. Conversely, alterations in the electrical characteristics of a sensor component are quantitated as an outcome of interactions with the analyte [10].

9.2.5 Chemical Detection

Here, the receptor experiences a chemical change, such as a chemical reaction or a physical adsorption, when the analyte is present, and can provide information about the chemical composition. In this process, first the analyte conditioning is needed (the monitoring of environmental parameters) and then the recognition of the required chemical property occurs. Subsequently, the signal transitions from a chemical to a quantifiable state that is more simplified, commonly mechanical, optical, or electrical. And finally, a quantification and interpretation of the signal is provided.

9.2.6 Chemiresistors

Chemiresistive sensors identify changes in the electrical properties of a substance. The variation in conductivity of the operational layer is attributable to the interaction involving the active layer and the analyte, which impacts the conductivity and can manifest as variability. This variability is indicative of analyte existence and encompasses effects such as responses at grain boundaries, physisorption, bulk reactions, catalytic responses, or chemisorption. Presence of an analyte has been observed to instigate a reduction or oxidation reaction within the active layer, or alternatively to induce a charge transfer process. The predominant functional layers in chemiresistors include mainly thin films as two-dimensional (2D), conducting polymer, carbon based, and metal oxide [20].

Despite the wide array of sensor detection principles that have been previously outlined, all sensors are characterized by a set of common performance parameters. These parameters include:

Selectivity, which has the capacity to detect differences in environments with multiple similar stimuli; it is also linked to the ability to avoid interference between signals. The accuracy and precision that encompasses two elements; first, the ability to reproduce results with consistent accuracy; and second, the capacity to discern minuscule variations in signal strength. Other important parameters to consider include the limits of detection, operational range, and response time,these factors contribute to the development of sensors that exhibit fast response and operational capabilities across a broad detection range. The reusability is intimately associated with the sensor's capacity to regain its original state prior to detection without undergoing alterations, and finally, it is important to consider the power consumption and autonomy as performance parameters of sensors [10].

As previously mentioned, sensors can be designed to detect the presence of an analyte or to quantify it. The importance of sensors in this regard is directly related to their applications. Sensors have become an integral component of modern life, facilitating numerous labors and enhancing efficiency. In addition to their current role in facilitating the performance of daily tasks, they play an essential part in enhancing human health, well-being, and quality of life. This is particularly evident in the domains of environmental detection, as well as biological sensing. The following discussion will mention the principles of environmental sensors, encompassing those that contribute to air and water quality, as well as biological sensors, or else known as biosensors, which have the potential to enhance human health. At the conclusion of each section, emphasis will be placed on sensors employing thin films that are considered to be making significant contributions to the advancement of this field, offering novel approaches and advantages over other types of sensors.

9.3 Environmental Sensors

9.3.1 Air Quality

Air pollution constitutes a serious environmental concern, having inflicted substantial damage on the planet, including the greenhouse effect. Moreover, it has emerged as a cautionary indicator due to its prevalent impact on the health and existence of all living species. Air pollutants have been identified as a primary contributing factor to the escalation in respiratory infections and cardiovascular diseases.

A variety of factors are implicated in the generation of air pollutants, including but not limited to motor vehicles, industrial facilities, and agricultural sources; these sources can be classified in primary and secondary. The primary contaminants are the direct emission of toxic gases, such as SO_x, from factories, and NO_x and CO, from vehicles. Secondary contaminants are produced in the atmosphere when primary contaminants interact, resulting in their non-direct emission [21].

The predominant atmospheric pollutants include carbon dioxide and monoxide (CO_2, CO), sulfur dioxide (SO_2), nitrogen oxides (NO_x), hydrocarbons, hydrogen sulfide (H_2S), ammonia (NH_3), and diverse chemical compounds. In addition to the aforementioned pollutants, the detection of hydrogen (H_2) gas is imperative, as it serves as a fuel source for spacecraft. With regard to the subject of CO_2, there has been a considerable increase in its concentrations. In the preindustrial era, the CO_2 concentration was recorded at 280 ppm. Presently, there has been a substantial increase in this figure, with a documented concentration of 420 ppm [22].

The contaminants classified as volatile organic compounds (VOCs) are the principal indoor air pollutants and are predominantly alcohols, including ethanol, isopropanol, and acetone. Evidence has emerged linking volatile organic compounds to leukemia and lymphoma diagnoses. These compounds have also been identified as potential biomarkers for various health conditions, including lung cancer and type 2 diabetes [23–25].

One of the best strategies to improve air quality is to monitor the presence of these pollutants that damage the environment, which is possible with gas sensors. Additionally, the manifestation of hazardous gases in minimal concentrations often remains undetectable by the human nose, necessitating the use of sensors for effective detection.

The study of gas sensors for environmental applications commenced in the 1960s with the documented zinc oxide (ZnO) thin film sensors measuring less than 100 nm. These sensors were utilized for the recognition of carbon dioxide, hydrocarbons, and various volatile organics at temperatures exceeding 400 °C, as reported by Seiyama et al. [26]. In the context of the stringent environmental pollution norms that have been instituted to regulate the emission of various toxic gases, there has been an exponential proliferation in research related to gas sensors in recent years.

Gas Sensors

A gas sensor provides information about the composition and concentration of gases and could be defined as a transducer that converts the interaction between an analyte (gas) and a sensing material into a form that can be processed (usually an electrical signal). A comparison of thin film sensors with other types of sensors reveals that thin film sensors have certain advantages over commonly used bulk sensors. Thin film sensors boast a high surface area-to-volume ratio, which has been demonstrated to enhance sensing capability, reduce power consumption, facilitate miniaturization of gas sensing systems, ensure durability, and simplify fabrication [27]. This combination of attributes is further complemented by their low gas detection limit.

Gas sensors can be classified into a variety of categories, with different classification standards applied to each category. However, the most prevalent classification system currently in use is based on working principles. The following classifications are included:

Electrochemical Gas Sensors

In this particular type of sensor, the gas experiences oxidation or reduction. For instance, the presence of carbon monoxide (CO) is oxidized to carbon dioxide (CO_2). This electrochemical reaction, which is the conversion of chemical energy into electrical energy, results in a change in the current signal. The components of the system under consideration are as follows: a gas-permeable membrane, an anode (the working electrode in which the chemical reaction takes place), a cathode (where the opposite electrochemical reaction occurs), an electrolyte (which allows for the transfer of charged ions between the electrodes), and a reference electrode (which is used to maintain a stable potential) [28]. These sensors demonstrate a high degree of sensitivity to toxic gases, including carbon monoxide, H_2S, hydrogen, ammonia, chlorine dioxide, and nitric oxide among others [29].

Electrochemical gas sensors offer several advantages, including high selectivity, which is a crucial parameter in sensor performance. These sensors also exhibit high sensitivity and low power consumption. However, there have been reports of slow response times [30]. In order to provide a comprehensive overview of the initiatives assumed to enhance response times, it is imperative to recognize the efforts that have been made in this regard.

Currie et al. [31] proposed the transition from the utilization of bulk sensors to the employment of thin films through micromachining techniques. This modification led to an enhancement in the performance of CO_2 sensors, characterized by reduced time response and recovery times. However, the utilization of thin films in this particular type of sensors does not originate with this research but rather dates to the early 1950s. During this period, the initial studies on electrochemical cells and thin film solid electrolytes were conducted on various silver and lead halogenides [32].

Catalytic Gas Sensors

This sensor employs the oxidation principle, similar to the electrochemical sensor. In this process, a high-temperature-resistant catalyst layer is deposited on the surface of a platinum resistor. At a specific temperature, the fuel gas is catalytically burned, resulting in a change in the electrical resistance. This device is generally employed for the detection of predominant gases in oxygen-rich environments or air. The gases detected include hydrocarbons such as butane (C_4H_{10}), propane (C_3H_8), methane (CH_4), ethanol, acetone and hydrogen (H_2) [33].

In recent developments, micro-hotplates have emerged as a prevalent alternative to platinum coils in the field of gas sensors, determined by the pressing need to minimize energy consumption in these applications. This variety of sensors generally comprises a platinum (Pt) resistor-coated catalytic surface on a hot plate. The Pt resistor functions to apply heat to the catalyst, thereby elevating its temperature to a remarkably high level. At this elevated temperature, any flammable gas molecules present in the environment are susceptible to ignition. The presence of gases in each environment can be determined through the observation of a change in the resistance of platinum, which occurs as result of temperature increase. This model was developed by Lei Xu et al. [34], who designed and fabricated a two-beam microplate for catalytic gas sensors introducing Pd–Pt as the catalytic material. The micro-hotplate design demonstrated a power per active area that was approximately 30% less than that of conventional micro-hotplates and this sensor demonstrated a high sensitivity to methane.

Thermal Conductivity Gas Sensors

Based on the difference in thermal conductivity between different gases and air, the change in thermal conductivity is interpreted into a change in resistance through the circuit. The sensing element of a thermal conductive sensor is typically composed of a metal wire or film with high thermal conductivity, such as platinum or tungsten. These sensors are constituted of two components: a detector and a compensator. The detector is a microheater that is exposed to the objective gas, while the compensator is a microheater that is located within a chamber containing air. The detector´s exposure to a gaseous medium results in a substantial discrepancy in thermal conductivity when compared with that of air. This phenomenon, in turn, gives rise to alterations in its resistance, attributable to the microheater's rate of heat dissipation. Subsequently, a comparison is made between the observed change and the values obtained from the compensator. The utilization of thermal conductivity sensors is primarily focused on the detection of gases characterized by low molecular weights [35, 36]. It is important to note that it is impractical to quantify specific gases, including oxygen, nitrogen, and carbon monoxide, due to their similar thermal conductivity to air [37]. These sensors are commonly used for detecting VOCs, hydrogen, and methane.

In addition to their limited selectivity, these sensors exhibit drawbacks such as substantial power consumption due to their considerable size and the incorporation of a heater, which restricts their versatility. Consequently, further miniaturization is imperative. As some of the promising alternatives to miniaturize the device and solve these problems, there is the work of gas sensor developed by Cho et al. The sensor in question utilizes a suspended 1D nano-heater, a design that facilitates gas detection while exhibiting an exceptionally ultralow power consumption rate [38].

Magnetic Gas Sensors

Magnetic gas sensors are devices that are capable of measuring changes in the magnetic properties of active materials. These sensors utilize various approaches to achieve this objective, including the magneto-optical Kerr effect, the Hall effect, spin orientation, magnetization, and ferromagnetic resonance. The interaction of magnetic materials with gas molecules, resulting in reduction or oxidation, leads to surface absorption of ions or molecules and this phenomenon constitutes the fundamental principle of operation. This, in turn, leads to alterations in the concentration of vacancies and defects within the materials. These alterations are fundamental in ascertaining the variation in the saturation magnetization, remanence magnetization, and coercivity of the materials. The modifications of the global magnetic performance of a material are attributable to an increase in the number of uncompensated surface spins. This phenomenon is the result of adsorbed molecules or ions inducing alterations in the electronic structure at the surface [39].

The utilization of magnetic gas sensors is a promising alternative due to their superior temporal response, which is significantly faster compared to chemical sensors. Moreover, the potential for explosion when the device is utilized in the proximity of reactive chemicals or hydrogen. The operational temperature of the sensors can be adjusted to the ambient temperature or to a lower or higher temperature by selecting magnetic materials with different Curie temperatures [40].

These sensors are particularly useful for the identification of gases, such as hydrogen and oxygen, and have the potential to affect the magnetic characteristics of materials.

Recent advancements in thin films have yielded notable progress in magnetic gas sensors field. Among these advancements, the utilization of extraordinary Hall effect (EHE) which is a phenomenon that arises from spin-orbit coupling and asymmetric electron scattering, has emerged as a highly effective solution for hydrogen gas sensing applications in ferromagnetic Co/Pd films. The palladium lattice undergoes substantial expansion upon the absorption of hydrogen due to the high solubility of hydrogen in palladium, resulting in the creation of palladium hydride. This effect is generally implemented in the construction of H_2 sensors that are composed of Co–Pd alloys and multi-layered films of Pd with transition metals. As demonstrated by the detailed H_2 sensing studies, the improved EHE response of the Co/Pd films can be achieved according to the film thickness, composition, and operational field. The utilization of these materials has emerged as a promising candidate for the development of magnetic gas sensors [40, 41].

Optical Gas Sensor

Optical sensors are devices that detect the interaction of light and gas molecules. A variety of optical sensors have been designed using diverse techniques, including ultraviolet–visible spectroscopy, infrared spectroscopy, and surface plasmon resonance. The initial two techniques cited are principally based on the principle of absorption spectroscopy, which entails the measurement of the quantity of photons that gas molecules absorb at specific wavelengths. Infrared absorption spectroscopy is a widely utilized technique for the analysis of gases such as carbon dioxide and methane. In the domain of ultraviolet–visible spectroscopy, the phenomenon of absorption can be attributed to electronic transitions. These sensors are employed for the purpose of monitoring various air pollutants, including ozone, nitrogen oxides, sulfur dioxide, and a wide range of VOCs [42].

The implementation of the surface plasmon resonance (SPR) technique for the detection of gas is only feasible when a sensitive layer is attached to the surface of a noble metal. This layer is indispensable for enabling the transmission of surface plasmon waves at interface involving the dielectric and metal layer. The existence of gas in the sensitive layer will induce a change in its refractive index. This modification in the refractive index will consequently lead to a corresponding alteration in the surface plasmon dispersion relation [43]. The surface plasmon dispersion relationship is directly related to the concentration of gas molecules that interact with the sensitive layer. Therefore, in order to achieve a highly sensitive and selective sensor, it is imperative to optimize the sensing film. The most promising developments with thin films using this technique include those reported by Ayushi Paliwal et al. [44].

These advances include the development of a highly sensitive and efficient surface plasmon resonance gas sensor based on Au and ZnO as the sensing layer (prism/Au/ZnO). The optimized ZnO thin film with a thickness of 200 nm presents an improved and consistent sensing response across a broad range of CO concentrations, ranging from 0.5 to 100 parts per million (ppm), with a rapid response time of 1 s. Besides, the selectivity studies further suggest that the sensor demonstrates a high degree of sensitivity, exhibiting minimal interference with other gases, including CO_2, NH_3, NOx, H_2 and LPG.

In a different category of optical detection, such as gasochromic, a WO_3 thin film gas sensor developed by researchers led by Michał Mazur, the sensor demonstrated an exceptional response to hydrogen concentrations as low as 25 ppm, exhibiting a response value of 6.3. This outcome positions the sensor as one of the most advanced WO_3 optical gas sensors operating based on the gasochromic effect. WO_3 exhibits chromogenic properties, indicating its capacity to undergo a color change, in this case from yellow to blue, in reaction to diverse physical factors, comprising exposure to hydrogen, light irradiation, temperature, and electric field. The material exhibits high coloring capacity and fast transition, making it suitable for use in smart windows and gas sensors applications [45]. Recent progress in the field of wearable technology has yielded notable developments, as evidenced by the work of Zheng et al. Their study introduces a light-activated gas sensor, which is fabricated with a thin film of In_2Se_3 on a transparent polyimide substrate, employing indium tin oxide (ITO) electrodes. The defining characteristics of sensor are its flexibility and transparency, which collectively enable superior performance in detecting C_2H_2 gas at ambient temperature. This property is further highlighted in the following reference: [46].

Metal Oxide Semiconductor Gas Sensors

In the context of gas sensors, the metal oxide semiconductor (MOS) type exhibits a change in conductivity when they are exposed to test gases of different concentrations. The following is a description of their specific working principle:

At operation temperature (usually high), the sensing film is exposed to oxidizing gas and the surface of sensor presents active sites for chemisorption of oxygen. As a consequence, electron depletion layer (EDL) is created on the sensor surface caused by the absorption of oxygen and reception of electrons from the conduction band of the element used for sensing. For a "n"-type sensing material, the EDL inhibits the electron migration between grains augmenting the surface resistance, conversely for a p-type semiconductor the opposite occurs.

In cases where the sensor is exposed to a reductant gas, oxygen ions that are adsorbed by sensing material liberate electrons into the conduction band. This phenomenon leads to the narrowing of the EDL band, consequently decreasing the resistance of devices. Conversely, if sensing material is p-type, the oxidation reaction of the test gas (which, as previously mentioned, is reducing) and the oxygen ions results in a decrease in the hole concentration, thereby increasing its resistance. The quantity of gas absorbed is determined by the properties of gas sensing, pressure of gas and operating temperature [47, 48].

In the case of the selection of oxides for MOS gas sensors, the most common are illustrated in Fig. 9.1. MOS gas sensors based on tin dioxide (SnO_2) have the capacity to detect carbon monoxide, methane, nitrogen dioxide, and VOCs. Tungsten dioxide (WO_2) can sense a wide variety of gases, including ammonia and ozone. ZnO is sensitive to nitrogen dioxide and hydrogen; for titanium dioxide (TiO_2), this is used for gases like hydrogen, ethanol (C_2H_5OH) and methanol (CH_3OH). Finally, indium oxide (In_2O_3) can detect carbon monoxide and NO_2 [49].

As mentioned in the operating principle of MOS gas sensors, they generally operate at temperatures above the ambient temperature, which is a major disadvantage, in addition to their poor selectivity. To overcome these obstacles, hybrid sensors that combine graphene (a two-dimensional sp2 bonded allotropy of carbon) have been investigated, for example with a MOS structure to enhance gas sensing performance, particularly for detecting gases like NO_2, NH_3, and CO at room temperature. As example of this type of innovative structure, a gas sensor was created for Seekaew

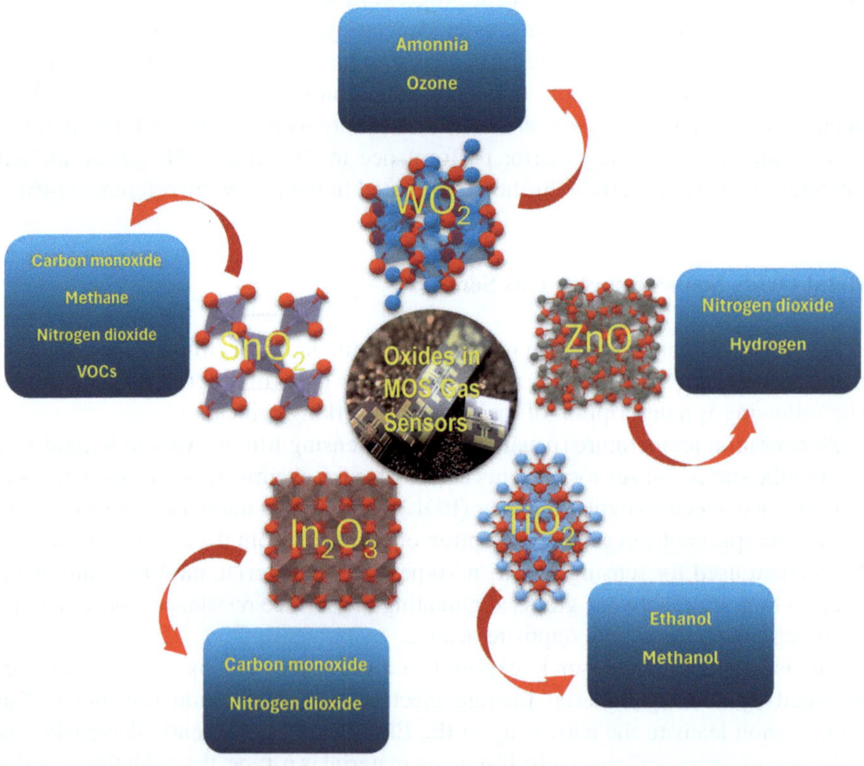

Fig. 9.1 Relevant thin film oxides for implementation in the creation of MOS gas sensors and the main gases that these materials can detect

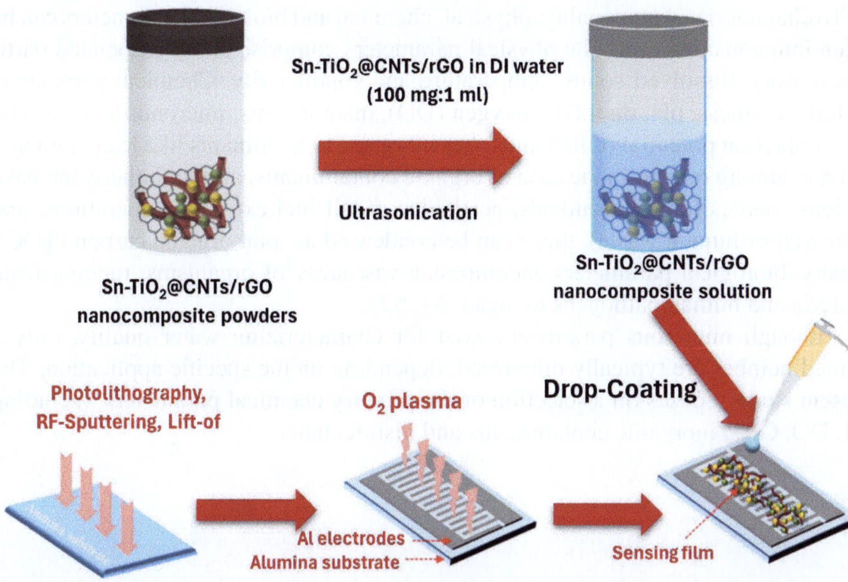

Fig. 9.2 Representation of the fabrication process for Sn − TiO2@rGO/CNT nanocomposite gas sensor [50]. Copyright 2019. Reproduced with permission from American Chemical Society

et Al. employing Sn-TiO_2 with reduced graphene and carbon nanotubes (Sn-TiO_2-rGO/CNT) as is illustrated in Fig. 9.2, demonstrating excellent levels of selectivity and high sensitivity [50].

9.3.2 *Water Quality*

Water Quality Parameters

Despite the abundant amount of water available on the planet, only a miniscule percentage is appropriate for human consumption (approximately 0.5% by 2025). Industrial and human activity, which pollutes the water, has contributed to this small percentage of water available for consumption, and, evidently, access to clean water is essential for a human survivor. This problem has motivated efforts to meet water purification and management requirements. In this context, sensors that provide information on water quality are essential to solve this problem.

Water quality sensors have a variety of applications, including industrial and municipal wastewater treatment, agriculture, and real-time monitoring of process water in industry, among other processes.

To characterize water quality, physical, chemical and biological parameters can be taken into consideration. The physical parameters comprise color, suspended particles density, dissolved solids, temperature and conductivity. Chemical parameters include hardness, pH, dissolved oxygen (DO), disinfectants, micronutrients, oxidation–reduction potential (ORP) and other inorganic contaminants like lead, mercury, arsenic, among others. In the case of organic contaminants, although these can have various sources (pharmaceuticals, pesticides, fossil fuel extraction operations, and industrial or human waste), these can be condensed as total organic carbon (TOC). Finally, biological parameters encompass a vast array of organisms, ranging from bacteria and human pathogens to algae [51, 52].

Although numerous parameters exist for characterizing water quality, only a limited number are typically monitored, depending on the specific application. The present study focuses on a selection of the primary chemical parameters, including pH, DO, ORP, inorganic contaminants and disinfectants.

pH

The sustained release of carbon dioxide into the atmosphere has been shown to result in an increase in hydrogen ion concentration in oceanic environments, accompanied by a concurrent decrease in the concentration of carbonate ions. Consequently, the process called "ocean acidification," in which the pH of seawater decreases occurs and impacts negatively in marine life and ecosystems. In this context, the $-\log [H_3O^+]$ or potential of hydrogen also known as the pH it is an important parameter to monitor. The pH is a fundamental concept in the field of water quality analysis given its role in influencing a range of biological and chemical processes. To provide a few examples, this phenomenon exerts a substantial influence on the survival rate of organisms, as evidenced by the critical pH range for life, which falls between 4 and 11. Furthermore, the pH level is a critical factor in corrosion processes, where its decline results in a more corrosive environment. In addition, the pH level has been demonstrated to influence the efficacy of disinfectants, with the efficiency of these agents being enhanced at lower pH values [53, 54].

The predominant pH sensor configuration consists of an electrode based on glass. This electrode contains an electrolyte-filled glass bubble and an internal silver chloride electrode. This configuration is frequently employed for precise pH measurements. Despite these advantages, glass electrodes possess several problems, including slow response times, elevated impedance, fragility, and difficulty in miniaturization. Consequently, these materials are not appropriate in circumstances involving acidity, alkalinity elevated temperatures and pressures [55].

In order to overcome the obstacles presented, there is a growing interest in the substitution of glass electrodes with thin films of sensing material for the creation of electrochemical pH sensors. This approach has emerged as a promising solution. To mention some of the most important developments in thin film pH sensors, Chien et al. manufactured a field-effect transistor with extended-gate employing a thin film of carbon nanotubes with improved sensing and electrical properties [56]. Another

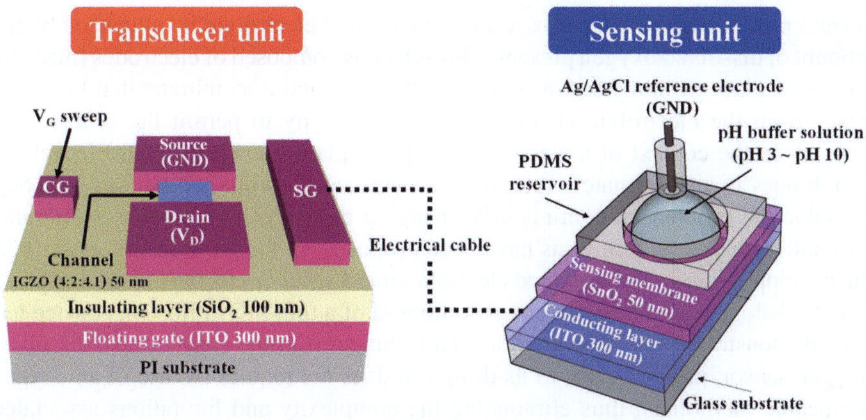

Fig. 9.3 Representation of coplanar dual-gate TFT transducer comprising an a-IGZO layer as channel, in conjunction with a based SnO_2 sensing units [58]. Copyright 2023. Reproduced with permission from MDPI

material employed for pH detection based on thin film transistor (TFT) is $InGaZnO_4$. In the research of Lu et al., this type of sensor demonstrated high sensitivity and low operating voltages in a pH range of 3 to 11 [57]. Additionally, in the research conducted by Hyun and Cho [58], this same material was used in an amorphous form, where in this case highly sensitive detection were obtained with 359.28 mV/pH (highly superior to the Nernstian limit) in a fully transparent pH sensor. The sensor created is a TFT transducer (with a dual-gate on flexible polyimide substrate) coupled to an extended-gate sensing unit, as shown in Fig. 9.3.

Complementing the creation of thin film flexible pH detectors, there is research by Chou et al. where noteworthy sensitivity (63 mV/pH) is observed in a pH range of 1 to 13 using nickel oxide as the detector layer. [59]. Finally, as an example of another novel development based on thin films, there is that of Iken et al. who created a thin film sensor for simultaneous detection of redox potential and pH [60].

Dissolved Oxygen

In context of the wastewater treatment industry, the process of degradation in organic and biological pollution species is predominantly driven by the action of aerobic bacteria, which consume substantial quantities of oxygen. This phenomenon is further exacerbated by the high solubility of oxygen in water at ambient temperatures. Consequently, the implementation of precise monitoring and control mechanisms for dissolved oxygen (DO) levels during the wastewater treatment process has been identified as a pivotal strategy to enhance its efficiency and reduce operational costs. Clark-type DO sensor is the typical DO sensor and works measuring the oxygen concentration in water through the application of a voltage between electrodes; the current produced during a redox reaction in a liquid electrolyte is influenced by the amount of dissolved oxygen present. This sensor is composed of electrodes (measure the redox current) immersed at a liquid electrolyte and a membrane that separates water from the electrolyte characterized by its ability to permit the diffusion of oxygen. In the context of a wastewater disposal plant, the operational lifecycle of membranes is approximately one month, even under optimal conditions [61, 62]. To solve the problem of continuously changing the oxygen permeable membrane, alternatives based on thin films have been proposed. In the work of Lee et al. [63], the developed sensor incorporated electrodes and a solid electrolyte, as well as polydimethylsiloxane and a sensor cover composed of a thin SiO_2 film. This device has been demonstrated to exhibit significant advantages over the conventional dissolved oxygen sensor, primarily due to its design. It does not require the use of an oxygen permeable membrane, thus eliminating the complexity and limitations associated with this traditional configuration. Additionally, it employs a solid electrolyte rather than a liquid electrolyte, a substitution that has been shown to enhance the reliability, stability, and precision of the device. Finally, this type of sensor can measure dissolved oxygen in water without the need for replacement of the membrane [64].

On the other hand, using optical techniques, Gillanders et al. developed a DO sensor composed of thin films. The sensor demonstrated an intense phosphorescence response when immersed in contaminated water and subjected to DO interaction. Additionally, the sensor exhibited remarkable mechanical strength and fast response times [65]. Recently Wijesooriya et al. propose a DO sensor based on electrodeposited Cu_2O thin film exhibited optimal performance for DO sensing, exhibiting a significant response. This advancement in thin film sensor technology underscores their capacity to substitute or supplement prevailing methodologies [66].

Oxidation–Reduction Potential

The oxidation–reduction potential (ORP) of water is a quantification of its ability to transfer electrons; in water subjected to treatment processes, the ORP is predominantly influenced by the composition and concentration of the introduced disinfectants. Positive ORP values indicate an oxidizing environment, suggesting strong disinfectant properties, while negative values suggest reducing environment conducive to contaminant and microbial growth [67]. The relationship between ORP and disinfectant efficiency is well-documented; therefore, it is recommended that ORP be utilized as the fundamental criterion in the context of water treatment processes in drinking, industrial and wastewater plants [10]. Among the ORP sensors developed with thin films, there is, for example, the work of Iken et al. who developed a thin film sensor that can detect changes in both pH and redox levels based on a bi-electrode which consists of Ti/Pt and Cr/Au with thickness of 20–200 nm and 20–200 nm, respectively. The fabricated sensor redox potential was successfully tested in a wide range of solutions comprising real test samples pool water and soda [60].

Disinfectants and Other Inorganic Contaminants

Disinfectants in water treatment have been demonstrated to eliminate pathogens through a reaction with diverse components of bacterial cells, thereby impeding essential the cellular processes that support their proliferation. Among the most commonly used disinfectants are free chlorine, chlorine dioxide, hydrogen peroxide, potassium permanganate, monochloramine, and hypobromous acid. The efficacy of disinfectants in preserving water quality (pathogen-free) relies on the concentration of the disinfectant agent in the water, consequently necessitating regular monitoring. The prevailing standard methodologies for the quantification of disinfectant concentrations are predominantly colorimetric and titrimetric; however, the utilization of such methods requires the employment of spectroscopic readout devices, which have been observed to be susceptible to interference from oxidizing species and organic constituents commonly found in water. Consequently, titrimetric and colorimetric techniques are not adequate for constant disinfectant monitoring [68]. A good alternative is thin film chemiresistive sensors, within this field those that incorporate graphene or carbon nanotubes stand out as demonstrated by recent research by Akbar et al., Zubiarrain et al. and Aryasomayajula et al. [68, 69].

In the case of other inorganic pollutants, such as lead, tin, arsenic, mercury, cyanide, chromium, nickel, barium and strontium, these chemicals are typically toxic even at low levels, emphasizing the critical importance of their monitoring for the preservation of human health. For the detection of these types of contaminants, as mentioned for the detection of disinfectants, carbon-based sensor developments are showing great interest from the scientific community. In the context of arsenic analysis, research has focused on the development of thin film carbon-based sensors. This includes the design of an advanced electrochemical sensor which is constructed from a combination of double-walled carbon nanotubes and graphene-hybrid thin film, in which the sensor demonstrates the capacity for linear detection of arsenic (V) with a minimal limit of detection of 0.287 ppb within the range of detection of 1 to 10 parts per billion [70]. For lead detection, the development of a fluorescent chemical sensor, derived from carbon quantum dots and dithizone, has been achieved through its integration within the polymeric films composed of chitosan. This sensor was successfully tested in various natural water resources, showing excellent potential for real-time monitoring of Pb^{2+} in real water samples [71]. While in the case of mercury, there is a research work developed by Vyas and Joshi [72]. In their study, a dual mode sensor (colorimetric and fluorescent) based on diphenyl carbazone and a thin film of carbon quantum dots was reported as a novel detection method for mercury ions. Employing a fluorometric method, this sensor has been demonstrated to reach a detection limit of 1.8 ppb within the scale of 0 to 100 μM, with a sensing time of two minutes.

9.4 Biological Sensors (Biosensors)

The evolution of biological detection can be dated back to 1960, when invention of the first biosensor by Clark and Lyon marked significant progress in the field. This biosensor was designed to detect hydrogen peroxide, facilitating the measurement of glucose levels using electrochemically immobilized glucose oxidase electrodes. Since then, the evolution of these devices has been marked by continuous advancement and innovation. This is attributable to the innumerable applications they present, which have significantly enhanced our health and life. The evolution of biosensor technology has yielded a wide range of applications, including medicine, agriculture, pharmaceutical, and mobile-based healthcare services. Recent trends in biosensor development have been characterized by the proliferation of compact and user-friendly sensor technologies that can be incorporated into our daily lives. Within this panorama, the main trends in biosensors are miniaturization, wearable biosensors, and the implementation of "lab-on-a-chip" technology, that integrates multiple laboratory functions, like chemical analysis or synthesis, onto a single small chip in real time [73].

A biosensor is defined as an analytical device with the capacity to produce information using a biological detection element (sensitive element). This recognition element could be an antibody, enzyme, cell, or nucleic acid integrated with a physicochemical transducer that processes the signal and identifies the characteristics of biochemical substance and its concentration.

The classification of biosensors is typically established on the characteristics of the bioactive elements involved. These include the bioaffinity sensors that are devices that employ the selective interaction between the analyte of interest and a modified surface, which has been prepared through the attachment of a specific receptor (antibodies or nucleic acids). The most advanced bioaffinity sensors are of the immunoreceptor variety, which are based on antigen–antibody interactions. Other types of biosensors in this classification include those who utilize an immobilized biocatalyst, which can be cells, enzymes (single or multiple), or organisms. This biocatalyst identifies, connects, and then transforms the objective analyte. The enzyme sensors have been the focus of exhaustive research, as their application in the field of biosensors is extensive. These sensors have the capacity to detect a wide range of analytes, which are crucial for numerous practical applications [74, 75].

Based on the aforementioned classification, a large number of biosensors subclassifications emerge, with diverse applications, operating principles, and recent technological developments. Therefore, below we will focus exclusively on glucose sensors, immunosensors, piezoelectric, and optical biosensors and their most relevant developments based on the application of thin films.

9.4.1 Glucose Sensors

Diabetes mellitus is a disorder of the endocrine system and metabolism that is characterized by insufficient production of insulin, leading to elevated blood glucose levels. This is one of the fastest-growing chronic affections, currently affecting 1 in 9 people. Periodic glucose monitoring is an extremely useful tool that can help combat the disease or prevent it from progressing to a critical point. The development of various sensing technologies (categorized into enzymatic and non-enzymatic) has enabled the precise quantification of glucose concentration in the bloodstream. Enzymatic reactions involve the use of glucose oxidase and glucose dehydrogenase as the transducing elements, facilitating the conversion of glucose into other chemical compounds. These types of sensors are very popular and have commercially accessible devices (sensing strips); however, their fabrication is an exacting process that demands meticulous adherence to specific manufacturing conditions. These conditions include, but are not limited to, sterilization, enzyme immobilization, precise control of humidity and temperature, and maintenance of a precise pH level for the enzyme environment. As an alternative, non-enzymatic glucose sensors present certain advantages for implementation, including stability under extreme environmental conditions, cost-effectiveness in production, enhanced sensitivity, prolonged durability, and high specificity [76]. The non-enzymatic glucose sensors facilitate direct oxidation of glucose on the surface of electrode, and this technology is the fourth generation of glucose sensors. As a recent advance in this type of sensor developed with nanostructures, we have the work of Bano et al. who created an ultrasensitive glucose sensor with real-time health-monitoring capabilities, for which they used 2D-MoS_2 nanocatalyst by means of a one-step hydrothermal technique and was immobilized on a glassy carbon electrode (GCE) to detect glucose sensitivity. This method yielded a superior sensitivity of 1194 $\mu A\ \mu M^{-1}\ cm^{-2}$ and detection range of $0.01 - 0.20$ mM [77].

In case of the non-enzymatic glucose sensors fabricated with thin films, film thickness plays a very important role since thickness affects the conductivity, grain size, and film-substrate interface. The ideal working electrode is characterized by an optimal film thickness and a substantial effective area. Singh and Ayyub worked with ultra-thin films of NiS, which were deposited through the mechanism of self-limiting surface reactions involving H_2S as the S precursor and Ni(acac)2 as the Ni precursor using atomic layer deposition, these films were used to create superior sensing glucose sensor, and the films demonstrate a minimal detection limit of 0.052 μM, an outstanding sensitivity (5.78 $\mu A\ mM^{-1}\ cm^{-2}$), and an excellent sensing time of < 2 s [78]. Among the most studied thin films for this type of sensors are those based on copper and nickel, since they show very promising responses, being simple and cost-efficient. [79–81]

Finally, as one of the most important advances in non-invasive glucose detection, there are developments such as that of Wang et al. who created an electrochemical sensor using thin films of polyethylene terephthalate (PET) with gold electrode for detection of glucose in sweat (fabrication and working principle is illustrated in

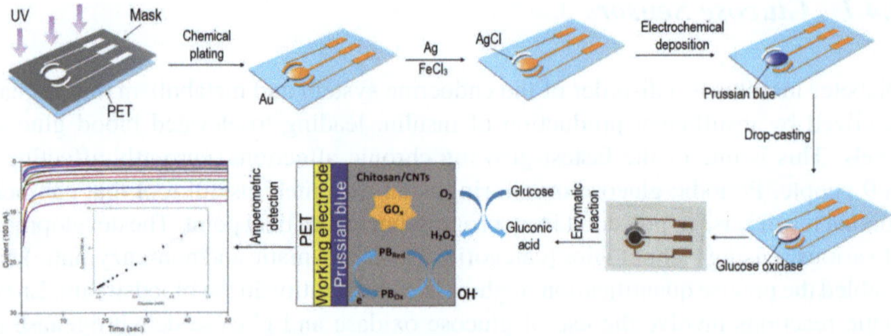

Fig. 9.4 Fabrication of the PET-based gold electrode glucose sensor and working principle of amperometric determination of glucose [82]. Copyright 2019. Reproduced with permission from Elsevier

Fig. 9.4). Long-term stability is shown by this sensor, and good selectivity is also presented. Additionally, interference by urea, lactic acid, acetaminophen, ascorbic acid and dopamine were not observed [82]. In the work of Sharma et al., an extremely sensitive glucose detector in saliva was created. This detector is non-invasive and employs metal oxide transistors, specifically using a tri-channel In_2O_3/ZnO heterojunction thin film transistors. The glucose in artificial and real saliva can be selectively detected by this sensor over a large range of concentrations. The limits of detection are 416 nM for real saliva and 365 pM in artificial saliva and this sensing is achieved in less than a minute [83].

9.4.2 *Immunosensors*

Immunosensors are a type of biosensor that uses antibodies to recognize and bind to antigens. This reaction produces further changes in the potential of the biosensitive membrane, enabling the measurement of high-molecular-weight compounds with high selectivity. Immunosensors can be classified by immune response type as either labeled or unlabeled. The immune substance functions as a sensitive element and is attached to the surface of the electrode in the case of a labeled immunosensors. After the antigen–antibody reaction reaches an equilibrium state, it transforms the signal of the biological reaction into an electrical signal, which is then output through the electrodes. Non-labeled immunosensors work based on the principle that the combination of antigens and antibodies produces electrical or electrochemical changes due to the large amount of charge present in each one. These alterations can be measured as changes in ion concentrations, membrane potential, and conductivity. The immunosensor can detect an immune response based on these changes without using a label, so they are also called direct immunoelectrodes. Within the immunosensors with thin films, we have those based on organic thin film transistors, used for the creation of devices with a flexible design for use in wearable and reusable sensing systems for daily health management. Minamiki et al. present a design for an immunosensor made on an ultra-thin film with a dual-gate configuration showed excellent operation stability and achieved the quantitative detection of proteins [84].

In the healthcare field, there have also been important advances with thin film immunosensors, such as the one reported by Aydın et al., where a sensor was created for the recognition of tumor necrosis factor α (TNFα) in saliva and serum. This sensor was manufactured using ITO thin films covered with the conjugated semiconductive polymer (Poly (3-thiophene acetic acid)) [85]. Or also in the rapid detection of viruses as dangerous as SARS-CoV-2, where in the work of Kim et al., a thin film transistor with high-performance and dual-gate oxide semiconductor was used to create an immunosensor that can perform a fast SARS-CoV-2 detection. This sensor precisely measures electrical alterations originating from SARS-CoV-2, demonstrating high antigen selectivity and sensitivity. It eliminates the requirements of polymerase chain response tests and labeling [86].

9.4.3 Piezoelectric Biosensors

In recent times, piezoelectric biosensors have seen a surge in applications, particularly in inspection of food hygiene, clinical diagnosis (e.g., detection of intestinal bacteria, HIV, insulin, and immunoglobulin), biotechnology, and other fields. This increased utilization can be attributed to the unique characteristics of piezoelectric biosensors, which offer distinct advantages over other biosensor types. These advantages include simple structure, short detection time, the absence of labeling, and selectivity. The piezoelectric biosensor is a technological innovation that integrates the specificity of biological responses with the sensitivity of piezoelectric sensors. The operation principle is based on the piezoelectric crystal selectively adsorbing biologically active substances in the gas or liquid phase. The mass load and viscous coupling then origin a change in frequency for piezoelectric quartz crystal. The measurement of the alteration in the vibration frequency of the crystal following the adsorption of foreign substances enables the determination of the mass of the adsorbed substance. This, in turn, facilitates the fabrication of a piezoelectric biosensor based on the aforementioned principle.

In the field of biosensor development, there has been a notable focus on the utilization of inorganic-based flexible piezoelectric thin films for biomedical applications. These films have garnered significant attention due to their advantageous properties, including high piezoelectric performance, flexibility, slim design, lightweight characteristics, and biocompatibility. Among the most important advantages of using these types of films is the possibility of implementing self-powered energy storage systems for in vivo diagnosis systems and bio-implantable medical devices which are intended to eliminate the need for batteries. The most studied thin films for this type of biosensors are $Pb(Zr_x,Ti_{1-x})O_3$ (PZT) and $(1\text{-}x)Pb(Mg_{1/3}Nb_{2/3})O_3 - xPbTiO_3$ (PMN-PT) [87]. In the case of PZT, biosensors with novel medical applications have been reported, such as the work of Lee et al. [88] who report an acoustic-piezoelectric nanosensor for biomimetic artificial hair cells using thin films of flexible PZT, achieving restore sensorineural hearing loss, through the adoption of a polyvinylidene fluoride positioned on the basilar membrane of the cochlea to sense vibrations (in response to sound waves) of the basilar membrane and transduce the stimulus into electric signals to mimic the role of auditory hair cells.

For the PMN-PT a direct stimulation of a living heart was reported, without the need for supplementary power sources. This was achieved by applying temporary electric output energy from a flexible PMN-PT thin film harvesting device. The investigation was developed by Hwang et al. [89].

9.4.4 Optical Biosensor

An optical biosensor can be defined as a device capable of converting the quantitative variation of a specific biological analyte into a visible light signal. This optical signal is subsequently converted into an electrical signal through the use of photoelectric elements. The classification of biosensors according to the method of light signal generation within the sensing layer includes three types:

Passive sensors, lacking an excitation source, employ optical emission produced by their individual biochemical reactions subsequent to the combination of the material to be analyzed with the sensing or active layer. Following this conversion, the optical signals are transformed into electrical signals without any associated energy excitation.

The second variety of biosensors utilizes a light signal as the excitation source and is referred to as a photo-induced optical biosensor. Typically, these biosensors employ an external excitation light source. This enables them to guarantee a stable incident light signal and to select the wavelength of the excitation photons within a considerable range. Furthermore, the primary light signals (reflected, transmitted, refracted, and scattered light signals) and the secondary light signals (phosphorescence and fluorescence) that are produced subsequent to the excitation light interact with the detection layer. These light signals can be utilized as fundamental detection parameters.

The final type of biosensor utilizes the electrical signal as the excitation source. This biosensor is known as the electro-optical biosensor. The construction of the electro-optical biosensor is typically based on electrochemiluminescence principle. This biosensor frequently necessitates the incorporation of a working electrode within its sensing layer. When a specific voltage is applied to the working electrode, a particular electrochemical reaction arises. The transitional material produced by this reaction subsequently interacts with the measured substance, resulting in the release of a quantifiable light signal.

Some of the developments created for thin film optical biosensors include the use of materials such as TiO_x and Au, which operate based on the principle of surface plasmon resonance, as demonstrated in [90, 91]. One of the most interesting applications in the field of optical thin film sensors is the integration of photodetectors and light emitters into a single device for biological detection, also known as a "lab on a chip." These generally use thin films of hydrogenated amorphous silicon and transparent conducting oxides. Additionally, the use of organic photodiodes in this technology has yielded promising results for implementation [92].

9.5 Conclusions and Outlook

This book chapter explores the application of thin film sensors to environmental and biological sensing. While these topics have numerous variants and countless applications, offering diverse alternatives to overcome the challenges that each sensor type may present, some examples demonstrate the importance and, above all, the contributions that the use of thin films makes to each of the potential applications. Among the approaches that provide novel properties to the sensors analyzed, we can mention the use of functionalizable materials such as carbon-based materials, the use of devices such as thin film transistors forming heterostructures with different materials, the incorporation of nanostructured thin films, and the variation in the thickness of thin films to manipulate their properties and characteristics.

Despite the need for continued improvement in the field of thin film sensors, significant and rapid progress has been made, offering a promising outlook for their large-scale implementation.

References

1. Greene, J. E. (2014). Tracing the 5000-year recorded history of inorganic thin films from~ 3000 BC to the early 1900s AD. *Applied Physics Reviews, 1*(4). https://doi.org/10.1063/1.4902760
2. Memon, A. G., Giwa, A. S., & Ahmad, J. (2025). Nanotechnology and thin films: explores nanoscale thin films, nanocomposites, nanoparticles, and their application. In *Thin Films and Coatings: Engineering Applications* (pp. 187–215). Singapore: Springer Nature Singapore. https://doi.org/10.1007/978-981-96-0005-2
3. Krishnan, R. M., & Meyer III, F. O. (1974). Quantum size effect in thin films. *Thin Solid Films, 23*(1), 7-13. https://doi.org/10.1016/0040-6090(74)90212-0
4. Spaepen, F. (2000). Interfaces and stresses in thin films. *Acta materialia, 48*(1), 31-42. https://doi.org/10.1016/S1359-6454(99)00286-4
5. Li, W., Shi, J., Zhang, K. H., & MacManus-Driscoll, J. L. (2020). Defects in complex oxide thin films for electronics and energy applications: challenges and opportunities. *Materials Horizons, 7*(11), 2832-2859. https://doi.org/10.1039/D0MH00899K
6. Mao, L. F. (2010). The quantum size effects on the surface potential of nanocrystalline silicon thin film transistors. *Thin Solid Films, 518*(12), 3396-3401. https://doi.org/10.1016/j.tsf.2009.10.143
7. Chopra, K. (2012). *Thin film device applications.* Springer Science & Business Media. https://doi.org/10.1007/978-1-4613-3682-2
8. Javaid, M., Haleem, A., Rab, S., Singh, R. P., & Suman, R. (2021). Sensors for daily life: A review. *Sensors International*, 2, 100121. https://doi.org/10.1016/j.sintl.2021.100121

9. VS, A. P., Joseph, P., SCG, K. D., Lakshmanan, S., Kinoshita, T., & Muthusamy, S. (2017). Colorimetric sensors for rapid detection of various analytes. *Materials Science and Engineering: C, 78*, 1231–1245. https://doi.org/10.1016/j.msec.2017.05.018
10. Kruse, P. (2018). Review on water quality sensors. *Journal of Physics D: Applied Physics, 51*(20), 203002. https://doi.org/10.1088/1361-6463/aabb93
11. Ng, S. M., Koneswaran, M., & Narayanaswamy, R. (2016). A review on fluorescent inorganic nanoparticles for optical sensing applications. *RSC advances, 6*(26), 21624-21661. https://doi.org/10.1039/C5RA24987B
12. Kats, M. A., & Capasso, F. (2016). Optical absorbers based on strong interference in ultra-thin films. *Laser & Photonics Reviews, 10*(5), 735-749. https://doi.org/10.1002/lpor.201600098
13. Bae, G., Seo, M., Lee, S., Bae, D., & Lee, M. (2021). Colorimetric Detection of Mechanical Deformation in Metals using Thin-Film Mechanochromic Sensor. *Advanced Materials Technologies, 6*(10), 2100479. https://doi.org/10.1002/admt.202100479
14. Privett, B. J., Shin, J. H., & Schoenfisch, M. H. (2010). Electrochemical sensors. *Analytical Chemistry*, 82(12), 4723-4741. https://doi.org/10.1021/ac101075n
15. Bakker, E., & Telting-Diaz, M. (2002). Electrochemical sensors. *Analytical chemistry, 74*(12), 2781-2800. https://doi.org/10.1021/ac0202278
16. Yoon, D., Baggett, A., Phongikaroon, S., King, J. A., & Marsden, K. (2019). Fundamental data acquisition toward silver-silver chloride reference electrode. *Journal of The Electrochemical Society*, 166(6), E159. https://doi.org/10.1149/2.0721906jes
17. Zhou, J., Zhang, L., & Tian, Y. (2016). Micro electrochemical pH sensor applicable for real-time ratiometric monitoring of pH values in rat brains. Analytical chemistry, 88(4), 2113-2118. https://doi.org/10.1021/acs.analchem.5b03634
18. Rajapaksha, R. D. A. A., Hashim, U., Gopinath, S. C. B., & Fernando, C. A. N. (2018). Sensitive pH detection on gold interdigitated electrodes as an electrochemical sensor. *Microsystem Technologies, 24*(4), 1965-1974. https://doi.org/10.1007/s00542-017-3592-5
19. Mazzara, F., Patella, B., D'Agostino, C., Bruno, M. G., Carbone, S., Lopresti, F.,... & Inguanta, R. (2021). PANI-based wearable electrochemical sensor for pH sweat monitoring. *Chemosensors*, 9(7), 169. https://doi.org/10.3390/chemosensors9070169
20. Wang, J., Wei, J. L., Cao, Q., Cheng, X. F., Chen, Z. K., & He, J. H. (2023). Chemresistive detection of NO2 of ppb level in humid air at 350 K using azo-spaced polycroconamide. *ACS sensors, 9*(1), 236-243. https://doi.org/10.1021/acssensors.3c01869
21. Sitaras, I. E., & Siskos, P. A. (2008). The role of primary and secondary air pollutants in atmospheric pollution: Athens urban area as a case study. *Environmental Chemistry Letters, 6*(2), 59-69. https://doi.org/10.1007/s10311-007-0123-0
22. Hashimoto, K. (2019). Global temperature and atmospheric carbon dioxide concentration. In *Global carbon dioxide recycling: for global sustainable development by renewable energy* (pp. 5–17). Singapore: Springer Singapore. https://doi.org/10.1007/978-981-13-8584-1
23. Dutta, D., Chong, N. S., & Lim, S. H. (2018). Endogenous volatile organic compounds in acute myeloid leukemia: origins and potential clinical applications. *Journal of Breath Research, 12*(3), 034002. https://doi.org/10.1088/1752-7163/aab108
24. Duan, X., Chen, Z., Xia, C., Zhong, R., Liu, L., & Long, L. (2024). Increased levels of urine volatile organic compounds are associated with diabetes risk and impaired glucose homeostasis. *The Journal of Clinical Endocrinology & Metabolism, 109*(2), e531-e542. https://doi.org/10.1210/clinem/dgad584
25. Tang, H., Lu, Y., Zhang, L., Wu, Z., Hou, X., & Xia, H. (2017). Determination of volatile organic compounds exhaled by cell lines derived from hematological malignancies. *Bioscience reports, 37*(3), BSR20170106. https://doi.org/10.1042/BSR20170106
26. Seiyama, T., & Kagawa, S. (1966). Study on a Detector for Gaseous Components Using Semiconductive Thin Films. *Analytical chemistry, 38*(8), 1069-1073. https://doi.org/10.1021/ac60240a031
27. Panžić, I., Bafti, A., Radovanović-Perić, F., Gašparić, D., Shi, Z., Borenstein, A., & Mandić, V. (2025). Advancements in Nanostructured Functional Constituent Materials for Gas Sensing Applications: A Comprehensive Review. *Applied Sciences, 15*(5), 2522. https://doi.org/10.3390/app15052522

28. Okamura, K., Ishiji, T., Iwaki, M., Suzuki, Y., & Takahashi, K. (2007). Electrochemical gas sensor using a novel gas permeable electrode modified by ion implantation. *Surface and Coatings Technology*, *201*(19-20), 8116-8119. https://doi.org/10.1016/j.surfcoat.2006.02.065
29. Khan, M. A., Qazi, F., Hussain, Z., Idrees, M. U., Soomro, S., & Soomro, S. (2017). Recent trends in electrochemical detection of NH3, H2S and NOx gases. *International Journal of Electrochemical Science, 12*(3), 1711–1733. https://doi.org/10.20964/2017.03.76
30. Wan, H., Yin, H., & Mason, A. J. (2017). Rapid measurement of room temperature ionic liquid electrochemical gas sensor using transient double potential amperometry. *Sensors and Actuators B: Chemical, 242*, 658-666. https://doi.org/10.1016/j.snb.2016.11.103
31. Currie, J. F., Essalik, A., & Marusic, J. C. (1999). Micromachined thin film solid state electrochemical CO2, NO2 and SO2 gas sensors. *Sensors and Actuators B: Chemical, 59*(2-3), 235-241. https://doi.org/10.1016/S0925-4005(99)00227-0
32. Janz, G. J., & Taniguchi, H. (1953). The Silver-Silver Halide Electrodes. Preparation, Stability, and Standard Potentials in Aqueous and non-Aqueous Media. *Chemical Reviews, 53*(3), 397–437. https://doi.org/10.1021/cr60166a002
33. Ivanov, I. I., Baranov, A. M., Spirjakin, D. N., Akbari, S., Mironov, S. M., Karami, H., & Gharehpetian, G. B. (2022). Expanding catalytic sensor capabilities to combustible gas mixtures monitoring. *Measurement, 194*, 111103. https://doi.org/10.1016/j.measurement.2022.111103
34. Xu, L., Li, T., Gao, X., & Wang, Y. (2012, March). A high heating efficiency two-beam microhotplate for catalytic gas sensors. In *2012 7th IEEE International Conference on Nano/ Micro Engineered and Molecular Systems (NEMS)* (pp. 65–68). IEEE. https://doi.org/10.1109/ NEMS.2012.6196724
35. Tardy, P., Coulon, J. R., Lucat, C., & Menil, F. (2004). Dynamic thermal conductivity sensor for gas detection. *Sensors and Actuators B: Chemical*, *98*(1), 63-68. https://doi.org/10.1016/j. snb.2003.09.019
36. Wang, J., Liu, Y., Zhou, H., Wang, Y., Wu, M., Huang, G., & Li, T. (2022). Thermal conductivity gas sensor with enhanced flow-rate independence. *Sensors, 22*(4), 1308. https://doi.org/10. 3390/s22041308
37. Samotaev, N., Podlepetsky, B., Mashinin, M., Ivanov, I., Obraztsov, I., Oblov, K., & Dzhumaev, P. (2024). Thermal conductivity gas sensors for high-temperature applications. *Micromachines, 15*(1), 138. https://doi.org/10.3390/mi15010138
38. Cho, W., Kim, T., & Shin, H. (2022). Thermal conductivity detector (TCD)-type gas sensor based on a batch-fabricated 1D nanoheater for ultra-low power consumption. *Sensors and Actuators B: Chemical, 371*, 132541. https://doi.org/10.1016/j.snb.2022.132541
39. Punnoose, A., Reddy, K. M., Thurber, A., Hays, J., & Engelhard, M. H. (2007). Novel magnetic hydrogen sensing: a case study using antiferromagnetic haematitenanoparticles. *Nanotechnology, 18*(16), 165502. https://doi.org/10.1088/0957-4484/18/16/165502
40. Shinde, P. V., & Rout, C. S. (2021). Magnetic gas sensing: Working principles and recent developments. *Nanoscale advances, 3*(6), 1551-1568. https://doi.org/10.1039/D0NA00826E
41. Das, S. S., Kopnov, G., & Gerber, A. (2018). Detection of hydrogen by the extraordinary Hall effect in CoPd alloys. *Journal of Applied Physics, 124*(10). https://doi.org/10.1063/1.5049647
42. Hodgkinson, J., & Tatam, R. P. (2012). Optical gas sensing: a review. *Measurement science and technology, 24*(1), 012004. https://doi.org/10.1088/0957-0233/24/1/012004
43. Tabassum, R., & Kant, R. (2020). Recent trends in surface plasmon resonance based fiber–optic gas sensors utilizing metal oxides and carbon nanomaterials as functional entities. *Sensors and Actuators B: Chemical, 310*, 127813. https://doi.org/10.1016/j.snb.2020.127813
44. Paliwal, A., Sharma, A., Tomar, M., & Gupta, V. (2017). Carbon monoxide (CO) optical gas sensor based on ZnO thin films. *Sensors and Actuators B: Chemical, 250*, 679-685. https://doi. org/10.1016/j.snb.2017.05.064
45. Mazur, M., Kapuścik, P., Weichbrodt, W., Domaradzki, J., Mazur, P., Kot, M., & Flege, J. I. (2023). WO3 thin-film optical gas sensors based on gasochromic effect towards low hydrogen concentrations. *Materials, 16*(10), 3831. https://doi.org/10.3390/ma16103831
46. Zheng, Z., Yao, J., Wang, B., & Yang, G. (2017). A flexible, transparent and high-performance gas sensor based on layer-materials for wearable technology. *Nanotechnology, 28*(41), 415501. https://doi.org/10.1088/1361-6528/aa8317

47. Nadargi, D. Y., Umar, A., Nadargi, J. D., Lokare, S. A., Akbar, S., Mulla, I. S.,... & Chaskar, M. G. (2023). Gas sensors and factors influencing sensing mechanism with a special focus on MOS sensors. *Journal of Materials Science*, *58*(2), 559-582. https://doi.org/10.1007/s10853-022-08072-0
48. Xue, S., Cao, S., Huang, Z., Yang, D., & Zhang, G. (2021). Improving gas-sensing performance based on MOS nanomaterials: a review. *Materials*, *14*(15), 4263. https://doi.org/10.3390/ma14154263
49. Goel, N., Kunal, K., Kushwaha, A., & Kumar, M. (2023). Metal oxide semiconductors for gas sensing. *Engineering Reports*, *5*(6), e12604. https://doi.org/10.1002/eng2.12604
50. Seekaew, Y., Pon-On, W., & Wongchoosuk, C. (2019). Ultrahigh selective room-temperature ammonia gas sensor based on tin–titanium dioxide/reduced graphene/carbon nanotube nanocomposites by the solvothermal method. *ACS omega*, *4*(16), 16916-16924. https://doi.org/10.1021/acsomega.9b02185
51. Gorde, S. P., & Jadhav, M. V. (2013). Assessment of water quality parameters: a review. *International Journal of Engineering Research and Applications*, *3*(6), 2029–2035. ISSN: 2248–9622
52. Omer, N. H. (2019). Water quality parameters. *Water quality-science, assessments and policy*, *18*, 1-34. https://doi.org/10.5772/intechopen.77531
53. Kennedy, L. C., Miller, S. E., Kantor, R. S., & Nelson, K. L. (2021). Effect of disinfectant residual, pH, and temperature on microbial abundance in disinfected drinking water distribution systems. *Environmental Science: Water Research & Technology*, *7*(1), 78-92. https://doi.org/10.1039/D0EW00809E
54. Boyd, C. E. (2019). Carbon dioxide, pH, and alkalinity. In *Water quality: an introduction* (pp. 177–203). Cham: Springer International Publishing. https://doi.org/10.1007/978-3-030-23335-8
55. Vonau, W., & Guth, U. (2006). pH monitoring: a review. *Journal of Solid State Electrochemistry*, *10*(9), 746-752. https://doi.org/10.1007/s10008-006-0120-4
56. Chien, Y. S., Yang, P. Y., Tsai, W. L., Li, Y. R., Chou, C. H., Chou, J. C., & Cheng, H. C. (2012). The pH sensing characteristics of the extended-gate field-effect transistors of multi-walled carbon-nanotube thin film using low-temperature ultrasonic spray method. *Journal of Nanoscience and Nanotechnology*, 12(7), 5423-5428. https://doi.org/10.1166/jnn.2012.6358
57. Lu, C. H., Hou, T. H., & Pan, T. M. (2016). Low-voltage InGaZnO ion-sensitive thin-film transistors fabricated by low-temperature process. *IEEE Transactions on Electron Devices*, *63*(12), 5060-5063. https://doi.org/10.1109/TED.2016.2614959
58. Hyun, T. H., & Cho, W. J. (2023). Fully transparent and highly sensitive pH sensor based on an a-IGZO thin-film transistor with coplanar dual-gate on flexible polyimide substrates. *Chemosensors*, *11*(1), 46. https://doi.org/10.3390/chemosensors11010046
59. Chou, J. C., Yan, S. J., Liao, Y. H., Lai, C. H., Chen, J. S., Chen, H. Y.,... & Wu, T. Y. (2017). Characterization of flexible arrayed pH sensor based on nickel oxide films. *IEEE Sensors Journal*, *18*(2), 605-612. https://doi.org/10.1109/JSEN.2017.2773143
60. Iken, H., Bronder, T. S., Goretzki, A., Kriesel, J., Ahlborn, K., Gerlach, F.,... & Schöning, M. J. (2019). Development of a Combined pH-and Redox-Sensitive Bi-Electrode Glass Thin-Film Sensor. *Physica status solidi (a)*, *216*(12), 1900114. https://doi.org/10.1002/pssa.201900114
61. Elanjeitsenni, V. P., Vadivu, K. S., & Prasanth, B. M. (2022). A review on thin films, conducting polymers as sensor devices. *Materials Research Express*, *9*(2), 022001. https://doi.org/10.1088/2053-1591/ac4aa1
62. Wei, Y., Jiao, Y., An, D., Li, D., Li, W., & Wei, Q. (2019). Review of dissolved oxygen detection technology: From laboratory analysis to online intelligent detection. *Sensors*, *19*(18), 3995. https://doi.org/10.3390/s19183995
63. Lee, H. J., Kim, H. M., Park, J. H., & Lee, S. K. (2017). Fabrication and characterization of micro dissolved oxygen sensor activated on demand using electrolysis. *Sensors and Actuators B: Chemical*, *241*, 923-930. https://doi.org/10.1016/j.snb.2016.10.145
64. Wang, Q., Zhang, J. M., & Li, S. (2019). Minreview: Recent advances in the development of gaseous and dissolved oxygen sensors. *Instrumentation Science & Technology*, *47*(1), 19-50. https://doi.org/10.1080/10739149.2018.1453835

65. Gillanders, R. N., Tedford, M. C., Crilly, P. J., & Bailey, R. T. (2005). A composite thin film optical sensor for dissolved oxygen in contaminated aqueous environments. *Analytica chimica acta, 545*(2), 189-194. https://doi.org/10.1016/j.aca.2005.04.086
66. Wijesooriya, H. E., Seneviratne, J. A., Jayathilaka, K. M. D. C., Fernando, W. T. R. S., Piyumal, P. L. A. K., Ranaweera, A. L. A. K.,... & Wijesundera, R. P. (2025). Cost-Effective Method for Dissolved Oxygen Sensing with Electrodeposited n-Cu2O Thin-Film Semiconductors. *Physchem, 5*(1), 6. https://doi.org/10.3390/physchem5010006
67. LeBaron, T. W., & Sharpe, R. (2022). ORP should not be used to estimate or compare concentrations of aqueous H2: An in silico analysis and narrative synopsis. *Frontiers in Food Science and Technology, 2*, 1007001. https://doi.org/10.3389/frfst.2022.1007001
68. Akbar, M. A., Sharif, O., Selvaganapathy, P. R., & Kruse, P. (2023). Identification and quantification of aqueous disinfectants using an array of carbon nanotube-based chemiresistors. *ACS Applied Engineering Materials, 1*(11), 3040-3052. https://doi.org/10.1021/acsaenm.3c00505
69. Aryasomayajula, A., Wojnas, C., Divigalpitiya, R., Selvaganapathy, P. R., & Kruse, P. (2018). Robust chemiresistive sensor for continuous monitoring of free chlorine using graphene-like carbon. *ACS sensors, 3*(2), 451-457. https://doi.org/10.1021/acssensors.7b00884
70. Duoc, P. N. D., Binh, N. H., Van Hau, T., Thanh, C. T., Van Trinh, P., Tuyen, N. V.,... & Van Chuc, N. (2020). A novel electrochemical sensor based on double-walled carbon nanotubes and graphene hybrid thin film for arsenic (V) detection. *Journal of Hazardous Materials, 400*, 123185. https://doi.org/10.1016/j.jhazmat.2020.123185
71. Vyas, T., Kumar, H., Choudhary, S., & Joshi, A. (2024). Carbon quantum dot (CQD)-dithizone-based thin-film chemical sensors for the specific detection of lead ions in water resources. *Environmental Science: Water Research & Technology, 10*(11), 2858-2868. https://doi.org/10.1039/D4EW00452C
72. Vyas, T., & Joshi, A. (2024). Carbon Quantum Dots (CQDs)-Diphenyl Carbazone (DPCO) loaded thin film sensors for fluorescent and colorimetric dual mode detection of mercury in various water resources. *Optical Materials, 154*, 115700. https://doi.org/10.1016/j.optmat.2024.115700
73. Chadha, U., Bhardwaj, P., Agarwal, R., Rawat, P., Agarwal, R., Gupta, I.,... & Chakravorty, A. (2022). Recent progress and growth in biosensors technology: A critical review. *Journal of Industrial and Engineering Chemistry, 109*, 21-51. https://doi.org/10.1016/j.jiec.2022.02.010
74. Andreescu, S., & Sadik, O. A. (2004). Trends and challenges in biochemical sensors for clinical and environmental monitoring. *Pure and applied chemistry, 76*(4), 861-878. https://doi.org/10.1351/pac200476040861
75. Wang, D., Liu, Z., Li, J., Tang, W., Huang, Y., Yu, J.,... & Ma, H. (2022). Thin-film transistor arrays for biological sensing systems. *Flexible and Printed Electronics, 7*(2), 023004. https://doi.org/10.1088/2058-8585/ac7928uns
76. Martínez-Saucedo, G., Ugalde-Reygadas, M., Peña, J. A., Lastra-Medina, G., Márquez-Marín, J., Torres-Delgado, G.,... & Chávez-Urbiola, I. R. (2023). Non-enzymatic glucose sensor using nanostructured copper oxide thin films deposited by spray pyrolysis. *Surfaces and Interfaces, 37*, 102702. https://doi.org/10.1016/j.surfin.2023.102702
77. Bano, M., Naikoo, G. A., BaOmar, F., Rather, J. A., Hassan, I. U., Sheikh, R. A.,... & Tambuwala, M. M. (2024). Revolutionizing glucose monitoring: Enzyme-free 2D-MoS2 nanostructures for ultra-sensitive glucose sensors with real-time health-monitoring capabilities. *ACS omega, 9*(18), 20021-20029. https://doi.org/10.1021/acsomega.3c10117
78. Singh, R., & Ayyub, M. M. (2021). Atomic layer deposition of crystalline β-NiS for superior sensing in thin-film non-enzymatic electrochemical glucose sensors. *ACS Applied Electronic Materials, 3*(4), 1912-1919. https://doi.org/10.1021/acsaelm.1c00145
79. Rinaldi, A. L., & Carballo, R. (2016). Impedimetric non-enzymatic glucose sensor based on nickel hydroxide thin film onto gold electrode. *Sensors and Actuators B: Chemical, 228*, 43-52. https://doi.org/10.1016/j.snb.2015.12.101
80. Dai, Y., Molazemhosseini, A., Abbasi, K., & Liu, C. C. (2018). A cuprous oxide thin film non-enzymatic glucose sensor using differential pulse voltammetry and other voltammetry methods and a comparison to different thin film electrodes on the detection of glucose in an alkaline solution. *Biosensors, 8*(1), 4. https://doi.org/10.3390/bios8010004

81. Pötzelberger, I., Mardare, A. I., & Hassel, A. W. (2017). Non-enzymatic glucose sensing on copper-nickel thin film alloy. *Applied Surface Science*, *417*, 48–53. https://doi.org/10.1016/j.apsusc.2016.12.193
82. Wang, Y., Wang, X., Lu, W., Yuan, Q., Zheng, Y., & Yao, B. (2019). A thin film polyethylene terephthalate (PET) electrochemical sensor for detection of glucose in sweat. *Talanta*, *198*, 86-92. https://doi.org/10.1016/j.talanta.2019.01.104
83. Sharma, A., AlGhamdi, W. S., Faber, H., Lin, Y. H., Liu, C. H., Hsu, E. K.,... & Anthopoulos, T. D. (2023). Non-invasive, ultrasensitive detection of glucose in saliva using metal oxide transistors. *Biosensors and Bioelectronics*, *237*, 115448. https://doi.org/10.1016/j.bios.2023.115448
84. Minamiki, T., Minami, T., Chen, Y. P., Mano, T., Takeda, Y., Fukuda, K., & Tokito, S. (2021). Flexible organic thin-film transistor immunosensor printed on a one-micron-thick film. *Communications Materials*, *2*(1), 8. https://doi.org/10.1038/s43246-020-00112-z
85. Aydın, E. B., Aydın, M., & Sezgintürk, M. K. (2017). A highly sensitive immunosensor based on ITO thin films covered by a new semi-conductive conjugated polymer for the determination of TNFα in human saliva and serum samples. *Biosensors and Bioelectronics*, *97*, 169-176. https://doi.org/10.1016/j.bios.2017.05.056
86. Kim, J., Jeong, S., Sarawut, S., Kim, H., Son, S. U., Lee, S.,... & Park, S. H. K. (2022). An immunosensor based on a high performance dual-gate oxide semiconductor thin-film transistor for rapid detection of SARS-CoV-2. *Lab on a Chip*, *22*(5), 899-907. https://doi.org/10.1039/D1LC01116B
87. Hwang, G. T., Byun, M., Jeong, C. K., & Lee, K. J. (2015). Flexible piezoelectric thin-film energy harvesters and nanosensors for biomedical applications. *Advanced healthcare materials*, *4*(5), 646-658. https://doi.org/10.1002/adhm.201400642
88. Lee, H. S., Chung, J., Hwang, G. T., Jeong, C. K., Jung, Y., Kwak, J. H.,... & Lee, K. J. (2014). Flexible inorganic piezoelectric acoustic nanosensors for biomimetic artificial hair cells. *Advanced Functional Materials*, *24*(44), 6914-6921. https://doi.org/10.1002/adfm.201402270
89. Hwang, G. T., Park, H., Lee, J. H., Oh, S., Park, K. I., Byun, M.,... & Lee, K. J. (2014). Self-powered cardiac pacemaker enabled by flexible single crystalline PMN-PT piezoelectric energy harvester. *Advanced Materials (Deerfield Beach, Fla.)*, *26*(28), 4880–4887. https://doi.org/10.1002/adma.201400562
90. Dominik, M., Leśniewski, A., Janczuk, M., Niedziółka-Jönsson, J., Hołdyński, M., Godlewski, M.,... & Śmietana, M. (2017). Titanium oxide thin films obtained with physical and chemical vapour deposition methods for optical biosensing purposes. *Biosensors and Bioelectronics*, *93*, 102-109. https://doi.org/10.1016/j.bios.2016.09.079
91. Handoyo, T., & Kondoh, J. (2020, May). Development of gold thin-film for optical-based biosensor. In *AIP Conference Proceedings* (Vol. 2230, No. 1, p. 020008). AIP Publishing LLC. https://doi.org/10.1063/5.0006687
92. Pires, N. M. M., Dong, T., Hanke, U., & Hoivik, N. (2014). Recent developments in optical detection technologies in lab-on-a-chip devices for biosensing applications. *Sensors*, *14*(8), 15458-15479. https://doi.org/10.3390/s140815458

Zaira Jocelyn Hernández Simón was born in Puebla, Mexico, in 1991. She obtained her bachelor's degree in mechanical and electrical engineering and subsequently completed her master's and doctoral degrees in Semiconductor Devices at the Autonomous University of Puebla. She is currently affiliated with Mexico's National Research System, a distinguished peer-reviewed recognition that symbolises the quality and prestige of scientific contributions. Her research interests include: The development of devices based on nanostructured materials obtained with cost-effective and scalable techniques (ultrasonic spray pyrolysis, spin coating, HFCVD, etc.) has applications in the energy sector, specifically sustainable solar photovoltaic energy. Technological development focused on optoelectronic applications (photodetectors and light-emitting devices).

Ferroelectric and multiferroic materials for application in the development of solid-state electronic devices.

Chapter 10
Porous Anodic Aluminum Oxide: A Versatile Template for the Fabrication of Metafilms with Tunable Plasmonic Response

R. González-Campuzano, D. E. Martínez-Lara, and D. Mendoza

10.1 Introduction

The PAAO is described as a vertical arrangement of Al oxide cells oriented perpendicular to the Al surface (Fig. 10.1). The cell array forms a minimum-energy arrangement known as hexagonal compact packing. Every cell includes a pore at its center, which is sealed at the base by a hemispherical anodic oxide barrier layer. The cell growth and formation mechanism has been described in detail in [1–3].

The structure is defined by various structural parameters, which include interpore distance (D_{int}), pore diameter (D_{p}), barrier layer (t_{b}), pore wall thickness (t_{w}), layer thickness (t_{p}), pore density (δ_{p}), and porosity (P). These features are known to be dependent on various anodizing conditions, including electrolyte type and concentration, anodizing potential, current density, and temperature, among others [2, 4].

PAAO templates synthesized via electrochemical anodization have gained significant popularity in recent years, serving as versatile molds for fabricating various nanoscale structures [4]. Recent research efforts have focused on expanding the range

R. González-Campuzano (✉) · D. E. Martínez-Lara (✉) · D. Mendoza
Instituto de Investigaciones en Materiales, Universidad Nacional Autónoma de México, Ciudad de México, México
e-mail: naedra@ciencias.unam.mx

D. E. Martínez-Lara
e-mail: davideduardo@ciencias.unam.mx

D. Mendoza
e-mail: doroteo@unam.mx

D. E. Martínez-Lara
Escuela Nacional Preparatoria No. 7 "Ezequiel A. Chávez", Universidad Nacional Autónoma de México, Ciudad de México, México

S. Sharma (ed.), *Multifunctional Heterostructure Thin Films Handbook*,
https://doi.org/10.1007/978-3-032-18692-8_10

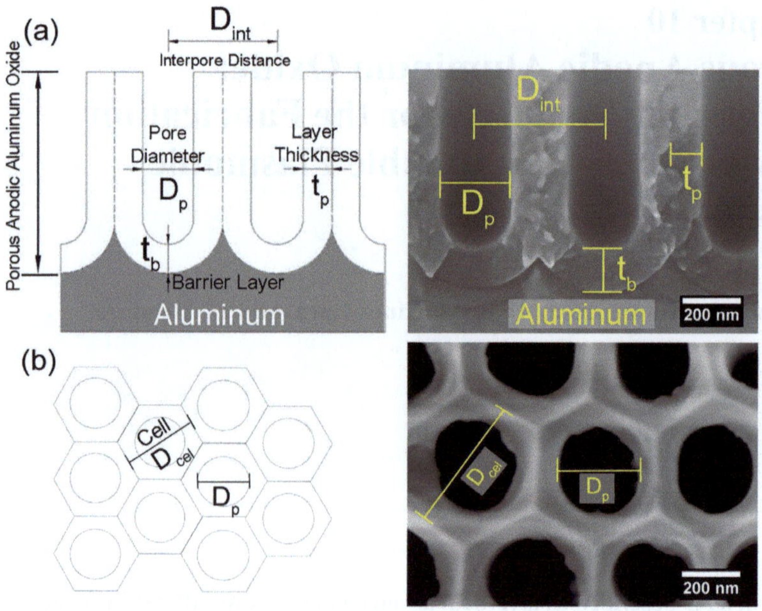

Fig. 10.1 Representative schematic diagrams (left) and real SEM micrographs (right) of a PAAO template. **a** Vertical cross-section and **b** top view illustrating all geometric parameters that characterize it

in D_{int} or increasing the D_{p} while minimizing changes to the order and organization of the nanopore arrangement. This is essential for specific applications that require a high degree of regularity and uniformity [5, 6]. The distribution of D_{p} and D_{int} can be adjusted according to various experimental parameters in the electrochemical anodizing process of Al. Moreover, research indicates that the thickness of the t_{b} determines both cell and pore size. Additionally, D_{p} and D_{int} exhibit a direct proportionality to the anodizing potential, while the concentration and type of electrolyte have minimal impact on D_{p}. Conventional anodizing processes (15–200 V) yield D_{int} within a limited range of 10–500 nm [7–11]. In contrast, employing high voltages (exceeding 400 V) enables D_{int} to be achieved between 884–2036 nm [12–14].

While most of the PAAO templates synthesized at high voltages (more than 200 V) demonstrate good reproducibility and reliability, specific problems persist: anodization processes generally require prolonged times (over 24 h) to achieve a highly ordered pore arrangement. Complex cooling systems are necessary to maintain low temperatures (below 0 °C) and ensure stable anodizing conditions. High voltages exceeding 200 V applied during anodizing generate heat due to intense electric fields, resulting in macroscopic burns or cracks on the surface of the PAAO because of the chemical reactions occurring under these conditions. Moreover, increased anodizing voltage leads to greater disorder in pore arrangements. The rapid and mechanically stable fabrication of PAAO films using high voltages presents a significant challenge.

Commonly used electrolytes for achieving D_{int} values between 50 and 600 nm with anodizing potentials up to 400 V include phosphoric acid (H_3PO_4) [15–18], malonic acid ($C_4H_6O_6$) [19–21], citric acid ($C_6H_8O_7$) [13, 14, 22], acetylenedicarboxylic acid ($C_4H_2O_4$) [23], selenic acid (H_2SeO_4) [24–28], and etidronic acid [29, 30], among others. Increasing the anodizing potential above 400 V requires the use of electrolytes consisting of acid and alcohol mixtures, which function as heat diffusers in the electrochemical cell [14, 31]. González-Campuzano et al. [32, 33] synthesized regular arrays of nanopores and nanodomes at different anodizing voltages, employing mixtures of acids such as oxalic acid ($C_2H_2O_4$), H_3PO_4, and citric acid ($C_6H_8O_7$) in combination with ethylene glycol (EG). Figure 10.2 indicates that the dimensions of these nanostructures (D_{int} and D_p) are dependent on the applied voltage, exhibiting an impressive linear correlation between size and anodization voltage. Electrochemical anodization offers a viable alternative to more expensive techniques like lithography, as it enables the achievement of specific dimensions through the selection of appropriate electrolytes and anodization voltages.

The potential applications of a PAAO template with a pore size modulation range over 600 nm are of importance in materials science and engineering. This interest stems in part from its exceptional physical and chemical properties, particularly its optical properties. The interaction of light with PAAO yields various optical properties significant to both research and practical applications. Leung et al. [34] found that appropriately modified PAAO templates, with pore dimensions of approximately 700 nm, perform effectively as light harvesters in the 700 nm wavelength range. Pore size modulation is crucial for specific applications.

Figure 10.3 illustrates a standard process for obtaining a PAAO template on the surface of Al. Figure 10.3a shows the Al surface post-cleaning and pretreatments; Fig. 10.3b illustrates the Al surface after electrochemical polishing. Figure 10.3c shows the upper surface of the PAAO template after the initial anodization, revealing

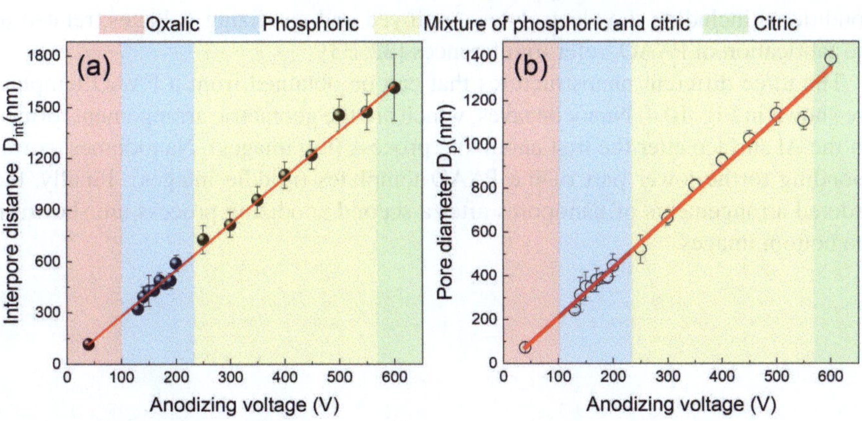

Fig. 10.2 Relationship between **a** interpore distance (D_{int}) and **b** pore diameter (D_p) as a function of anodizing voltage in regular arrays of nanopores for different acidic electrolytes

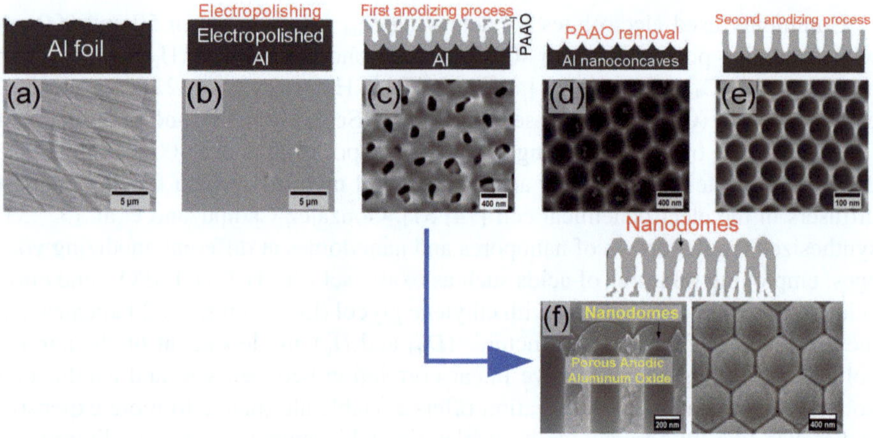

Fig. 10.3 Stages of the typical double anodizing process. The upper images show a schematic representation of the experimental procedure for the synthesis of PAAO templates, while the lower images show SEM micrographs of the real material obtained

an evident disordered arrangement of nanopores; Fig. 10.3d illustrates the Al surface after the selective removal of the PAAO. This metallic surface exhibits a pattern of nanoconcaves that function as nucleation sites for the development of a new PAAO template characterized by ordered pores. Figure 10.3e illustrates the PAAO template following a second anodization. Figure 10.3f displays the cross-section (left) and bottom (right) views of the PAAO template after the selective removal of Al, revealing the base of the pores arranged in structures known as nanodomes. The second anodizing process enhances the pore arrangement of PAAO templates throughout initial growth stages and reduces mechanical stresses or strains in the cell arrangement within a defined geometric space. For additional information about the conditions, including the electrolytes employed and anodizing voltages, related to the fabrication of PAAO, refer to references [32, 33].

The three different nanostructures that can be obtained from a PAAO template are shown in Fig. 10.4. Nanoconcaves, which are the geometric arrangement formed on the Al surface after the first anodizing process (top images). Nanodomes, corresponding to the lower part of the PAAO templates (middle images). Finally, the ordered arrangements of nanopores after a second anodizing process are shown in the bottom images.

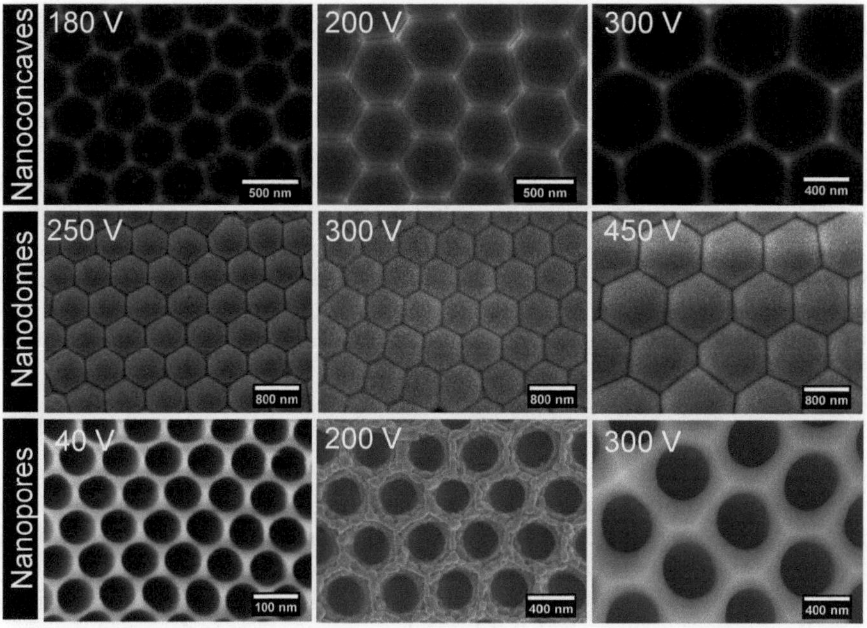

Fig. 10.4 SEM micrographs showing the highly arranged nanoconcaves, nanodomes, and nanopores obtained by the electrochemical anodization of Al

10.2 Plasmonic Response of PAAO Templates and Potential Applications

PAAO templates offer remarkable versatility for the fabrication of multifunctional nanostructured materials [35]. By depositing metals onto the surface of the nanopores array (see bottom images in Fig. 10.4), it is possible to generate nanohole systems with unique physical and optical properties that are highly valuable for technological applications. Unlike continuous metallic films, nanohole arrays exhibit phenomena such as quantum effects in electrical conductivity [36, 37], enhanced optical transmission [38, 39], and distinct plasmonic responses [40, 41].

Surface plasmons (SPs) are coherent oscillations of free electrons at the interface between a metal and a dielectric and can be supported when the following dispersion relation is satisfied [42, 43]:

$$K_{SPR} = K_0\sqrt{\frac{\varepsilon_c \varepsilon_d}{\varepsilon_c + \varepsilon_d}} \tag{10.1}$$

where K_{SPR} is the in-plane wave vector of the surface plasmon (SP), $K_0 = \omega/c$ is the wave vector in vacuum, and ε_d and ε_c are the dielectric functions of the dielectric and the conductor, respectively. The condition for SP excitation requires that the real parts of the dielectric functions have opposite signs, with the conducting material exhibiting a negative real component.

On flat surfaces, direct SP excitation is not possible due to the requirement of momentum conservation for the incident photon [44, 45]. However, this limitation can be overcome through the use of scattering centers, evanescent waves [43], or introducing periodic surface textures [42, 44]. Nanodomes and nanocavities provide the additional momentum required to fulfill the resonance condition.

Using an idealized theoretical model as a first approximation, surface plasmon resonances (SPRs) can also be related to the geometry and periodicity of the nanostructured surface. In the case of a two-dimensional hexagonal array, the SPR conditions are described by the following Equation [46]:

$$\frac{\varepsilon_m(\lambda)\varepsilon_d}{\varepsilon_m(\lambda)+\varepsilon_d} = \sin^2\theta + \frac{2}{\sqrt{3}}\frac{\lambda}{D_P}(2i+j)\sin\theta + \frac{4}{3}\frac{\lambda^2}{D_P^2}(i^2+j^2+ij) \tag{10.2}$$

where θ is the angle of incidence of the electromagnetic wave, i and j are the diffraction orders, and D_P is the period of the array, in our case, corresponding to interdome (D_D) or interconcavity (D_I) distances (see Fig. 10.5).

In Fig. 10.5, metafilms composed of Al, Ag, Bi, Pb, and Nb with hexagonal periodic geometries, deposited by thermal evaporation on both nanodomes and nanocaves, are shown. In these structures, the minimum observed in the reflectance spectra, associated with surface plasmon resonances (l_{SPR}) [42] (see Fig. 10.6), can be tuned to specific wavelengths by adjusting the D_D or D_I distances. This tunability follows the predictions of the ideal theoretical model described in Eq. (10.2). Based on this model, analytical SPR curves were calculated using experimental dielectric function values of Al [47], Ag [48], Bi [49, 50], Pb [51], and Nb [52]. These theoretical results were then compared with experimental data obtained for the reflectance

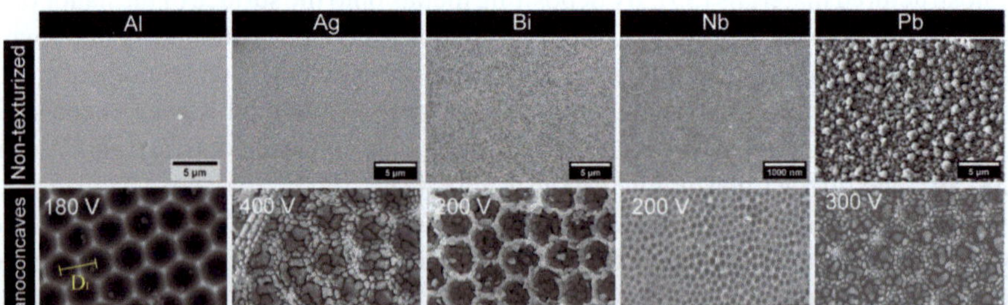

Fig. 10.5 SEM micrographs of untextured thin films deposited on glass substrates, and metafilms of Al, Ag, Bi, Nb, and Pb on nanoconcaves and nanodomes. The shown voltage corresponds to that applied during the substrate anodization process

spectra corresponding to metafilms (nanodomes and nanoconcaves) [53–57], as illustrated in Fig. 10.7.

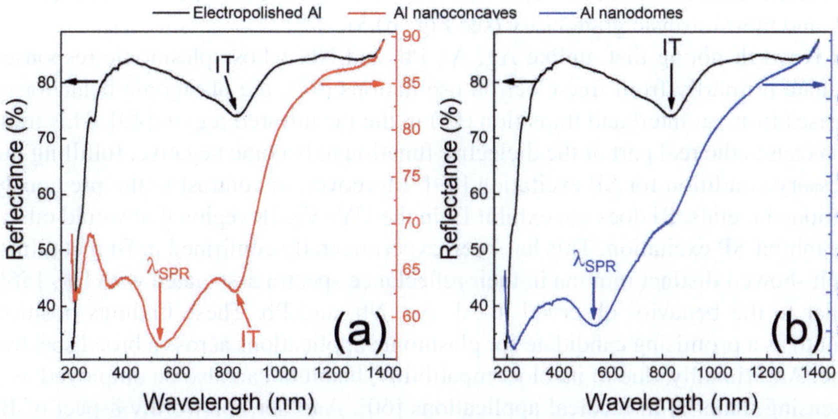

Fig. 10.6 Reflectance spectra of **a** Al nanoconcaves and **b** Al nanodomes, showing minima related to interband transition (IT) and surface plasmonic resonance (SPR). Results for electropolished Al foil are shown for comparison

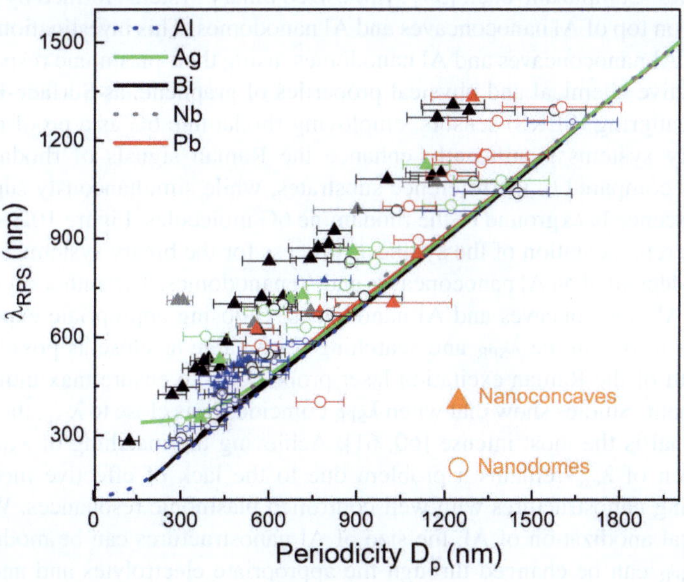

Fig. 10.7 Analytical curves obtained by means of Eq. (10.2), using experimental values of the dielectric functions of each metal [47–52]. The experimental points corresponding to the nanoconcaves (hollow circles) and nanodomes (filled triangles), for Al (blue), Ag (green), Bi (black), Nb (grey) and Pb (red)

As shown in Fig. 10.7, the analytical curves generally follow the overall trend of the experimental data. In most cases, the experimental points are redshifted toward longer wavelengths compared to the theoretical curves. This deviation is attributed to surface roughness-induced dispersion effects [58], the finite thickness of the films [59], and their intrinsic granularity (see Fig. 10.5).

It is worth noting that, unlike Ag, Al, Pb, and Nb, whose plasmonic responses originate primarily from free-electron oscillations [42], the plasmonic behavior of Bi arises from an interband transition (IT) in the far-infrared region [49]. This transition causes the real part of the dielectric function to become negative, fulfilling the necessary condition for SP excitation [49]. Moreover, in contrast to the previously mentioned metals, Bi does not exhibit IT in the UV–Vis-IR region that would otherwise inhibit SP excitation. This has been experimentally confirmed in Bi metafilms, which showed distinct minima in their reflectance spectra associated with l_{SPR} [56], similar to the behavior observed in Al, Ag, Nb, and Pb. These findings position bismuth as a promising candidate for plasmonic applications across a broad spectral range. Additionally, due to its biocompatibility, bismuth can also be employed as a biosensing material in medical applications [60]. Another noteworthy aspect of Bi is that, in thin film form, it exhibits a critical thickness at which quantum confinement effects emerge, leading to a semimetal-to-semiconductor transition [61]. This phenomenon opens new possibilities for the development of advanced optoelectronic devices.

González-Campuzano et al. [59] synthesized binary systems formed by graphene deposited on top of Al nanoconcaves and Al nanodomes. This investigation explores the use of Al nanoconcaves and Al nanodomes, using their plasmonic response with the distinctive chemical and physical properties of graphene, as Surface-Enhanced Raman Scattering (SERS) sensors, employing rhodamine 6G as a proof molecule. The binary systems significantly enhance the Raman signals of rhodamine 6G molecules compared to the reference substrates, while simultaneously suppressing the fluorescence background of the rhodamine 6G molecules. Figure 10.8 shows the schematic representation of the synthesis process for the binary systems formed by graphene deposited on Al nanoconcaves and Al nanodomes. The authors synthesized arrays of Al nanoconcaves and Al nanodomes, choosing appropriate experimental conditions to obtain the λ_{SPR} and searching for a value as close as possible to the wavelength of the Raman excitation laser probe (λ_{exc}) to ensure maximum Raman enhancement. Studies show that when λ_{SPR} coincides or is close to λ_{exc}, the obtained SERS signal is the most intense [60, 61]. Achieving the matching of λ_{SPR} around the position of λ_{exc} remains a problem due to the lack of effective methods for synthesizing nanostructures with well-controlled plasmonic resonances. With electrochemical anodization of Al, the size of Al nanostructures can be modified, and in turn λ_{SPR} can be changed through the appropriate electrolytes and anodization voltages.

Figure 10.9 a, b show Raman spectra of rhodamine 6G molecules in solution ($1X10^{-6}$ mol/L) detected in graphene on Al nanoconcaves and Al nanodomes, respectively, as well as on substrates used as references (electropolished Al and SiO_2(306 nm)/Si. For the case of rhodamine 6G molecules deposited on graphene

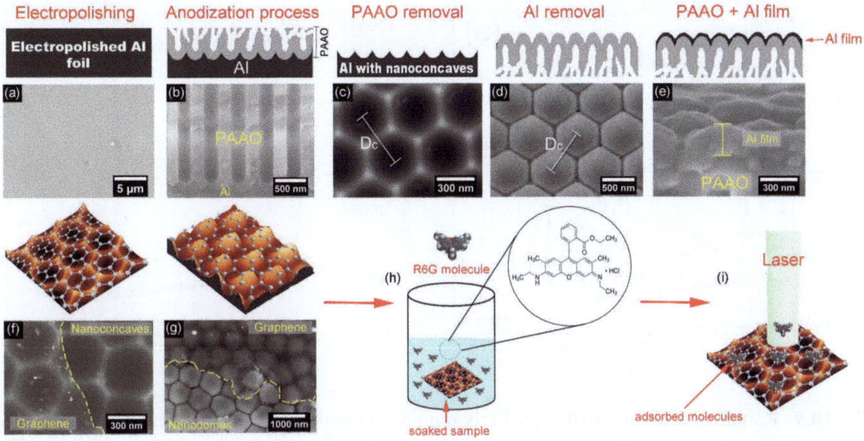

Fig. 10.8 Schematic representation of design of binary systems formed by graphene on Al nanoconcaves and graphene on Al nanodomes. **a** Electropolished Al foil surface. **b** PAAO after anodization. **c** Al nanoconcaves on top surface of Al foil after removal PAAO. **d** Nanodomes on top surface of PAAO after removal Al foil. **e** Nanodomes after deposition of Al film. **f**, **g** Top surface of Al nanoconcaves and Al nanodomes after deposition of graphene, respectively. **h**, **i** Schematic representation of the sample preparation for the Raman measurements

Al nanoconcaves (see Fig. 10.9a), the enhancement factors of the band located at 1647 cm^{-1} (peak labeled 9) was ~14-fold with respect to graphene on electropolished Al and ~5-fold with respect to graphene on SiO_2 (306 nm)/Si; while for rhodamine 6G molecules on graphene on Al nanodomes (see Fig. 10.9b), the enhancement factors of the band located at 1570 cm^{-1} was ~7-fold with respect to graphene on electropolished Al, and ~13-fold with respect to graphene on SiO_2 (306 nm)/Si.

Using the plasmonic response of Al nanoconcaves and Al nanodomes, it is possible to design SERS substrates formed by the combination of graphene and metal nanostructures called graphene-mediated SERS substrates (G-SERS substrates) with superior characteristics compared to other traditional SERS substrates.

10.3 Conclusion and Outlook

We have demonstrated the remarkable versatility of a simple method for producing substrates that enable the synthesis of metafilms, thereby tuning their plasmonic resonances across a broad wavelength range spanning the UV–Vis-IR spectrum. The dimensions of the substrate structures depend mainly on the type of electrolyte and anodization voltage. The most stringent condition for this technique to have a narrow plasmonic resonance is to have a long-range pore order, which is obtained using high-purity aluminum, including annealing processes of the starting material and more than one anodizing stage.

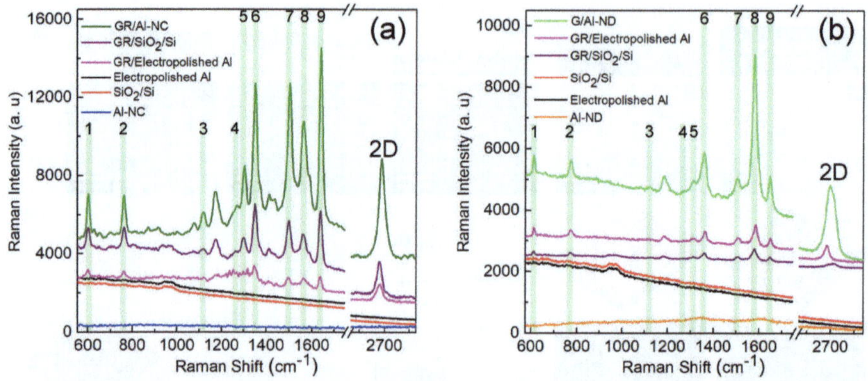

Fig. 10.9 Raman spectra of $1X10^{-6}$ mol/l rhodamine 6G molecules (peaks labeled 1–9) deposited on top of **a** graphene on Al nanoconcaves (GR/Al-NC) and **b** graphene on Al nanodomes (GR/Al-ND). Graphene (GR) spectra on electropolished Al and on SiO_2 (306 nm)/Si are shown for comparison

The use of PAAO templates may be extended to produce a great variety of metamaterials, for example, nanowires (filling the pores with other material), besides the metafilm deposited on the nanoconcaves and nanodomes. Depending on the physical property to explore, the limitation of this approach is to have a lower dimension of the pores; which is restricted to a minimum dimension of the order of 10 nm using sulfuric acid [7]. To have nanostructures with lower dimensions, more sophisticated lithographic techniques must be used.

Acknowledgements Omar Novelo, Josué Esaú, and Lourdes Bazán from the University Laboratory of Electron Microscopy (LUME-UNAM) for their assistance with the SEM imaging. Miguel Ángel Canseco for his support with the UV-Vis-IR spectrophotometer, Carlos Ramos for his help with vacuum equipment, and Caín González for his support with computational equipment, all of them affiliated with the IIM-UNAM. We are also grateful to Prof. J. M. Saniger from ICAT-UNAM and the University Laboratory of Spectroscopic Characterization (LUCE-ICAT-UNAM) for providing access to the facilities for Raman measurements. R. González-Campuzano would like to recognize SECIHTI for its support with the scholarship "Estancias posdoctorales por México 2023 (1)."

References

1. Abdel-Karim, R., & El-Raghy, S. M. (2016). Fabrication of nanoporous alumina. In *Nanofabrication using nanomaterials*. One Central Press (OCP).
2. Lee, W., & Park, S. (2014). Porous anodic aluminum oxide: Anodization and templated synthesis of functional nanostructures. *Chemical Reviews, 114*(13), 7487–7556. https://doi.org/10.1021/cr500002z
3. Runge, J. M. (2018). A brief history of anodizing aluminum. In *The metallurgy of anodizing aluminum: Connecting science to practice*. Springer International Publishing. https://doi.org/10.1007/978-3-319-72177-4_2

4. Sulka, G. D. (2008). Highly ordered anodic porous alumina formation by self-organized anodizing. In *Nanostructured materials in electrochemistry* (Vol. 1, pp. 1–116). Wiley-VCH. https://doi.org/10.1002/9783527621507.ch1
5. Masuda, H. (2005). Highly ordered nanohole arrays in anodic porous alumina. In *Ordered porous nanostructures and applications*. Springer International Publishing. https://doi.org/10.1007/0-387-25193-6_3
6. Wehrspohn, R. B. (2005). *Ordered porous nanostructures and applications*. Springer International Publishing. https://doi.org/10.1007/b106900
7. O'Sullivan, J. P., & Wood, G. C. (1970). The morphology and mechanism of formation of porous anodic films on aluminium. *Proceedings of the Royal Society of London. Series A, Mathematical and Physical Sciences, 317*(1531), 511–543. https://doi.org/10.1098/rspa.1970.0129
8. Li, A. P., Müller, F., Birner, A., Nielsch, K., & Gösele, U. (1998). Hexagonal pore arrays with a 50–420 nm interpore distance formed by self-organization in anodic alumina. *Journal of Applied Physics, 84*(11), 6023–6026. https://doi.org/10.1063/1.368911
9. Jessensky, O., Müller, F., & Gösele, U. (1998). Self-organized formation of hexagonal pore arrays in anodic alumina. *Applied Physics Letters, 72*(10), 1173–1175. https://doi.org/10.1063/1.121004
10. Nielsch, K., Choi, J., Schwirn, K., Wehrspohn, R. B., & Gösele, U. (2002). Self-ordering regimes of porous alumina: The 10 porosity rule. *Nano Letters, 2*(7), 677–680. https://doi.org/10.1021/nl025537k
11. Zaraska, L., Stępniowski, W. J., Jaskuła, M., & Sulka, G. D. (2014). Analysis of nanopore arrangement of porous alumina layers formed by anodizing in oxalic acid at relatively high temperatures. *Applied Surface Science, 305*, 650–657. https://doi.org/10.1016/j.apsusc.2014.03.154
12. Chen, X., Yu, D., Cao, L., Zhu, X., Song, Y., Huang, H., Lu, L., & Chen, X. (2014). Fabrication of ordered porous anodic alumina with ultra-large interpore distances using ultrahigh voltages. *Materials Research Bulletin, 57*, 116–120. https://doi.org/10.1016/j.materresbull.2014.05.037
13. Ma, Y., Wen, Y., Li, J., Li, Y., Zhang, Z., Feng, C., & Sun, R. (2016). Fabrication of self-ordered alumina films with large interpore distance by Janus anodization in citric acid. *Scientific Reports, 6*, 39165. https://doi.org/10.1038/srep39165
14. Wang, Q., Long, Y., & Sun, B. (2013). Fabrication of highly ordered porous anodic alumina membrane with ultra-large pore intervals in ethylene glycol-modified citric acid solution. *Journal of Porous Materials, 20*(3), 785–788. https://doi.org/10.1007/s10934-012-9653-3
15. Li, Y., Zheng, M., Ma, L., & Shen, W. (2006). Fabrication of highly ordered nanoporous alumina films by stable high-field anodization. *Nanotechnology, 17*(20), 5101. https://doi.org/10.1088/0957-4484/17/20/010
16. Masuda, H., Yada, K., & Osaka, A. (1998). Self-ordering of cell configuration of anodic porous alumina with large-size pores in phosphoric acid solution. *Japanese Journal of Applied Physics, 37*(11B), L1340–L1342. https://doi.org/10.1143/JJAP.37.L1340
17. Xu, Y., Liu, H., Li, X., Kang, W., Cheng, B., & Li, X. (2015). A novel method for fabricating self-ordered porous anodic alumina with wide interpore distance using phosphoric/oxalic acid mixed electrolyte. *Materials Letters, 151*, 79–81. https://doi.org/10.1016/j.matlet.2015.03.049
18. Zhang, R., Jiang, K., & Ding, G. (2010). Surface morphology control on porous anodic alumina in phosphoric acid. *Thin Solid Films, 518*(14), 3797–3800. https://doi.org/10.1016/j.tsf.2010.01.004
19. Knörnschild, G., Poznyak, A. A., Karoza, A. G., & Mozalev, A. (2015). Effect of the anodization conditions on the growth and volume expansion of porous alumina films in malonic acid electrolyte. *Surface and Coatings Technology, 275*, 17–25. https://doi.org/10.1016/j.surfcoat.2015.04.030
20. Santos, A., Alba, M., Rahman, M. M., Formentín, P., Ferré-Borrull, J., Pallarès, J., & Marsal, L. F. (2012). Structural tuning of photoluminescence in nanoporous anodic alumina by hard anodization in oxalic and malonic acids. *Nanoscale Research Letters, 7*, 228. https://doi.org/10.1186/1556-276X-7-228

21. Sun, B., Li, J., Jin, X., Zhou, C., Hao, Q., & Gao, X. (2013). Self-ordered hard anodization in malonic acid and its application in tailoring alumina taper-nanopores with continuously tunable periods in the range of 290–490 nm. *Electrochimica Acta, 112*, 327-332. https://doi.org/10.1016/j.electacta.2013.08.147
22. Kikuchi, T., Nishinaga, O., S., Natsui, & Suzuki, R. O. (2014). Fabrication of anodic nanoporous alumina via acetylenedicarboxylic acid anodizing. *ECS Electrochemistry Letters, 3*, C25. https://doi.org/10.1149/2.0071407eel
23. Akiya, S., Kikuchi, T., Natsui, S., & Suzuki, R. O. (2015). Optimum exploration for the self-ordering of anodic porous alumina formed via selenic acid anodizing. *Journal of The Electrochemical Society, 162*, E244. https://doi.org/10.1149/2.0391510jes
24. Kikuchi, T., Nishinaga, O., Natsui, S., & Suzuki, R. O. (2014). Self-ordering behavior of anodic porous alumina via selenic acid anodizing. *Electrochimica Acta, 137*, 728-735. https://doi.org/10.1016/j.electacta.2014.06.078
25. Nazarkina, Y., Gavrilov, S., Terryn, H., Petrova, M., & Ustarroz, J. (2015). Investigation of the ordering of porous anodic alumina formed by anodization of aluminum in selenic acid. *Journal of The Electrochemical Society, 16*, E166. https://doi.org/10.1149/2.0571509jes
26. Nazarkina, Y., Kamnev, K., Dronov, A., Dudin, A., Pavlov, A., & Gavrilov, S. (2017). Features of porous anodic alumina growth in galvanostatic regime in selenic acid based electrolyte. *Electrochimica Acta, 231*, 327-335. https://doi.org/10.1016/j.electacta.2017.02.049
27. Nishinaga, O., Kikuchi, T., Natsui, S., & Suzuki, R. O. (2013). Rapid fabrication of self-ordered porous alumina with 10-/sub-10-nm-scale nanostructures by selenic acid anodizing. *Scientific Reports, 3*, 2748. https://doi.org/10.1038/srep02748
28. Kikuchi, T., Nishinaga, O., Natsui, S., & Suzuki, R. O. (2015). Fabrication of self-ordered porous alumina via etidronic acid anodizing and structural color generation from submicrometer-scale dimple array. *Electrochimica Acta, 156*, 235-243. https://doi.org/10.1016/j.electacta.2014.12.171
29. Kikuchi, T., Takenaga, A., Natsui, S., & Suzuki, R. O. (2017). Advanced hard anodic alumina coatings via etidronic acid anodizing. *Surface and Coatings Technology, 326*, 72-78. https://doi.org/10.1016/j.surfcoat.2017.07.043
30. González-Campuzano, R., Saniger, J. M., & Mendoza, D. (2017). Plasmonic resonances in hybrid systems of aluminum nanostructured arrays and few layer graphene within the UV–IR spectral range. *Nanotechnology, 28*, 465704. https://doi.org/10.1088/1361-6528/aa8ce4
31. González-Campuzano, R., Mata-Zamora, M. E., López-Romero, S., & Mendoza, D. (2018). Excitation of plasmonic resonances within UV-Vis wavelength range using low-purity aluminum nanoconcave arrays. *Applied Physics Letters, 113*, 221604. https://doi.org/10.1063/1.5059556
32. Leung, S., Yu, M., Q., Lin, Kwon, K., Ching, K., Gu, L., Yu, K., & Fan, Z. (2012). Efficient photon capturing with ordered three-dimensional nanowell arrays. *Nano Letters, 12*, 3682–3689. https://doi.org/10.1021/nl3014567
33. Sousa, C. T., Leitao, D. C., Proenca, M. P., Ventura, J., Pereira, A. M., & Araujo, J. P. (2014). Nanoporous alumina as templates for multifunctional applications. *Applied Physics Reviews, 1*, 031102. https://doi.org/10.1063/1.4893546
34. Nakanishi, T., & Ando, T. (1996). Quantum interference effects in antidot lattices in magnetic fields. *Physical Review B, 54*, 8021. https://doi.org/10.1103/PhysRevB.54.8021
35. Uryu, S., & Ando, T. (1998). Numerical study of localization in antidot lattices. *Physical Review B, 58*, 10583. https://doi.org/10.1103/PhysRevB.58.10583
36. Ebbesen, T. W., Lezec, H. J., Ghaemi, H. F., Thio, T., & Wolff, P. A. (1998). Extraordinary optical transmission through sub-wavelength hole arrays. *Nature, 391*, 667–669. https://doi.org/10.1038/35570
37. Ruan, Z., & Qiu, M. (2006). Enhanced transmission through periodic arrays of subwavelength holes: The role of localized waveguide resonances. *Physical Review Letters, 96*, 233901. https://doi.org/10.1103/PhysRevLett.96.233901
38. Schmidt, T. M., Bochenkov, V. E., Espinoza, J. A., Smits, E. C. P., Muzafarov, A. M., Kononevich, Y. N., & Sutherland, D. S. (2014). Plasmonic fluorescence enhancement of

DBMBF2 monomers and DBMBF2–toluene exciplexes using Al-hole arrays. *The Journal of Physical Chemistry C, 118*, 2138–2145. https://doi.org/10.1021/jp4110823
39. Schwind, M., Kasemo, B., & Zorić, I. (2014). Localized and propagating plasmons in metal films with nanoholes. *Nano Letters, 13*, 743–1750. https://doi.org/10.1021/nl400328x
40. Maier, S. A. (2006). Plasmonics: Fundamentals and applications. Springer-Verlag.
41. Sun, G., & Zheng, J. (2005). Investigation of some electromagnetic modes in the metal-dielectric system by the absorption spectrum. *Science and Technology of Advanced Materials, 6*, 848–854 https://doi.org/10.1016/j.stam.2005.05.012
42. Raether, H. (1988). Surface plasmons on smooth and rough surfaces and on gratings. Springer-Verlag.
43. Khurgin, J. B., & Boltasseva, A. (2012). Reflecting upon the losses in plasmonics and metamaterials. *MRS Bulletin, 37*, 768–779. https://doi.org/10.1557/mrs.2012.173
44. Papaioannou, E. T., Kapaklis, V., Melander, E., Hjörvarsson, B., Pappas, S. D., Patoka, P., Giersig, M., Fumagalli, P., Garcia-Martin, A., & Ctistis, G. (2007). Surface plasmons and magneto-optic activity in hexagonal Ni anti-dot arrays. *Optics Express, 19*, 23867. https://doi.org/10.1364/OE.19.023867
45. Palik, E. D. (1985). Handbook of optical constants of solids (Vol. 2). Academic Press.
46. González-Campuzano, R., & Mendoza, D. (2019). Excitation of plasmons in self-ordered arrays of aluminum and silver nanoconcaves within UV–IR spectral range. *Journal of Physics: Conference Series, 1221*, 012001. https://doi.org/10.1088/1742-6596/1221/1/012001
47. Toudert, J., Serna, R., Camps, I., Wojcik, J., Mascher, P., Rebollar, E., & Ezquerra, T. A. (2017). Unveiling the far infrared-to-ultraviolet optical properties of bismuth for applications in plasmonics and nanophotonics. *The Journal of Physical Chemistry C, 121*, 3511–3521. https://doi.org/10.1021/acs.jpcc.6b10331
48. Werner, W. S., Glantschnig, K., & Ambrosch-Draxl, C. (2009). Optical constants and inelastic electron-scattering data for 17 elemental metals. *Journal of Physical and Chemical Reference Data, 38*, 1013. https://doi.org/10.1063/1.3243762
49. Mathewson, A. G., & Myers, H. P. (1971). Absolute values of the optical constants of some pure metals. *Physica Scripta, 4*, 291. https://doi.org/10.1088/0031-8949/4/6/009
50. Windt, D. L., Cash, W. C., Scott, M., Prendt, P., Newnam, B., Fisher, R. F., & Swartzlander, A. B. (1988). Optical constants for thin films of Ti, Zr, Nb, Mo, Ru, Rh, Pd, Ag, Hf, Ta, W, Re, Ir, Os, Pt, and Au from 24 A to 1216 A. *Applied Optics, 27*, 246–278. https://doi.org/10.1364/AO.27.000246
51. Martínez-Lara, D. E., González-Campuzano, R., Benítez, J. L., & Mendoza, D. (2021). Localized overheating on aluminum metafilms mediated by surface plasmons. *Revista Mexicana de Física, 67*, 1–6. https://doi.org/10.31349/revmexfis.67.051001
52. González-Campuzano, R., Martínez-Lara, D. E., & Mendoza, D. (2020). Lead plasmonics on texturized substrates: Pb metafilms. *Applied Physics Letters, 117*, 031603. https://doi.org/10.1063/5.0016131
53. Martínez-Lara, D. E., González-Campuzano, R., & Mendoza, D. (2022). Bismuth plasmonics in the visible spectrum using texturized films. *Photonics and Nanostructures - Fundamentals and Applications, 52*, 101058. https://doi.org/10.1016/j.photonics.2022.101058
54. Martínez Lara, D. E. (2024). Resonancias plasmónicas y resistividad eléctrica de metapelículas de Bi; búsqueda del grosor crítico para el confinamiento cuántico en películas delgadas (Tesis doctoral). *Universidad Nacional Autónoma de México.*
55. Barnes, W. L., Dereux, A., & Ebbesen, T. W. (2003). Surface plasmon subwavelength optics. *Nature, 424*, 824–830. https://doi.org/10.1038/nature01937
56. Golovashkin, A. I., & Motulevich, G. P. (2005). Investigation of some electromagnetic modes in the metal-dielectric system by the absorption spectrum. *Science and Technology of Advanced Materials, 6*, 848–854. https://doi.org/10.1016/j.stam.2005.05.012
57. Beliaev, L. Y., Lavrinenko, A. V., & Takayama, O. (2024). Alternative plasmonic materials for biochemical sensing: A review. *Progress In Electromagnetics Research, 180*, 25-53. https://doi.org/10.2528/PIER24080104

58. Martínez-Lara, D. E., González-Campuzano, R., & Mendoza, D. (2025). Quantum confinement effect in thin bismuth films: In search of the critical thickness for the semimetal–semiconductor transition. *Thin Solid Films, 821*, 140678. https://doi.org/10.1016/j.tsf.2025.140678
59. González-Campuzano, R., Sato-Berrú, R. Y., & Mendoza, D. (2021). Surface-enhanced Raman scattering effect in binary systems formed by graphene on aluminum plasmonic nanostructures. *Nano Express, 2*, 010029. https://doi.org/10.1088/2632-959X/abe991
60. González-Campuzano, R., Mata-Zamora, M. E., López-Romero, S., & Mendoza, D. (2018). Excitation of plasmonic resonances within UV–vis wavelength range using low-purity aluminum nanoconcave arrays. *Applied Physics Letters, 113*, 221604. https://doi.org/10.1063/1.5059556
61. Kahraman, M., Daggumati, P., Kurtulus, O., Seker, E., & Wachsmann-Hogiu, S. (2013). Fabrication and characterization of flexible and tunable plasmonic nanostructures. *Scientific Reports, 3*, 3396. https://doi.org/10.1038/srep03396

R. González-Campuzano is a Doctor in Science and Engineering of Materials from the Instituto de Investigaciones en Materiales (IIM) at the Universidad Nacional Autónoma de México (UNAM). His main research lines focus on the synthesis and the study of optical and electronic properties of photonic nanostructures, two-dimensional materials such as graphene and related systems, as well as thin films for environmental remediation, particularly in the photocatalytic degradation of pollutant molecules. He is currently a postdoctoral researcher at the IIM and teaches physics and mathematics courses at the Facultad de Ciencias and the Escuela Nacional de Ciencias de la Tierra (ENCiT) of the UNAM.

D. E. Martínez-Lara is a Doctor in Science and Engineering of Materials from the Instituto de Investigaciones en Materiales (IIM) at the Universidad Nacional Autónoma de México (UNAM). His research focuses on plasmonic, thin-film nanostructures and quantum confinement effects in bismuth and related materials. He has also developed Ag–Cu nanocoatings (SakCu®) with antimicrobial properties for medical devices. He is currently a full-time professor at the Escuela Nacional Preparatoria No. 7 "Ezequiel A. Chávez," UNAM, where he integrates experimental nanoscience into high-school education through innovative pedagogical projects. His work combines scientific research, hands-on instrumentation, and science communication to bridge the gap between advanced materials research and public understanding of science.

D. Mendoza is Doctor in physics from the Universidad Nacional Autónoma de México (UNAM), Mexico City, Mexico, 1989. He is currently a Researcher with the Instituto de Investigaciones en Materiales, UNAM, where he is involved in the synthesis and the study of physical properties of different carbon structures (amorphous, fullerenes, nanotubes, graphene, and related materials); also, he works in the study of optical and electronic properties of photonic nanostructures based on graphene, C60, aluminum, silver, and bismuth. He teaches Solid State Physics courses at Facultad de Ciencias UNAM.

Chapter 11
Integration of Heterostructure Thin Films in Flexible and Wearable Devices and Sensors

Anjali Varshney and Sunil Chauhan

11.1 Introduction

Wearable devices or sensors have been an area of scientific interest for the past forty years, but recently, research in this field has grown rapidly. This increase is mainly due to technological advancements that have helped overcome challenges related to materials, device construction, power sources, and how information is stored and processed. As illustrated in Fig. 11.1, wearable sensors can be grouped based on several key factors, and one of the most important is how the sensor is worn on the body. The word "wearable" is generally used in two different ways. In the first sense, it refers to common commercial devices like smartwatches and pedometers. These devices are based on standard semiconductor technology and have benefited from the miniaturization trends in the electronics industry. While these devices are wearable, they are usually rigid and more accurately described as portable rather than fully wearable.

In the second usage, the term "wearable" refers to devices that are more seamlessly integrated with the human body or clothing. These sensors are designed to be more comfortable, flexible, and less noticeable, making them more truly wearable in everyday life. This type of wearable sensor is currently being actively researched for health monitoring and environmental sensing. Such body-integrated or fabric-embedded devices or sensors can be further divided into two categories. One category includes conformal sensors, which stay in close contact with the skin and can move with the body without causing discomfort. The other includes non-conformal sensors, which are not directly attached to the skin but are still placed in clothing

A. Varshney · S. Chauhan (✉)
Department of Physics & Environmental Sciences, Sharda School of Engineering & Science, Sharda University, Greater Noida, Uttar Pradesh, India
e-mail: sunil.chauhan@sharda.ac.in

Centre for Solar Cell and Renewable Energy, Department of Physics & Environmental Sciences, Sharda School of Engineering & Science, Sharda University, Greater Noida, Uttar Pradesh, India

S. Sharma (ed.), *Multifunctional Heterostructure Thin Films Handbook*,
https://doi.org/10.1007/978-3-032-18692-8_11

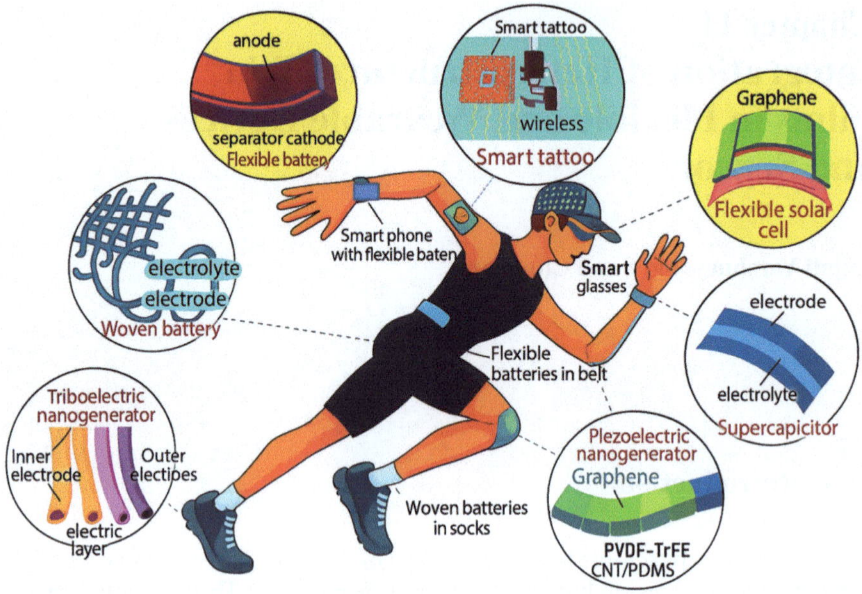

Fig. 11.1 Diagram illustrating various sources of wearable energy harvesting and their corresponding storage devices [1]

or accessories. Even for non-conformal sensors, flexibility and comfort are essential for successful use in wearable applications.

Apart from how a wearable sensor fits on the body, two other important things to consider are power and data handling. Some wearable devices can generate their own power using energy-harvesting methods, some come with built-in batteries, and others may need to be plugged into an external power source during use. When it comes to data, the key points include how data is stored, processed, and analyzed. Some sensors have built-in features to handle all of these tasks, while others send the data to a smartphone or computer, which processes and displays the information.

Wearable sensors can also be grouped by what they measure—for example, whether they detect physical or chemical signals and work actively or passively. Also, mobile phones, in-ear monitors or earbuds, smart bands or watches, rings, glasses, and even smart tattoos (Fig. 11.1) allow people to interact with the outside world and receive individualized information [1–8]. These wearable smart gadgets have several uses, but adhering to strict size restrictions presents a significant problem for their power supply components. As mentioned earlier, if the sensor is going to be truly wearable—meaning it blends in with the skin or clothing—it must be flexible and ideally unnoticeable to the person wearing it. This feature of wearability is the main focus of this paper. Here, we explore the types of materials needed to make wearable and flexible sensors. We don't limit our discussion to specific device or sensor types, configurations, power systems, or data setups. Instead, we focus on how different materials can be combined to make these sensors practical and comfortable to wear.

There are still many challenges to solve before wearable sensors can become common in everyday life. One of the most significant challenges is managing the power supply, data storage, data transfer, and real-time analysis. Many research papers have already explored these topics in detail, covering things like how sensors or devices work, how to make them stretchable, what problems still exist, and how to put all the parts together into a system. Several strategies have been tried to make devices/sensors both wearable and reliable, and many exciting breakthroughs have come from using new materials, creative designs, and smart combinations of properties to track human activity and health.

This chapter primarily emphasizes the stretchable designs, working principles, and specific material choices of sensors and devices, providing a comprehensive understanding beyond the basic types. Other research has already covered these areas. Instead, we focus on the need to combine many different types of materials, each with its thermal, mechanical, and electrical characteristics. This kind of material integration is necessary to build truly wearable device or sensors. In Part 2, we discuss the different methods being developed to combine these materials cost-effectively and reliably to create flexible and long-lasting wearable sensors. In Part 3, we present a brief overview of the wide range of materials needed with some of their applications. We don't list every possible material, but we provide enough examples to show the diversity of material properties and sensor designs used today (Fig. 11.2).

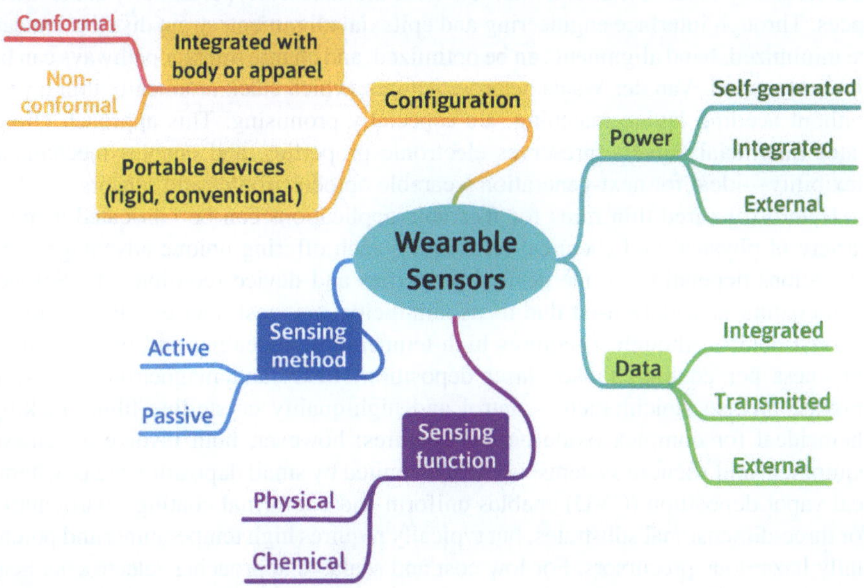

Fig. 11.2 Illustrates a conceptual classification of wearable sensors or devices. The sensor's body structure and the methodologies implemented for power and data pertain directly to wearability concerns. The sensing function and methodology are relevant to all sensors

11.2 Fundamentals of Heterostructured Thin Films for Wearable Devices

Heterostructured thin films are advanced material systems composed of two or more distinct layers with tailored physical, electrical, optical, or mechanical properties. These films are engineered at the nanoscale and stacked in a precise sequence to achieve synergistic properties not found in the individual components alone. In the context of wearable electronics, such thin films offer a powerful platform for multifunctional integration, allowing devices to remain lightweight, flexible, and mechanically durable while maintaining high performance in real-world environments. At the core of these heterostructures lies the ability to form well-defined interfaces between dissimilar materials—ranging from metals, semiconductors, dielectrics, polymers, and even emerging 2D materials like graphene or molybdenum disulfide (MoS_2). Each material layer contributes a specific function, such as conductivity, sensing, energy harvesting, or mechanical compliance. For instance, by integrating a rigid but piezoelectric thin film (e.g., ZnO or $BaTiO_3$) with a soft elastomeric substrate (e.g., PDMS), the composite structure can bend and conform to human skin while generating an electrical signal in response to mechanical stress—critical for wearable strain or pressure sensors.

A significant advantage of using heterostructured thin films in wearable devices is their ability to tailor material interactions at the nanoscale, particularly at the interfaces. Through interface engineering and epitaxial alignment, strain distribution can be minimized, band alignment can be optimized, and charge transfer pathways can be precisely tuned. Van der Waals heterostructures, which stack atomically thin layers without needing lattice matching, are especially promising. This approach eliminates interfacial defects, preserves electronic properties, and enables mechanical flexibility—ideal for next-generation wearable optoelectronics and sensors.

Heterostructured thin films for wearable applications can be fabricated using a variety of physical and chemical techniques, each offering unique advantages and limitations depending on the desired properties and device requirements. Sol–gel spin coating is widely used due to its simplicity, low cost, and excellent compositional control, though it requires high-temperature annealing and offers limited thickness per coating. Pulsed laser deposition (PLD) and magnetron sputtering provide precise stoichiometry control and high-quality crystalline films, making them ideal for complex oxide heterostructures; however, both involve expensive equipment and vacuum systems, with PLD limited by small deposition areas. Chemical vapor deposition (CVD) enables uniform and conformal coatings, particularly for three-dimensional substrates, but typically requires high temperatures and potentially hazardous precursors. For low cost and scalable approaches, electrochemical deposition (ECD) and doctor blade/drop casting are attractive, especially for flexible substrates, though they may suffer from poor crystallinity, limited uniformity, or restricted substrate compatibility. Atomic layer deposition (ALD) offers atomic-scale thickness control and excellent interface quality, critical for high-performance heterostructures, yet it is slow and costly. Inkjet printing, on the other hand, allows mask-less, direct patterning of functional inks onto flexible or wearable substrates, minimizing material waste, though it often requires post-processing and provides lower resolution. Understanding these merits and demerits allows researchers to strategically select fabrication techniques that balance performance, scalability, and flexibility, tailoring heterostructured thin films to specific wearable device applications (Table 11.1). Chemical vapor deposition (CVD) is commonly used for growing large-area graphene and other 2D materials. Atomic layer deposition (ALD) offers angstrom-level thickness control, ideal for constructing ultra-thin oxide layers with excellent uniformity. Pulsed laser deposition (PLD), sputtering, and solution-based methods like spin coating or inkjet printing are also widely used, especially when integrating hybrid organic–inorganic components onto flexible substrates. These techniques allow for scalable and low-temperature processing, essential for wearable device fabrication.

The mechanical properties of heterostructured films are critical. By designing films with controlled thickness, incorporating neutral mechanical plane structures, and embedding ductile layers, these systems can tolerate high degrees of bending, twisting, and even stretching without performance degradation. For example, Kim et al. (2015) demonstrated a flexible ferroelectric heterostructure that maintained its functionality even when bent to a 2.5 mm radius, showcasing the resilience of these films in wearable applications [9]. Functionally, heterostructured thin films serve multiple roles in wearable systems. They can be used for sensing (e.g., strain, pressure, temperature, or biochemical analytes), actuation (e.g., for haptic feedback), energy generation (e.g., piezoelectric or triboelectric nanogenerators), and energy storage (e.g., thin-film batteries or supercapacitors). The integration of multiple functionalities within a single-layered platform significantly reduces device size and weight while enhancing performance and user comfort [10–14]. In summary, heterostructured thin films are a cornerstone technology for the development of high performance, flexible, and wearable electronic systems. Their ability to integrate multiple material functions, conform to complex geometries, and operate reliably under mechanical stress makes them ideal for a wide range of wearable applications—from healthcare monitoring to smart textiles and human–machine interfaces. Ongoing advancements in material science, interfacial engineering, and deposition technologies will continue to expand the capabilities and practical adoption of these films in real-world wearable systems.

11.3 Materials for Wearable Sensors

Wearable sensors work based on the same fundamental principles as traditional measurement devices. Depending on what needs to be measured—such as temperature, pressure, chemical levels, or physical movement—different sensing mechanisms and materials are selected. These materials must be carefully combined to ensure the final sensor is not only accurate and functional but also soft, flexible, and durable enough for regular wear. The materials commonly used include

Table 11.1 Comparative heterostructured thin-film fabrication techniques suitable for wearable applications, including advantages and disadvantages

Process/technique	Principle/description	Merits (Advantages)	Demerits (Disadvantages)
Sol–gel spin coating	Chemical solution is deposited on substrate and spun at high speed to form a thin film	Simple, low cost; good compositional control; suitable for large-area deposition	Requires high-temperature annealing; limited film thickness per coating; solvent residues may affect film quality
Pulsed laser deposition (PLD)	High-energy laser ablates target material, which deposits as a thin film on substrate	Precise stoichiometry control; high-quality crystalline films; suitable for complex oxides	Expensive equipment; small deposition area; requires vacuum system
Magnetron sputtering	Target material bombarded with ions; atoms sputter onto substrate forming a thin film	Uniform, dense films; good adhesion; compatible with large substrates	High vacuum required; stoichiometry control challenging for complex oxides; slow deposition rate
Chemical vapor deposition (CVD)	Chemical precursors react/decompose on substrate to form thin film	Excellent uniformity; conformal coating over complex shapes; high-quality crystalline films	High temperature needed; toxic/flammable precursors; complex and costly equipment
Electrochemical deposition (ECD)	Films grown via electrochemical reactions from solution onto conductive substrate	Low cost, simple setup; can coat large areas; thickness controlled by deposition time	Limited to conductive substrates; poor crystallinity unless annealed; composition control can be difficult
Atomic layer deposition (ALD)	Films grow layer-by-layer via sequential self-limiting chemical reactions	Atomic-scale thickness control; excellent uniformity on 3D structures; high-quality interfaces	Very slow deposition rate; high cost of precursors/equipment; complex process optimization
Doctor blade/drop casting	Paste or solution spread onto substrate and dried to form films	Very simple, low cost; suitable for flexible substrates; easy for lab-scale testing	Poor uniformity; low control over thickness/morphology; limited scalability
Inkjet printing	Functional inks deposited in patterned manner on substrate to form films	Mask-less, direct patterning; compatible with flexible substrates; minimal material waste	Limited ink viscosity and particle size; post-processing often needed; lower resolution than lithography

metals, silicon-based semiconductors, flexible polymers, smart textile materials, and two-dimensional (2D) materials like graphene.

11.3.1 Metals

Metals play a key role in any electronic sensor, mainly to create electrical connections and also to serve sensing functions themselves. For example, metals such as gold and bismuth are used to detect heavy metals in human sweat. In some wearable sensors, a technique called voltammetry is used that relies on three electrodes—working, reference, and counter electrodes—made from metals. Selecting the right metal is crucial for accurate and stable results. Gold is widely used for its effectiveness and safety. In one study, researchers developed a sensor using gold and bismuth to measure metals like lead, mercury, and cadmium in sweat, along with skin temperature. Other metals like titanium and platinum have been used for temperature detection, and materials like gallium indium tin (GaInSn) have been sandwiched between soft layers to create strain sensors. Silver-silver chloride is another common electrode material, though alternatives like copper are being explored due to their lower cost and outstanding conductivity. However, copper needs better durability and adhesion to be viable. Research is also ongoing on using very thin metal films placed in soft materials or shaped into patterns like waves or serpentine lines to make sensors that can flex with the body [2, 15–17].

For instance, Gao et al. [15] constructed a microsensor or device array utilizing gold (Au) and bismuth (Bi) working electrodes, a silver (Ag) reference electrode, an Au counter electrode, and chromium/gold (Cr/Au) microlines for temperature measurement (Fig. 11.3). The Au/Bi working electrodes simultaneously detect amounts of some elements, like Pb, Hg, Cu, Zn, and Cd, in human sweat. They established the independence of these signals from the presence of various heavy metals and calibrated the measurement to the skin temperature recorded by the Cr/Au sensor. The array can be calibrated utilizing off-body measures and thereafter worn as a wristband for on-body monitoring in real time [17–24].

11.3.2 Semiconductor Materials

Traditional semiconductor materials like silicon, which are usually rigid, can also be adapted for flexible electronics if processed in very thin layers. For example, researchers have designed silicon circuits that can stretch and fold by building them on removable base layers. Once the base is removed, the flexible circuits can be transferred to soft materials like PDMS. These sheets can then be used as flexible electronic devices. Thin silicon layers, called nanomembranes, have been incorporated into flexible sensor designs, even though silicon is inherently brittle. These

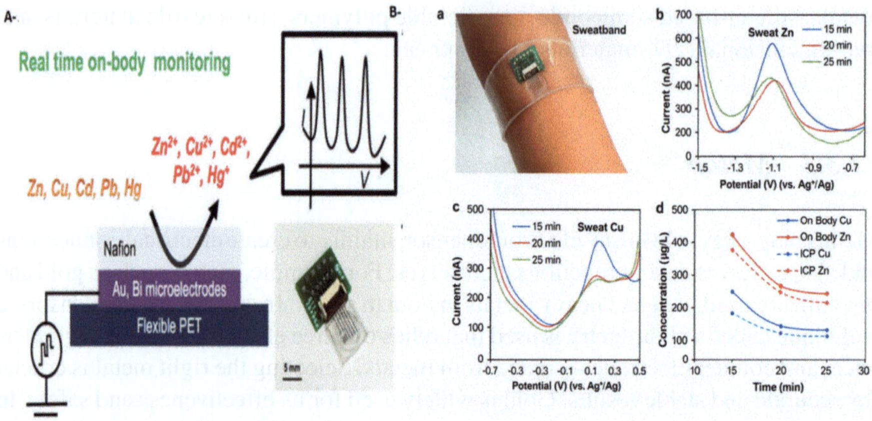

Fig. 11.3 **A** Schematic illustrating the process of metal ion deposition and stripping on microelectrodes. **B** Real-time trace metal detection on the human body using a multiplexed microsensor array during an indoor cycling exercise at a constant load of 150 W. **a** Photograph of the microsensor array embedded within a transparent wristband worn on the body. **b** Real-time analysis of zinc (Zn) levels during cycling, measured by a bismuth-based microsensor at various time points throughout the exercise. **c** a similar chart for sweat Cu, and **d** a chart comparing on-body and ICP concentrations of Cu and Zn over time. The charts display data trends without specific numerical details. [15]

designs make it possible to include traditional silicon-based electronics in wearable technology [17, 25].

11.3.3 Polymer

Polymers, which are flexible plastic materials, are another key component in wearable sensors. They are mostly used as soft, flexible base layers on which other components are added. PDMS and polyimide are two of the most commonly used polymers for this purpose [26–29]. Polymers are also used to make the electronics themselves, including transistors [30–44]. Some special polymers can directly perform sensing tasks. The ferro- and pyroelectric characteristics of PVDF and P(VDF-TrFE) facilitate their application as touch and temperature sensors. Xue et al. [41] illustrated the application of PVDF for energy harvesting. It functions as a self-sustaining temperature sensor for monitoring human respiration (Fig. 11.4).

Coating with a palladium sheet enables its usage as a hydrogen sensor [45]. A notable method for processing this polymer involves electrospinning, where a thin stream of polymer solution is sprayed onto a collector to form highly aligned fibers. These fibers show excellent piezoelectric behavior and can be woven into wearable bands or fabrics that detect motion and pressure. In addition to mechanical sensing, PVDF can also detect temperature and touch. When coated with palladium, it can even detect hydrogen gas.

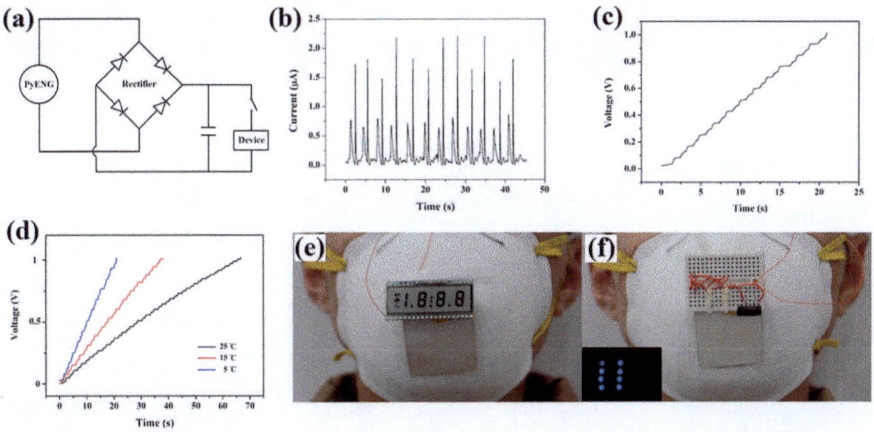

Fig. 11.4 Plots of the PyNG are influenced by human respiratory activity. **a** The circuit schematic illustrating the PyNG interfaced with a rectifier, a capacitor, and an electrical apparatus. **b** The output current generated by the PyNG after rectification. **c** The voltage measured across a capacitor charged by the rectified PyNG output when activated by human respiration at an ambient temperature of 5 °C. **d** The voltage profiles of the capacitor charged by the rectified PyNG under varying ambient temperatures. **e** and **f** Photographs showing an LCD and eight LEDs, respectively, illuminated by the rectified PyNG output [41]

11.3.4 Smart Textiles

Moving beyond traditional materials, smart textiles are being developed to turn everyday clothing into sensing devices. Despite ongoing development, researchers have made significant progress. For instance, researchers have successfully created fabric-based sensors using materials like polypyrrole, reduced graphene oxide, and zinc oxide. These sensors can be woven into clothing to monitor glucose levels, hydration, breathing patterns, and body movement [47–52]. Making these smart textiles involves complex steps such as coating fibers with sensing materials, heating and drying, and sometimes applying in-situ polymerization techniques. These smart fabrics are expected to be used in future wearable medical devices, including pressure-controlling medical braces. For instance, Wang et al. [46] fabricated fiber OECT devices using polypyrrole (PPy)-reduced graphene (rGO) composites on PA6 fiber filaments for glucose sensing (Fig. 11.5).

11.3.5 Two-Dimensional Materials

Two-dimensional materials, especially graphene, are of great interest due to their unique properties. These materials are only one atom thick and can act as metals, semiconductors, or insulators depending on how they are made [53, 54]. Their large surface area and strong chemical reactivity make them perfect for detecting gases.

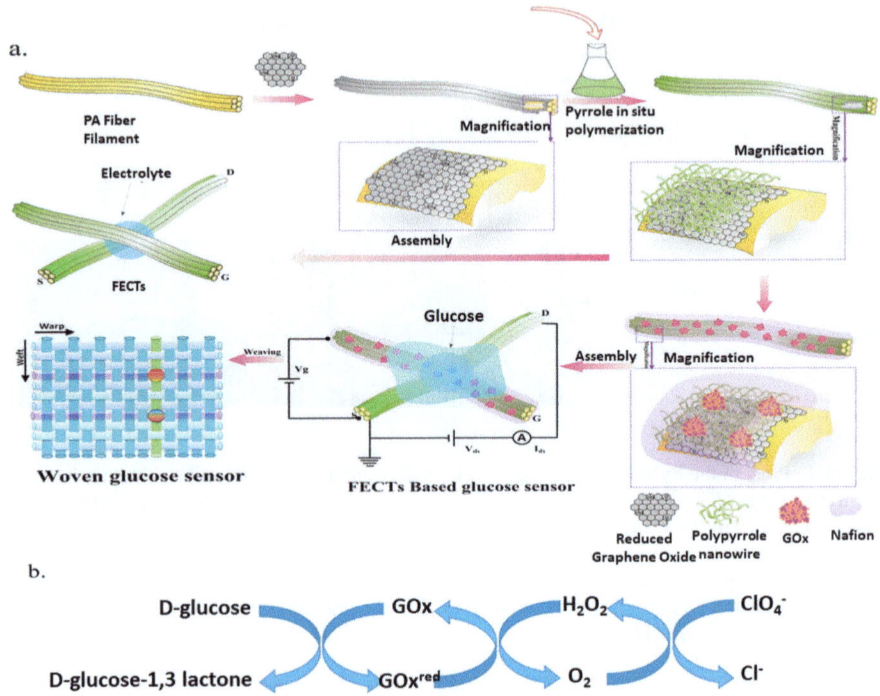

Fig. 11.5 **a** Diagram illustrates a glucose sensing device that utilizes fiber organic electrochemical transistors, featuring an active layer of PPy nanowires and rGO. **b** The reaction cycle of GO is involved in glucose sensing [46]

Graphene is particularly useful because it works well at room temperature, making it ideal for wearable gas sensors. These sensors can detect gases such as ammonia, hydrogen, carbon dioxide, and even chemical threats like explosives. They are built using different sensor designs, including resistors, transistors, and acoustic sensors [9, 55–57]. High-quality graphene is typically made using chemical vapor deposition (CVD), which allows researchers to grow large, uniform sheets of graphene that can be transferred onto flexible surfaces like polyimide films. These graphene-based sensors work well even under bending or in humid conditions. Some versions can even be printed onto paper [58–68]. Chemically modified graphene, known as reduced graphene oxide (rGO), is also used for sensing. It is made by oxidizing graphite and then reducing it using different chemicals [69, 70].

The choice of reducing agent—such as hydrazine, ascorbic acid, or sodium borohydride—affects the final sensor's sensitivity and response to gases. For instance, rGO sensors combined with silver nanoparticles can detect ammonia more effectively than pure rGO. Other researchers have made hybrid sensors by combining graphene with metal particles or metal oxides, like palladium or cuprous oxide [72]. These hybrids improve sensitivity and speed. Conductive polymers like PEDOT: PSS

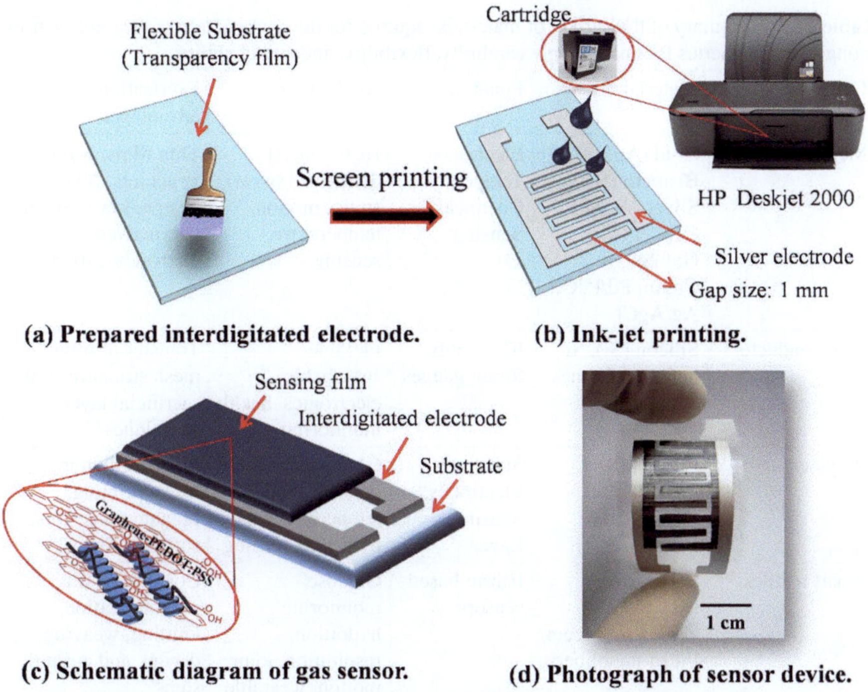

Fig. 11.6 **a-b** Schematic illustration of gas-sensor fabrication methods [71]

have also been mixed with graphene to make low cost, flexible sensors capable of detecting harmful gases even at low concentrations [71, 73] (Fig. 11.6).

Overall, integrating these various materials into a single flexible platform makes wearable sensors functional, comfortable, and effective for real-world applications (Table 11.2). These innovations open up many possibilities for future technologies in health monitoring, environmental sensing, and personal safety.

11.4 Design Strategies and Integration Techniques for Wearable Heterostructured Thin Films

The effective design and integration of heterostructured thin films into wearable devices requires a multi-faceted strategy that combines materials selection, interface control, mechanical architecture, and system-level packaging. Design strategies begin with identifying the key functionalities needed—such as sensing, energy harvesting, signal transmission, or actuation—and then selecting materials that provide these properties while remaining compatible with each other and with flexible substrates [15, 26, 35].

Table 11.2 Summary of the variety of materials required for the construction of wearable sensors, along with key tactics for enhancing wearability, flexibility, and stretchability

Category	Material	Function	Applications	Fabrication/ advantages
Metals	Gold (Au), Bismuth (Bi), Silver (Ag), Copper (Cu), GaInSn Cr/Au, Pd/PdO, Ag/AgCl	Electrodes, Temperature/ Chemical Sensing	Heavy metal detection (sweat), strain, motion, temperature sensing	Thin films, wavy structures, NMP design, low cost Cu alternatives, microfabrication
Semiconductors	Silicon (Si), Si nanomembranes	IC circuits, Strain gauges	Flexible/ stretchable electronics, health monitoring	Transfer printing, mesh structure, and sacrificial layer techniques
Polymers	PDMS, PI, PEDOT:PSS, P3AT, PVDF, P(VDF-TrFe)	Substrates, Electrodes, Sensing Layer	Glucose, pressure, motion, temperature, and hydrogen sensing	Electrospinning, inkjet printing, organic transistors, soft electronics
Smart textiles	Polypyrrole (PPy), rGO, ZnO, PA6 Fibers, Piezo-nanofibers	Fabric-based sensors	Glucose monitoring, hydration, respiration, joint motion, wearable strain	Coating, in-situ polymerization, knitting, weaving, drying, and curing steps
2D Materials	Graphene, rGO, MoS_2, WS_2, h-BN, Black Phosphorus, Cu_2O	Gas sensing, Flexible electronics	NH_3, NO_2, DNT, H_2, CO_2, DMPP detection, hazardous gas, and explosive sensing	CVD, Hummer's method, reduction agents, inkjet/screen printing, metal/oxide/ polymer composites
2D Materials	rGO + AgNPs/ PdNPs, rGO + PEDOT:PSS, rGO + Cu_2O	Gas sensing, Flexible electronics	NH_3, NO_2, DNT, H_2, CO_2, DMPP detection, hazardous gas, and explosive sensing	Room temperature operation, high sensitivity, flexible substrates (PET, PI, paper), printable sensors

One of the primary design strategies is the creation of ultra-thin, layered architectures where each layer is engineered for specific tasks. A typical heterostructure may comprise a superior active sensing layer (e.g., MoS_2 or ZnO), an intermediate charge transport layer (e.g., graphene or PEDOT: PSS), and a flexible encapsulating layer (e.g., PDMS or PI) to ensure mechanical stability and environmental protection. Stacking may be vertical or lateral, contingent upon the desired electrical paths and device configuration.

Geometric design also plays a significant function. Structures like serpentine interconnects, kirigami and origami-inspired folds, or undulating buckled films are utilized to enhance stretchability and durability. These designs enable rigid materials to stretch or contract without structural breakdown. Furthermore, positioning the functional thin films adjacent to the neutral mechanical plane reduces strain during flexion, thereby maintaining performance and prolonging device longevity.

Advanced fabrication techniques are utilized for the deposition, patterning, and transfer of thin films onto flexible substrates. Transfer printing is frequently employed to transfer fragile films from growing substrates, such as silicon wafers, to flexible platforms. Soft lithography, inkjet printing (Fig. 11.7), and laser patterning allow for precise control over spatial architecture and connectivity. Heterogeneous integration—joining different functional layers via thermal bonding, lamination, or adhesive films—ensures mechanical and electrical continuity across interfaces.

Furthermore, interface engineering techniques such as surface treatment (plasma, UV-ozone), functionalization (e.g., self-assembled monolayers), and graded interlayers are applied to enhance adhesion, reduce interfacial resistance, and prevent delamination under strain. Encapsulation strategies using parylene, elastomers, or hydrogels are employed to protect devices from moisture, oxidation, and biofluids, making them suitable for long-term use on human skin.

In the context of real-world deployment, system-level integration includes combining the heterostructured thin film with interconnects, power sources (e.g., printed batteries or nanogenerators), data transmission units (e.g., Bluetooth modules), and packaging—all on a wearable platform like a patch, textile, or band. These systems must maintain performance under dynamic body motion, be biocompatible, and operate efficiently under low power.

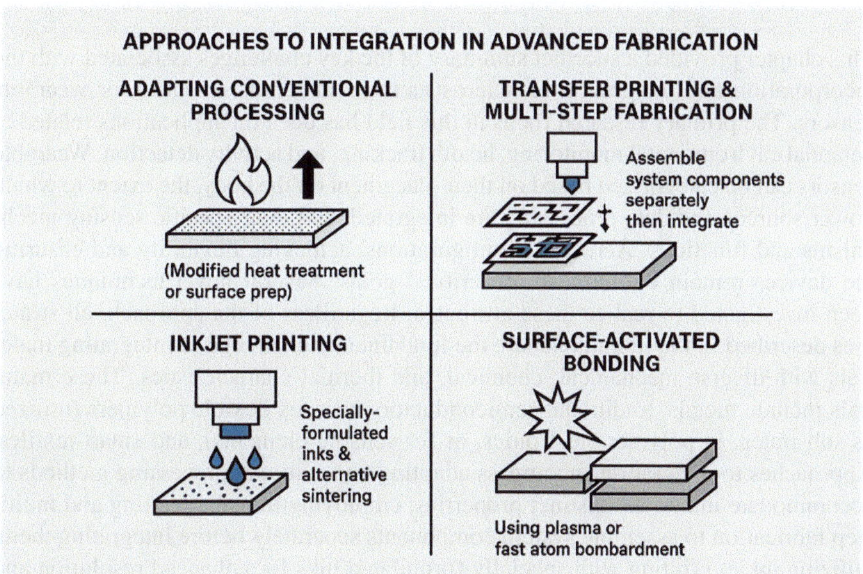

Fig. 11.7 Approaches to integration in advanced fabrication

Through the combination of thoughtful material design, architectural innovation, and scalable integration processes, heterostructured thin films are being transformed into robust, multifunctional components that can meet the rigorous demands of wearable electronics. As research progresses, these techniques will evolve to support increasingly complex, miniaturized, and intelligent wearable systems [71, 74–81].

11.5 Summary

The integration of diverse materials to create reliable and robust wearable sensors remains a major challenge and has attracted significant research interest. An effective integration process must enable high-resolution patterning and precise layer-to-layer alignment. It should be capable of accommodating heterogeneous materials exhibiting a wide variety of chemical, mechanical, and thermal characteristics while ensuring strong adhesion so that the overall device structure remains durable against mechanical stresses and environmental exposure. Moreover, the process must be compatible with high-throughput and scalable manufacturing. The comprehensive review by Gao and Haddara [15, 82] provides an in-depth examination of these integration challenges.

11.6 Conclusion

This chapter provided a succinct summary of the key challenges associated with the incorporation and integration of heterostructured materials essential for wearable sensors. The primary research focus in this field has been on applications related to personal environmental monitoring, health tracking, and activity detection. Wearable sensors can be categorized based on their placement on the body, the extent to which power sources and data processing are integrated, and their specific sensing mechanisms and functions. Across all configurations, achieving flexibility and ensuring the devices remain unobtrusive are critical goals. Several novel techniques have been investigated to realize these attributes. Regardless of the approach, all strategies described in the literature share the fundamental challenge of integrating materials with diverse mechanical, chemical, and thermal characteristics. These materials include metals, traditional semiconductors, various flexible polymers (utilized as substrates, in polymer electronics, or for sensing elements), and smart textiles. Approaches to integration encompass adapting conventional processing methods to accommodate materials' distinct properties, employing transfer printing and multi-step fabrication to assemble system components separately before integrating them, utilizing inkjet printing with specially formulated inks for enhanced resolution and adhesion along with alternative low-temperature sintering techniques, and applying surface-activated bonding processes.

References

1. Sekar, M.; Sriramprabha, R.; Sekhar, P. K.; Bhansali, S.; Ponpandian, N.; Pandiaraj, M.; & Viswanathan, C; Towards wearable sensor platforms for the electrochemical detection of cortisol. Journal of The Electrochemical Society, **2020**, 167(6), 067508, https://doi.org/10.1149/1945-7111/ab7e24.
2. Bandodkar, A.J.; Jeerapan, I.; Wang, J. Wearable Chemical Sensors: Present Challenges and Future Prospects. *ACS Sens.* **2016**, *1*, 464–482, https://doi.org/10.1021/acssensors.6b00250.
3. Trung, T.Q.; Lee, N.-E. Flexible and Stretchable Physical Sensor Integrated Platforms for Wearable Human-Activity Monitoring and Personal Healthcare. *Adv. Mater.* **2016**, *28*, 4338–4372, https://doi.org/10.1002/adma.201504244.
4. Rogers, J.A.; Someya, T.; Huang, Y. Materials and mechanics for stretchable electronics. *Science* **2010**, *327*, 1603–1607, https://doi.org/10.1126/science.1182383.
5. Someya, T.; Bauer, S.; Kaltenbrunner, M. Imperceptible organic electronics. *MRS Bull.* **2017**, *42*, 124–130, https://doi.org/10.1557/mrs.2017.1.
6. An, B.W.; Shin, J.H.; Kim, S.-Y.; Kim, J.; Ji, S.; Park, J.; Lee, Y.; Jang, J.; Park, Y.-G.; Cho, E.; et al. Smart Sensor Systems for Wearable Electronic Devices. *Polymers* **2017**, *9*, 303, https://doi.org/10.3390/polym9080303.
7. Thiyagarajan, K.; Jeong, U. Strategies for stretchable polymer semiconductor layers. *MRS Bull.* **2017**, *42*, 98–102, https://doi.org/10.1557/mrs.2017.2.
8. Dagdeviren, C.; Joe, P.; Tuzman, O.L.; Park, K.-I.; Lee, K.J.; Shi, Y.; Huang, Y.; Rogers, J.A. Recent progress in flexible and stretchable piezoelectric devices for mechanical energy harvesting, sensing, and actuation. *Extreme Mech. Lett.* **2016**, *9*, 269–281, https://doi.org/10.1016/j.eml.2016.05.015.
9. Kim, Y.H.; Kim, S.J.; Kim, Y.-J.; Shim, Y.-S.; Kim, S.Y.; Hong, B.H.; Jang, H.W. Self-Activated Transparent All-Graphene Gas Sensor with Endurance to Humidity and Mechanical Bending. *ACS Nano* **2015**, *9*, 10453–10460, https://doi.org/10.1021/acsnano.5b04680.
10. Sindhu, P.; Ballav, N; Thin Films of MOF-on-Guest@MOF: A Simple Strategy of Designing Electronic Heterostructures', Inorganic Chemistry, **2023,** vol. 62, no. 28, pp. 10887–10891, https://doi.org/10.1021/acs.inorgchem.3c01489.
11. Kang, K.; Lee, K.-H.; Han, Y.; Gao, H.; Xie, S.; Muller, D.A; Park, J.; ~~2017,~~ Layer-by-layer assembly of two-dimensional materials into wafer-scale heterostructures, **2017**, Nature, vol. 550, no. 7675, pp. 229-233, https://doi.org/10.1038/nature23905.
12. Vadla, S.S.; Guru, S.; Parida, T.; John, S.; Roy, S.C. and Rao, G.R.; Electrodeposited $NiFe_2O_4/Cu_2O$ heterostructure thin films with enhanced photocurrent generation. Journal of Photochemistry and Photobiology, **2023**, 15, p.100181, https://doi.org/10.1016/j.jpap.2023.100181.
13. Britnell, L.; Ribeiro, R.M.; Eckmann, A.; Jalil, R.; Belle, B.D.; Mishchenko, A.; Kim, Y.J.; Gorbachev, R.V.; Georgiou, T.; Morozov, S.V. and Grigorenko, A.N.; Strong light-matter interactions in heterostructures of atomically thin films. Science, **2013**, 340(6138), pp.1311-1314, https://doi.org/10.1126/science.1235547.
14. Cataldo, S.; Sartorio, C.; Giannazzo, F.; Scandurra, A. and Pignataro, B.; Self-organization and nanostructural control in thin film heterojunctions. *Nanoscale*, 2014, *6*(7), pp.3566-3575, https://doi.org/10.1039/C3NR05027K.
15. Gao, W.; Nyein, H.Y.Y.; Shahpar, Z.; Fahad, H.M.; Chen, K.; Emaminejad, S.; Gao, Y.; Tai, L.-C.; Ota, H.; Wu, E.; et al. Wearable Microsensor Array for Multiplexed Heavy Metal Monitoring of Body Fluids. *ACS Sens.* **2016**, *1*, 866–874, https://doi.org/10.1021/acssensors.6b00287.
16. Eom, T.H.; Han, J.I. The effect of the nickel and chromium concentration ratio on the temperature coefficient of the resistance of a Ni–Cr thin film-based temperature sensor. *Sens. Actuators Phys.* **2017**, *260*, 198–205, https://doi.org/10.1016/j.sna.2017.04.024.
17. Yeo, W.-H.; Kim, Y.-S.; Lee, J.; Ameen, A.; Shi, L.; Li, M.; Wang, S.; Ma, R.; Jin, S.H.; Kang, Z.; et al. Multifunctional Epidermal Electronics Printed Directly onto the Skin. *Adv. Mater.* **2013**, *25*, 2773–2778, https://doi.org/10.1002/adma.201204426.

18. Qin, Y.; Alam, A.U.; Pan, S.; Howlader, M.M.R.; Ghosh, R.; Hu, N.-X.; Jin, H.; Dong, S.; Chen, C.-H.; Deen, M.J. Integrated water quality monitoring system with pH, free chlorine, and temperature sensors. *Sens. Actuators B Chem.* **2017**, *255*, 781–790, https://doi.org/10.1016/j.snb.2017.07.188.
19. Matsuzaki, R.; Tabayashi, K. Highly Stretchable, Global, and Distributed Local Strain Sensing Line Using GaInSn Electrodes for Wearable Electronics. *Adv. Funct. Mater.* **2015**, *25*, 3806–3813, https://doi.org/10.1002/adfm.201501396.
20. Pei, X.; Kang, W.; Yue, W.; Bange, A.; Heineman, W.R.; Papautsky, I. Disposable Copper-Based Electrochemical Sensor for Anodic Stripping Voltammetry. *Anal. Chem.* **2014**, *86*, 4893–4900, https://doi.org/10.1021/ac500277j.
21. Redhwan, T.Z.; Alam, A.U.; Haddara, Y.M.; Howlader, M.M.R. Bonding mechanism and electrochemical impedance of directly bonded liquid crystal polymer and copper. In Proceedings of the 2017 5th International Workshop on Low Temperature Bonding for 3D Integration (LTB-3D), Tokyo, Japan, 16–18 May 2017; p. 40, https://doi.org/10.23919/LTB-3D.2017.7947436.
22. Howlader, M.M.R.; Suga, T.; Surface Activated Bonding Method for Flexible Lamination. In Proceedings of the 6th International Conference on Polymers and Adhesives in Microelectronics and Photonics, Tokyo, Japan, 16–18 January **2007**, pp. 271–276, https://doi.org/10.1109/POLYTR.2007.4339181.
23. Qin, Y.; Howlader, M.M.R.; Deen, M.J.; Haddara, Y.M.; Selvaganapathy, P.R. Polymer integration for packaging of implantable sensors. *Sens. Actuators B Chem.* **2014**, *202*, 758–778, https://doi.org/10.1016/j.snb.2014.05.063.
24. Yang, Y.; Ding, S.; Araki, T.; Jiu, J.; Sugahara, T.; Wang, J.; Vanfleteren, J.; Sekitani, T.; Suganuma, K. Facile fabrication of stretchable Ag nanowire/polyurethane electrodes using high intensity pulsed light. *Nano Res.* **2016**, *9*, 401–414, https://doi.org/10.1007/s12274-015-0921-9.
25. Kim, D.-H.; Ahn, J.-H.; Choi, W.M.; Kim, H.-S.; Kim, T.-H.; Song, J.; Huang, Y.Y.; Liu, Z.; Lu, C.; Rogers, J.A. Stretchable and Foldable Silicon Integrated Circuits. *Science* **2008**, *320*, 507–511, https://doi.org/10.1126/science.1154367.
26. Rezai, P.; Selvaganapathy, P.R.; Wohl, G.R.; Plasma enhanced bonding of polydimethylsiloxane with parylene and its optimization. *J. Micromech. Microeng.* **2011**, *21*, 065024, https://doi.org/10.1088/0960-1317/21/6/065024.
27. Su, Z.; Chen, X.; Chen, H.; Song, Y.; Cheng, X.; Meng, B.; Song, Z.; Zhang, H.; Bioinspired microporous elastomer with enhanced and tunable stretchability for strain sensing device. In Proceedings of the 2017 IEEE 30th International Conference on Micro Electro Mechanical Systems (MEMS), Las Vegas, NV, USA, 22–26 January 2017; pp. 1036–1039, https://doi.org/10.1109/MEMSYS.2017.7863589.
28. Carlson, J.A.; English, J.M.; Coe, D.J. A flexible, self-healing sensor skin. *Smart Mater. Struct.* **2006**, *15*, N129–N135, https://doi.org/10.1088/0964-1726/15/5/N05.
29. Meyer, J.-U.; Stieglitz, T.; Scholz, O.; Haberer, W.; Beutel, H. High density interconnects and flexible hybrid assemblies for active biomedical implants. *IEEE Trans. Adv. Packag.* **2001**, *24*, 366–374, https://doi.org/10.1109/6040.938305..
30. Liang, J.; Tong, K.; Sun, H.; Pei, Q. Intrinsically stretchable field-effect transistors. *MRS Bull.* **2017**, *42*, 131–137, https://doi.org/10.1557/mrs.2016.326.
31. Lee, Y.K.; Jang, K.-I.; Ma, Y.; Koh, A.; Chen, H.; Jung, H.N.; Kim, Y.; Kwak, J.W.; Wang, L.; Xue, Y.; et al. Chemical Sensing Systems that Utilize Soft Electronics on Thin Elastomeric Substrates with Open Cellular Designs. *Adv. Funct. Mater.* **2017**, *27*, 1605476, https://doi.org/10.1002/adfm.201605476.
32. Noh, J.-S. Conductive Elastomers for Stretchable Electronics, Sensors and Energy Harvesters. *Polymers* **2016**, *8*, 123, https://doi.org/10.3390/polym8040123.
33. Rogers, J.A.; Ghaffari, R.; Kim, D.-H. (Eds.) *Stretchable Bioelectronics for Medical Devices and Systems*; Microsystems and Nanosystems; Springer International Publishing: Cham, Switzerland, 2016; ISBN 978–3–319–28692–1, https://doi.org/10.1007/978-3-319-28694-5.

34. Root, S.E.; Savagatrup, S.; Printz, A.D.; Rodriquez, D.; Lipomi, D.J. Mechanical Properties of Organic Semiconductors for Stretchable, Highly Flexible, and Mechanically Robust Electronics. *Chem. Rev.* **2017**, *117*, 6467–6499, https://doi.org/10.1021/acs.chemrev.7b00003.
35. Zhang, S.; Hubis, E.; Tomasello, G.; Soliveri, G.; Kumar, P.; Cicoira, F. Patterning of Stretchable Organic Electrochemical Transistors. *Chem. Mater.* **2017**, *29*, 3126–3132, https://doi.org/10.1021/acs.chemmater.7b00181.
36. Macaya, D.J.; Nikolou, M.; Takamatsu, S.; Mabeck, J.T.; Owens, R.M.; Malliaras, G.G. Simple glucose sensors with micromolar sensitivity based on organic electrochemical transistors. *Sens. Actuators B Chem.* **2007**, *123*, 374–378, https://doi.org/10.1016/j.snb.2006.08.038.
37. Baek, J.-Y.; An, J.-H.; Choi, J.-M.; Park, K.-S.; Lee, S.-H. Flexible polymeric dry electrodes for the long-term monitoring of ECG. *Sens. Actuators Phys.* **2008**, *143*, 423–429, https://doi.org/10.1016/j.sna.2007.11.019.
38. Jung, H.C.; Moon, J.H.; Baek, D.H.; Lee, J.H.; Choi, Y.Y.; Hong, J.S.; Lee, S.H. CNT/PDMS Composite Flexible Dry Electrodes for Long-Term ECG Monitoring. *IEEE Trans. Biomed. Eng.* **2012**, *59*, 1472–1479, https://doi.org/10.1109/TBME.2012.2190288..
39. Lin, C.T.; Liao, L.D.; Liu, Y.H.; Wang, I.J.; Lin, B.S.; Chang, J.Y. Novel Dry Polymer Foam Electrodes for Long-Term EEG Measurement. *IEEE Trans. Biomed. Eng.* **2011**, *58*, 1200–1207, https://doi.org/10.1109/TBME.2010.2102353.
40. Persano, L.; Dagdeviren, C.; Su, Y.; Zhang, Y.; Girardo, S.; Pisignano, D.; Huang, Y.; Rogers, J.A. High performance piezoelectric devices based on aligned arrays of nanofibers of poly(vinylidenefluoride-*co*-trifluoroethylene). *Nat. Commun.* **2013**, *4*, 1633, https://doi.org/10.1038/ncomms2639.
41. Xue, H.; Yang, Q.; Wang, D.; Luo, W.; Wang, W.; Lin, M.; Liang, D.; Luo, Q. A wearable pyroelectric nanogenerator and self-powered breathing sensor. *Nano Energy* **2017**, *38*, 147–154, https://doi.org/10.1016/j.nanoen.2017.05.056.
42. Baumgärtel, K.H.; Zöllner, D.; Krieger, K.-L. Classification and Simulation Method for Piezoelectric PVDF Sensors. *Procedia Technol.* **2016**, *26*, 491–498, https://doi.org/10.1016/j.protcy.2016.08.062.
43. Zhang, B.; Zhang, L.; Deng, W.; Jin, L.; Chun, F.; Pan, H.; Gu, B.; Zhang, H.; Lv, Z.; Yang, W.; et al. Self-Powered Acceleration Sensor Based on Liquid Metal Triboelectric Nanogenerator for Vibration Monitoring. *ACS Nano* **2017**, *11*, 7440–7446, https://doi.org/10.1021/acsnano.7b03818.
44. Johnston, R.; Nakano, K.; Fujita, K.; Misaki, S.; Fujii, H.; Misaki, Y. A non-contact high resolution piezoelectric film-based sensor for monitoring breathing during sleep. *AIP Conf. Proc.* **2017**, *1865*, 070004, https://doi.org/10.1063/1.4993386.
45. Imai, Y.; Tadaki, D.; Ma, T.; Kimura, Y.; Hirano-Iwata, A.; Niwano, M. Response characteristics of hydrogen gas sensor with porous piezoelectric poly(vinylidene fluoride) film. *Sens. Actuators B Chem.* **2017**, *247*, 479–489, https://doi.org/10.1016/j.snb.2017.03.001.
46. Wang, Y.; Qing, X.; Zhou, Q.; Zhang, Y.; Liu, Q.; Liu, K.; Wang, W.; Li, M.; Lu, Z.; Chen, Y.; et al. The woven fiber organic electrochemical transistors based on polypyrrole nanowires/reduced graphene oxide composites for glucose sensing. *Biosens. Bioelectron.* **2017**, *95*, 138–145, https://doi.org/10.1016/j.bios.2017.04.018.
47. Soltanian, S.; Servati, A.; Rahmanian, R.; Ko, F.; Servati, P. Highly piezoresistive compliant nanofibrous sensors for tactile and epidermal electronic applications. *J. Mater. Res.* **2015**, *30*, 121–129, https://doi.org/10.1557/jmr.2014.361.
48. Badawy, D.; Soltanian, S.; Servati, P.; Ko, F. Piezoelectric zinc oxide woven fabric sensors for dynamic strain and damage sensing. In Proceedings of the 12th International Conference on Textile Composites, Raleigh, NC, USA, 26–29 May 2014.
49. Atalay, O.; Kennon, W. Knitted Strain Sensors: Impact of Design Parameters on Sensing Properties. *Sensors* **2014**, *14*, 4712–4730, https://doi.org/10.3390/s140304712.
50. Atalay, O.; Kennon, W.R.; Demirok, E. Weft-Knitted Strain Sensor for Monitoring Respiratory Rate and Its Electro-Mechanical Modeling. *IEEE Sens. J.* **2015**, *15*, 110–122, https://doi.org/10.1109/JSEN.2014.2339739.

51. Zhao, Z.; Yan, C.; Liu, Z.; Fu, X.; Peng, L.-M.; Hu, Y.; Zheng, Z. Machine-Washable Textile Triboelectric Nanogenerators for Effective Human Respiratory Monitoring through Loom Weaving of Metallic Yarns. *Adv. Mater.* **2016**, *28*, 10267–10274, https://doi.org/10.1002/adma.201603679.
52. Servati, A.; Zou, L.; Wang, Z.; Ko, F.; Servati, P. Novel Flexible Wearable Sensor Materials and Signal Processing for Vital Sign and Human Activity Monitoring. *Sensors* **2017**, *17*, 1622, https://doi.org/10.3390/s17071622.
53. Yang, S.; Jiang, C.; Wei, S. Gas sensing in 2D materials. *Appl. Phys. Rev.* **2017**, *4*, 021304, https://doi.org/10.1063/1.4983310.
54. Singh, E.; Meyyappan, M.; Nalwa, H.S. Flexible Graphene-Based Wearable Gas and Chemical Sensors. *ACS Appl. Mater. Interfaces* **2017**, *9*, 34544–34586, https://doi.org/10.1021/acsami.7b07063.
55. Dresselhaus, M.S.; Araujo, P.T. Perspectives on the 2010 Nobel Prize in Physics for Graphene. *ACS Nano* **2010**, *4*, 6297–6302, https://doi.org/10.1021/nn1029789.
56. Hwang, E.H.; Adam, S.; Sarma, S.D. Carrier Transport in Two-Dimensional Graphene Layers. *Phys. Rev. Lett.* **2007**, *98*, https://doi.org/10.1103/PhysRevLett.98.186806.
57. Lee, C.; Wei, X.; Kysar, J.W.; Hone, J. Measurement of the Elastic Properties and Intrinsic Strength of Monolayer Graphene. *Science* **2008**, *321*, 385–388, https://doi.org/10.1126/science.1157996.
58. Chen, G.; Paronyan, T.M.; Harutyunyan, A.R. Sub-ppt gas detection with pristine graphene. *Appl. Phys. Lett.* **2012**, *101*, 053119, https://doi.org/10.1063/1.4742327.
59. Kumar, S.; Kaushik, S.; Pratap, R.; Raghavan, S. Graphene on Paper: A Simple, Low-Cost Chemical Sensing Platform. *ACS Appl. Mater. Interfaces* **2015**, *7*, 2189–2194, https://doi.org/10.1021/am5084122.
60. Marcano, D.C.; Kosynkin, D.V.; Berlin, J.M.; Sinitskii, A.; Sun, Z.; Slesarev, A.; Alemany, L.B.; Lu, W.; Tour, J.M. Improved Synthesis of Graphene Oxide. *ACS Nano* **2010**, *4*, 4806–4814, https://doi.org/10.1021/nn1006368.
61. Bagri, A.; Mattevi, C.; Acik, M.; Chabal, Y.J.; Chhowalla, M.; Shenoy, V.B. Structural evolution during the reduction of chemically derived graphene oxide. *Nat. Chem.* **2010**, *2*, 581–587, https://doi.org/10.1038/nchem.686.
62. Fowler, J.D.; Allen, M.J.; Tung, V.C.; Yang, Y.; Kaner, R.B.; Weiller, B.H. Practical Chemical Sensors from Chemically Derived Graphene. *ACS Nano* **2009**, *3*, 301–306, https://doi.org/10.1021/nn800593m.
63. Dua, V.; Surwade, S.P.; Ammu, S.; Agnihotra, S.R.; Jain, S.; Roberts, K.E.; Park, S.; Ruoff, R.S.; Manohar, S.K. All-Organic Vapor Sensor Using Inkjet-Printed Reduced Graphene Oxide. *Angew. Chem. Int. Ed.* **2010**, *49*, 2154–2157, https://doi.org/10.1002/ange.200905089.
64. Hu, N.; Wang, Y.; Chai, J.; Gao, R.; Yang, Z.; Kong, E.S.-W.; Zhang, Y. Gas sensor based on p-phenylenediamine reduced graphene oxide. *Sens. Actuators B Chem.* **2012**, *163*, 107–114, https://doi.org/10.1016/j.snb.2012.01.016.
65. Ghosh, R.; Midya, A.; Santra, S.; Ray, S.K.; Guha, P.K. Chemically Reduced Graphene Oxide for Ammonia Detection at Room Temperature. *ACS Appl. Mater. Interfaces* **2013**, *5*, 7599–7603, https://doi.org/10.1021/am4019109.
66. Ghosh, R.; Singh, A.; Santra, S.; Ray, S.K.; Chandra, A.; Guha, P.K. Highly sensitive large-area multi-layered graphene-based flexible ammonia sensor. *Sens. Actuators B Chem.* **2014**, *205*, 67–73, https://doi.org/10.1016/j.snb.2014.08.044.
67. Zhang, H.; Fan, L.; Dong, H.; Zhang, P.; Nie, K.; Zhong, J.; Li, Y.; Guo, J.; Sun, X. Spectroscopic Investigation of Plasma-Fluorinated Monolayer Graphene and Application for Gas Sensing. *ACS Appl. Mater. Interfaces* **2016**, *8*, 8652–8661, https://doi.org/10.1021/acsami.5b11872.
68. Sysoev, V.I.; Okotrub, A.V.; Asanov, I.P.; Gevko, P.N.; Bulusheva, L.G. Advantage of graphene fluorination instead of oxygenation for restorable adsorption of gaseous ammonia and nitrogen dioxide. *Carbon* **2017**, *118*, 225–232, https://doi.org/10.1016/j.carbon.2017.03.026.
69. Tran, Q.T.; Hoa, H.T.M.; Yoo, D.-H.; Cuong, T.V.; Hur, S.H.; Chung, J.S.; Kim, E.J.; Kohl, P.A. Reduced graphene oxide as an over-coating layer on silver nanostructures for detecting

NH3 gas at room temperature. *Sens. Actuators B Chem.* **2014**, *194*, 45–50, https://doi.org/10.1016/j.snb.2013.12.062.

70. Chen, F.; Wan, P.; Xu, H.; Sun, X. Flexible Transparent Supercapacitors Based on Hierarchical Nanocomposite Films. *ACS Appl. Mater. Interfaces* **2017**, *9*, 17865–17871, https://doi.org/10.1021/acsami.7b02460.
71. Seekaew, Y.; Lokavee, S.; Phokharatkul, D.; Wisitsoraat, A.; Kerdcharoen, T.; Wongchoosuk, C. Low-cost and flexible printed graphene–PEDOT: PSS gas sensor for ammonia detection. *Org. Electron.* **2014**, *15*, 2971–2981, https://doi.org/10.1016/j.orgel.2014.08.044.
72. Shin, D.H.; Lee, J.S.; Jun, J.; An, J.H.; Kim, S.G.; Cho, K.H.; Jang, J. Flower-like Palladium Nanoclusters Decorated Graphene Electrodes for Ultrasensitive and Flexible Hydrogen Gas Sensing. *Sci. Rep.* **2015**, *5*, https://doi.org/10.1038/srep12294.
73. Deng, S.; Tjoa, V.; Fan, H.M.; Tan, H.R.; Sayle, D.C.; Olivo, M.; Mhaisalkar, S.; Wei, J.; Sow, C.H. Reduced Graphene Oxide Conjugated Cu_2 O Nanowire Mesocrystals for High-Performance NO_2 Gas Sensor. *J. Am. Chem. Soc.* **2012**, *134*, 4905–4917, https://doi.org/10.1021/ja211683m.
74. Xuan,X.; Park, J.Y. Miniaturized flexible sensor with reduced graphene oxide/carbon nano tube modified bismuth working electrode for heavy metal detection. In Proceedings of the 2017 IEEE 30th International Conference on Micro Electro Mechanical Systems (MEMS), Las Vegas, NV, USA, 22–26 January 2017; pp. 636–639, https://doi.org/10.1109/MEMSYS.2017.7863488.
75. Wang,Z.; Sun, X.; Li, C.; He, X.; Liu, G. On-site detection of heavy metals in agriculture land by a disposable sensor based virtual instrument. Comput. Electron. Agric. 2016, 123, 176–183, https://doi.org/10.1016/j.compag.2016.02.017.
76. McLaughlin, J.A.; Hyland, M.; McAdams, E.T.; Holder, D.; Zhou, D.-M. Thin film flexible electrodes based on gold/polytetrafluoroethylene layers. In Proceedings of the 18th Annual International Conference of the IEEE Engineering in Medicine and Biology Society, Bridging Disciplines for Biomedicine, Amsterdam, The Netherlands, 31 October–3 November 1996; pp. 108–109, https://doi.org/10.1109/IEMBS.1996.656871.
77. Howlader, M.M.R.; Doyle, T.E.; Mohtashami, S.; Kish, J.R. Charge transfer and stability of implantable electrodes on flexible substrate. Sens. Actuators B Chem. 2013, 178, 132–139, https://doi.org/10.1016/j.snb.2012.12.051.
78. Gupta, A.; Mandal, S.; Katiyar, M.; Mohapatra, Y.N. Low temperature solution process for fabrication of electrodes on flexible substrate using gold nanoparticles. Int. J. Nanosci. 2011, 10, 659–663, https://doi.org/10.1142/S0219581X11009192.
79. Khan, Y.; Pavinatto, F.J.; Lin, M.C.; Liao, A.; Swisher, S.L.; Mann, K.; Subramanian, V.; Maharbiz, M.M.;Arias, A.C. Inkjet-Printed Flexible Gold Electrode Arrays for Bioelectronic Interfaces. Adv. Funct. Mater. 2016, 26, 1004–1013, https://doi.org/10.1002/adfm.201503316.
80. Sultana, I.; Razaq, A.; Idrees, M.; Asif, M.H.; Ali, H.; Arshad, A.; Iqbal, S.; Ramay, S.M.; Hussain, S.Q. Electrodeposition of Gold on Lignocelluloses and Graphite-Based Composite Paper Electrodes for Superior Electrical Properties. J. Electron. Mater. 2016, 45, 5140–5145, https://doi.org/10.1007/s11664-016-4727-7.
81. Bartlett, M.D.; Markvicka, E.J.; Majidi, C. Rapid Fabrication of Soft, Multilayered Electronics for Wearable Biomonitoring. Adv. Funct. Mater. 2016, 26, 8496–8504, https://doi.org/10.1002/adfm.201602733.
82. Haddara, Y.M.; Howlader, M.M.R. Integration of Heterogeneous Materials for Wearable Sensors. Polymers 2018, https://doi.org/10.3390/polym10010060.
83. Reis, S.; Correia, V.; Martins, M.; Barbosa, G.; Sousa, R.M.; Minas, G.; Lanceros-Mendez, S.; Rocha, J.G. Touchscreen based on acoustic pulse recognition with piezoelectric polymer sensors. In Proceedings of the 2010 IEEE International Symposium on Industrial Electronics (ISIE), Bari, Italy, 4–7 July2010; pp. 516–520, https://doi.org/10.1109/ISIE.2010.5637672.
84. Emamian, S.; Narakathu, B.B.; Chlaihawi, A.A.; Bazuin, B.J.; Atashbar, M.Z. Screen printing of flexible piezoelectric based device on polyethylene terephthalate (PET) and paper for touch and force sensing applications. *Sens. Actuators Phys.* **2017**, *263*, 639–647, https://doi.org/10.1016/j.sna.2017.07.045.

Sunil Chauhan is an Associate Professor of Physics at Sharda University, Noida (India), with over nine years of teaching and fifteen years of research experience in nanoscience and materials physics. His multidisciplinary research focuses on the synthesis and functionalization of nanostructured materials for biomedical, photocatalytic, piezocatalytic, and solar cell applications. He has extensive expertise in chemical and mechanical synthesis methods, as well as in fabricating nanostructured films via electrospinning and spin coating, followed by detailed characterization of their structural, optical, and magnetic properties. Dr. Chauhan has authored more than seventy research papers, holds six granted patents (with three pending), and has presented at numerous national and international conferences. He has completed four funded research projects, including a SERB (TAR/2019/000210) project, and his scholarly work has garnered over 1,762 citations, with an h-index of 22. He actively mentors research students and contributes to scientific innovation through the design and development of instruments.

Chapter 12
Future Trends in Thin Film Device Applications

Piyali Maity

12.1 Introduction

Thin films are thin to thick layers of materials ranging from nanometers to micrometers. Thin films can also be hetero-layers of different type of materials in different layer that can extend along any two dimensions with third dimension less than 1 μm [1–3]. The materials are essential elements for technological advancements in the fields of optoelectronics, photonics, magnetic, biomedical, and sensing devices. These have been studied for electronic and energy storage device applications such as light emitting diodes, thin film solar cells, photoconductors, IR detectors, interference filters, superconducting films, anticorrosive and decorative coatings, semiconductor devices, and thin film batteries [3–6]. Various types of devices can be integrated into thin films by processing materials. The characteristics of materials are different in thin film and bulk. Functional materials in thin film form give specific optical, electronic, structural, and magnetic properties different from bulk materials. Atoms, ions, or molecules are the smallest building blocks of materials to fabricate layers in thin films by modifying material surfaces [7–9]. The high structural stability with large surface area and novel architectures gives the thin film excellent loading capabilities and topological properties [10–13]. Low resistivity, high conductivity, and large surface area are required for thin film sensing applications with post-synthetic modifications. Figure 12.1 depicts the thin film deposited on solid substrate and flexible thin film which different mechanical strength.

Thin film technology garnered extensive growth over the past few decades in the field of materials science. The notable chemical and physical properties are observed when materials are used in the nanoscale devices with enhanced robustness, durability, and efficiency. This helps to deposit thin film with different techniques. Among these techniques, thermal evaporation, sputtering, molecular beam epitaxy,

P. Maity (✉)
City University of New York, New York, NY, USA
e-mail: dr.piyalimaity@gmail.com

S. Sharma (ed.), *Multifunctional Heterostructure Thin Films Handbook*,
https://doi.org/10.1007/978-3-032-18692-8_12

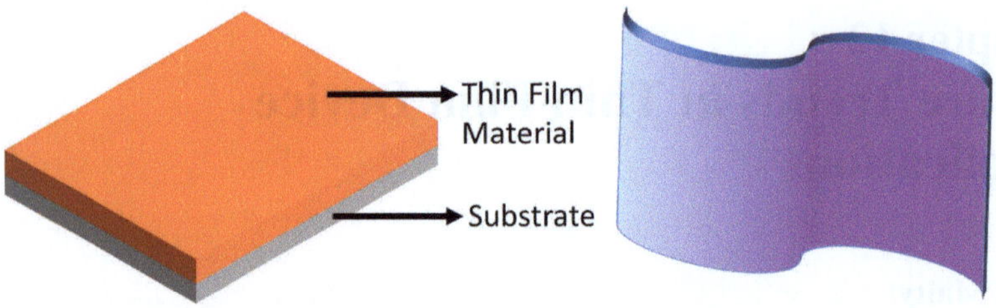

Fig. 12.1 Thin film deposited on substrate and flexible thin film

chemical vapor deposition, chemical bath deposition, atomic layer deposition, are widely used in research and commercial area. The thickness impacts the property of the thin film [14–19].

Large scale manufacturing is a challenge for thin films despite of its developments in multifunctional sensing devices. In order to overcome these obstacles, things which need to be managed are quality of materials, such as chemical composition stability, affinity, and shape dispersibility. Coatings of thin films are also used to enhance transmission, reflection, absorption, conductance, corrosion, electrical behavior, and penetration. Thin film technology has been also used in nano-, micro-, and opto-electronics. Thin film technology gives the evolution in dynamic area of research. It demonstrates the atom-by-atom construction of solid materials [20–23].

The purpose of this chapter is to provide most recent advancements and applications of the thin film. The basic theory of the three main modes of growth, Volmer-Weber, Frank and Van der Merwe, and Stranski-Krastanov plays crucial role to understand thin film fabrication. The chapter describes the trend in fabrication and application of thin films in research and industrial technology [24–29]. Figure 12.2 represents different thin film preparation techniques with various possible application field.

12.2 History and Background of Thin Film

During past decades, the thin film deposition technology has been advanced drastically. The improvement was controlled largely for the need of new devices and products in the optics and electronics industries. Solid-state electronic devices are the most important and primary application of thin film deposition processes with improved film qualities. These deposition techniques are still evolving to develop more complex and advanced future electronic and optoelectronic devices.

At first, thin films were prepared manually by gold sheathing. This gave the result of a thin deposition of gold layer on surfaces to beautify golds, bronzes, and other objects. Then Moche Indians developed oxidation-reduction coating process approximately 100 BC. This gave new direction of advanced fabrication of thin films. They also used the thin film coatings in different metals to improve the strength and beauty

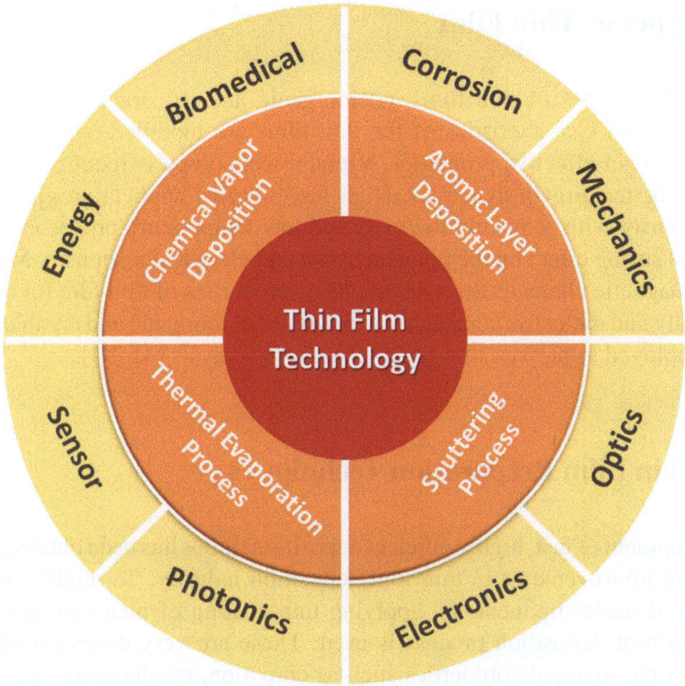

Fig. 12.2 Thin film preparation techniques with various possible application field

of the metal. This oxidation-reduction process is now known as electroless plating [30–32]. The first thin film coating was created in Venice in sixteenth century to prepare a reflective surface using silver (Ag) on a glass substrate to fabricate mirrors. Now-a-days semiconductor-based thin films are broadly used in numerous industries for their unique electrical properties. The properties are obtained by materials processing, and used in optical devices, antireflection coatings, electronic devices, interference filters, solar cells, flat panel displays, etc.

Vacuum technology gave invention of barometer which resulted in the thin film technology evolution. Scientists required clean surface to deposit thin films. Vacuum technology helped to get it. Third generation vacuum system was discovered by Otto von Guericke in 1652, which gave boost to the "vacuum science" [33–36]. In 1663, von Guericke discovered electrical power by converting mechanical energy to electrical energy. This discovery gave rise to physical vapor deposition (PVD) in mid-1800 century. Afterward, thin film technology grew rapidly and made a huge importance in semiconductor technology and energy devices manufacturing. Advanced microprocessors in electronic devices at nanoscale and photovoltaic cells are primary applications of this manufacturing technology [37–40].

12.3 Types of Thin Film

There are two types of thin films: (i) inorganic and (ii) organic thin films. The morphology and characteristics of the thin films are modified by growth process effectively to get effective properties. Vacuum technology is required to grow thin films with better growth dynamics. High flexibility of application gives thin film high popularity with relatively low cost and low temperature processes. Thin film surface has strong effect on the optoelectronic and electrical responses. Mechanical, thermal, magnetic characteristics are the basic properties of thin film for optimizing conductivity and dielectric constant. Various types of inorganic and organic materials can be employed to prepare thin films for precise analysis [41–44].

12.4 Thin Film Preparation Techniques

The development of fast, highly efficient deposition setups has been obtained by innovations and improvements in thin film deposition industry. To modify the surface properties of the components by applying thin coating of materials in nanometer range, thin film deposition process is used. These are very diverse modifications and impact the materials properties such as corrosion, conductivity, and hardness. Electronic displays, biomedical devices, solid surface coatings, data storage devices are key thin film applications. Thin film coatings are deposited onto substrate material. These coatings can be metals to different oxides or chalcogenides to complex compounds. The substrate elements can be transparent, some substrates can be scratch resistant, and some enhance or reduce the electrical conductivity or transmission of signals. Thin film technology produces optoelectronic, solid state, and biomedical devices and products, as well as semiconductor lasers, LED displays, fiber lasers, compound semiconductors. The deposition technique and configuration of the thin film affect the performance of the device application. The manufacturing of thin film is classified according to the source materials which will be used in thin film preparation, such as (i) gaseous state, (ii) solution state, (iii) semi-molten state. The mechanism of thin film preparation process is, first, (i) surface preparation, then (ii) atom and surface diffusion, followed by (iii) nucleation of atoms, (iv) interface formation, and finally (v) continuous film growth. Nanostructured thin films have been fabricated by several physical and chemical deposition techniques over last two decades. The deposition needs to be on transparent materials on conducting or non-conducting substrates [45–47]. Figure 12.3 shows different thin film Fabrication Techniques.

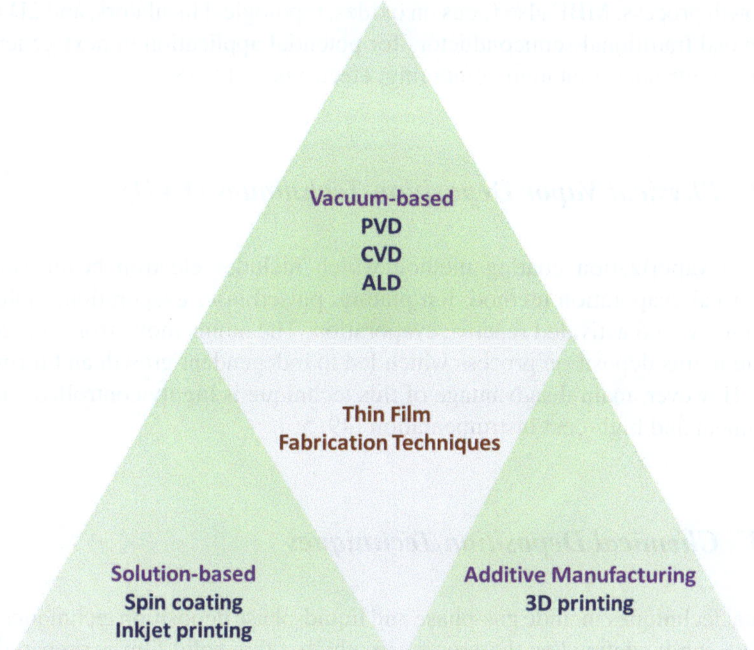

Fig. 12.3 Thin film fabrication techniques (i) vacuum-based, (ii) solution-based, and (iii) additive manufacturing

12.4.1 Molecular Beam Epitaxy (MBE)

Atomic-level control over fabrication of film is obtained by MBE technique. MBE is a sophisticated deposition technique where material is heated and evaporated under ultra-high vacuum in a direction toward substrate. As a result, it condenses in form of thin film. MBE is a high precision, ultra-high vacuum (UHV) technique for advanced semiconductor and quantum devices fabrication with high precisely controlled layer thickness and composition. This cutting-edge technique thin film preparation offers high control with high precision and elasticity in advanced material deposition. This method allows precise tuning of the composition, thickness, and crystallinity of the film materials. The source materials are highly pure elemental materials. They are heated to generate beams of atoms and molecules. Then they travel through ultra-high vacuum environment and condense on the surface of substrate. One monolayer is deposited per second. This makes MBE technique most suitable for preparing superlattices, quantum structures, superlattices, complex nano heterostructures. Since this technique is under UHV conditions, this helps to minimize the contamination allows highly pure film growth. These films grow with precisely sharp atomic interfaces which is key need for electronic and optoelectronic applications. Reflection high-energy electron diffraction (RHEED) helps to achieve real-time monitoring of the film growth process. MBE also focus on oxides, topological insulators, and 2D materials beyond traditional semiconductors for potential application in next-generation quantum spintronics, quantum computing, electronics [47, 48].

12.4.2 Physical Vapor Deposition Techniques (PVD)

PVD is a vaporization coating method which includes electron beam evaporation, thermal evaporation method, ion plating, pulsed laser evaporation, molecular beam epitaxy, and activated reactive evaporation. The atoms move from source to a substrate in this deposition process which led to independent growth and formation of film. However, main disadvantage of this technique is highly controlled vacuum environment and high-cost instrumentation [49, 50].

12.4.3 Chemical Deposition Techniques

Chemical techniques include gas-phase and liquid-phase deposition techniques. This technique can be defined as the process in which a thin solid film is prepared on a substrate of the adsorbed precursors of gas phase by the surface-mediated reaction. The CVD process in reactive in nature. This distinguishes them from physical processes (such as sputtering and evaporation). The surface-mediated term comes from the fact that heterogeneous reaction helps to form solid film which occurs at the substrate surface. Several types of CVD processed films are epitaxial compound semiconductors, polycrystalline silicon, epitaxial silicon, Dielectric thin films such as silicon dioxide, low-k dielectrics, aluminum oxide, titanium nitride, transition metal oxides, metals [44, 48, 51–55].

12.4.4 Evaporation Techniques

One of the most widely used thin film preparation techniques is thermal evaporation technique. The basic principle of this method is thermal evaporation and vaporization. Thermodynamics and molecular transport are few key concepts of thermal evaporation. Material is heated in controlled ambient atmosphere. Electron beam helps to achieve the required heating and this gives the optimum energy to cross the enthalpy of vaporization of material. The surface vapor pressure is directly related to its temperature. This is key point to predict the vapor pressure and evaporation rate of the material with varying temperatures. The chamber is maintained at low pressure to reduce the molecular collisions. The mean free path of these molecules is quite high enough for moving toward the substrate surface. This phenomenon is described by Knudsen diffusion model which mentions the molecular transport while deposition. The molecular vapor condenses and film is formed on the substrate. The substrate temperature, rate of deposition, film, and substrate interaction influences thin film growth process. Three primary models play the key role behind this: layer by layer deposition (Frank-Vander Merwe), island (Volmer-Weber), layer-plus-island (Stranski-Krastanov) growth.

The current focus is on enhancing film quality and homogeneity by the thermal evaporation technique. High-quality multilayer structures and semiconductor materials are produced by this method. The target material is vaporized by thermal evaporation technique at high temperatures. Then the material is transported to substrate through vacuum. This indicates condensation on the surface of the substrate and forms thin film. This technique is performed under high vacuum pressure for minimal interference. A high thermal vaporization rate is obtained by heating target material by tungsten wire coils. Semiconductor devices are manufactured by this method for transistors, conductors, cellular phones [35–37, 56, 57].

12.4.5 Sputtering

Sputtering is similar procedure to physical vapor deposition method to deposit films from target material onto substrates. When energetic ions in a plasma hit the target material, the ejection of atoms happens as a fundamental principle. This technique happens in a vacuum chamber when inert gases like argon are implemented at low pressure. The ions generated from plasma are then accelerated toward the target material. This causes ejection of atoms which subsequently get deposited onto substrate, which forms thin film. There are different types of sputtering techniques, (i) direct current (DC) sputtering, (ii) radiofrequency (RF) sputtering, and (iii) magnetron sputtering.

In DC sputtering, direct current is applied to conductive materials to maintain the plasma. In RF sputtering, radiofrequency power is used to make the target suitable for non-conductive materials. In magnetron sputtering, magnetic field is incorporated to confine the plasma near target. This increases the efficiency which leads the deposition rate high and improves the film property. Broad range of materials, such as metals, insulators, and semiconductors are used depositing materials which is a significant advantage of sputtering. Another advantage is its ability for producing uniform coatings in large area with different complex shapes. This technique gives precise control over composition and thickness of the film by adjusting the processing parameters. Recent advancements are to improve materials, as well as to integrate sputtering and deposition techniques. This reflects its dominant role in thin film preparation for different applications. This method involves the atoms removal from target due to high-energy ion bombardment. However, sputtering does not depend on thermal processes but employs ion impact on the target material. Performance improvements and flexibility are obtained by magnetron sputtering, reactive sputtering, bias sputtering, and ion-beam sputtering [51, 58].

12.4.6 CVD Technique

CVD technique is a widely used through chemical reactions involving gaseous precursors. The process involves introducing volatile precursors into a reaction chamber. The precursors participate chemical reactions to create a solid material that deposits onto a substrate to form thin film. This highly versatile technique is used to produce a broad range of materials. The substrate is kept in a reaction chamber for gaseous form exposure of precursors. To form a thin film, these precursors react on the substrate surface by heat or plasma. Film properties, such as, composition, thickness, and uniformity are controlled by pressure, temperature, and gas flow. There are different types of CVD, such as thermal CVD, PECVD, LPCVD, MOCVD.

PECVD utilizes plasma to increase the chemical reactions at lower temperature in compare to thermal CVD. This totally depends on external thermal energy. LPCVD prepares film under lower pressures. This can reduce the stress in film which results in improving the uniformity over large area. MOCVD is used to deposit compounds and various complex materials due to its handling ability toward metal-organic precursors. Most recent focus on CVD is to improve quality of films, deposition rates, and scalability.

12.4.7 PECVD

PECVD is a broadly used semiconductor thin film deposition technique, optical coatings, and photovoltaics. This technique is operated at comparatively lower temperatures than the traditional CVD techniques. This utilizes plasma to activate chemical reactions. This method maintains high quality of films and allows the materials deposition on temperature sensitive substrates. Plasma is generated from precursor gases using a RF or DC power in vacuum chamber. Electrons and ions of plasma enables the gas dissociation of gas molecules which interacts with the substrate and leads to film formation. The lower temperature allows the deposition on polymers or other materials which are degradable at high temperatures. Plasma increases the chemical reactions and gives result in thin film formation. Surface chemistry governs the PECVD reaction mechanism which includes surface diffusion, species reaction, and adsorption on the substrate. The high-energy environment of plasma provides additional energy to the gas molecules. This allows the precursors degradation and facilitates the thin film deposition process at lower temperatures compared to regular CVD techniques. Parameters such as radio frequency (RF) power, flow rate of gas, pressure, substrate temperature determine the rate of film deposition and quality in PECVD. To fabricate specific materials and applications, these parameters are optimized accordingly. This method is hugely used to produce silicon-based thin films in transistors and solar cells. Silicon nitride and dioxide dielectric layers are deposited through PECVD to prepare insulating layers in semiconductor devices. This technique is also capable of depositing complex materials like doped films or gradient composition of film for applications in microelectronics and nanotechnology.

12.4.8 LPCVD

LPCVD is most useful technique for microelectronics and semiconductor manufacturing thin film deposition. This method typically operates at lower pressure between 0.1 and 10 Torr in comparison to regular CVD technique. This gives better control in uniformity of thin film and reduces the particle formation defect occurrences. LPCVD is able to produce high-quality excellent step coverage and uniform films. These films are ideal for dielectric, semiconductor, and conductor thin film applications [52–55].

12.4.9 ALD

ALD method deposits films one layer at a time by self-limiting vapor-phase sequential chemical reactions with atomic level control. This high control allows the uniform ultra-thin film formation. This makes ALD remarkable in precise thickness-controlled applications in high-k dielectric materials for thin film transistors [53]. Nanotechnology and advanced materials are the fields where ALD gained significant attention for catalysts and protective coatings. It operates on chemical reaction principle between substrate and gaseous precursors. This makes ALD technique unique from other deposition techniques with precise thickness and uniformity control at high aspect ratio structures and complex geometries. There are two half-reactions in this deposition cycle. In each half reaction, one precursor gas is implemented to the substrate to form a monolayer. The second precursor reacts with the adsorbed monolayer which forms a thin film. Desired thickness is achieved by repeating the cycle multiple times. Each cycle deposits one atomic layer per cycle. This technique is widely used for modern transistors and capacitors fabrication by depositing thin high k-oxides dielectric layers. Moreover, this is used in protective coatings, energy storage devices, and biomedical devices. ALD produces high-quality uniform films and step coverage on intricate topography substrates. Additionally, spatial ALD has higher throughput capabilities which makes it industrial-scale suitable applications. ALD is integrated into thin film preparation techniques for 2D materials and high-performance transistor nanostructures. ALD helps to coat ultra-thin electrodes to enhance stability and performance of lithium-ion batteries. It also involves the precursor gases sequential exposure of a substrate, resulting thin film formation [59, 60].

12.4.10 CBD Technique

Chemical bath deposition (CBD) is a simple and cost-effective thin film deposition technique from aqueous solution. A substrate is dipped in a solution containing the precursor ions. Then a thin film is gradually deposited onto substrate by many chemical reactions in solution. This method is best suitable for metal oxides and chalcogenides deposition which includes CdS, ZnO, PbS, etc. These are extensively studied for optoelectronic devices such as solar cells, photodetectors, sensors. This depends on controlled precipitation of the target material on substrate surface from aqueous solution. The precipitation can be controlled by modulating temperature, concentration, pH. There are several advantages of CBD such as simple technique, low-cost equipment, large-area coating, complex shapes. Large-scale production is key attraction of thin films on substrates such as glass, metals, and polymers. However, the limitation is poor film adhesion, slow deposition rates, and poor control over stoichiometry and uniformity of film. Many researches have demonstrated the high-quality deposition with increased electrical and optical properties. CBD is simple, low cost, versatile deposition technique for large-area thin film applications from photovoltaics to biosensors.

This method is also known as the solution-based technique and is an established method for depositing films on substrates for metal oxides and chalcogenides films at lower temperatures. The deposition technique is facilitated by precursor solutions such as ammonia, triethanolamine, or ethylene-diamine-tetra acetic acid. Specific substrate positions allow the required film thickness by CBD [61–63]. Table 12.1 shows the comparative study of CVD, PVD, and ALD.

12.5 Thin Film Applications

Among the modern thin film applications, micro miniaturization has made the thin films highly on demand and imperative. Most of the thin film devices have been fabricated based on the physical properties. Table 12.2 shows the list of applications and future trend of thin films. Semiconductor devices, telecommunications, rectifiers, solar cells, photoconductors, magneto-optic memories, multilayer capacitors,

Table 12.1 Comparative study of CVD, PVD, and ALD

Property	CVD	PVD	ALD
Growth rate per minute	10–100 nm	100 nm–1 μm	0.1–1 nm
Materials used	Oxides, sulfides, nitrides, metals	Metals, alloys, oxides, nitrides	Oxides, sulfides, nitrides, metals
Temperature	High	Low	Wide range
Growth mechanism	Continuous	Continuous	Self-limiting

Table 12.2 Applications and future trend of thin films

Industry	Key applications	Future trends
Optics and photonic	Anti-reflective coatings, mirrors, low pass filters, high pass filters, bandpass filters	Ultra-thin, sensitive, flexible coatings for antireflecting devices
Semiconductor electronics	Flexible electronics, thin film light emitting transistors, printed circuit boards	Thin film devices for IoT and AI
Energy and energy storage	Thin film batteries, memory devices, Thin film solar cells	Memristors, neuromorphic devices, solar windows, energy-harvesting wearable films
Pharmacy and heath	Tissue regeneration, drug delivery, toxicology testing	Biodegradable films, real-time health monitoring
Airforce and defense	Anti-corrosive coatings, temperature sensitive coatings	Thin film sensors and NIR detectors
Electronic devices	OLED displays, foldable devices	Thin film sensors in wearable electronics for health monitoring
Structural coatings	Smart glass, sunglasses, protective coatings	Self-healing films, adaptive coatings

light emitting diodes, and transistors are emerging cutting-edge applications of thin films technology. Figure 12.4 demonstrates the classification of thin film devices for potential applications.

12.5.1 Sensors

Different types of chemical and biological sensors are fabricated on thin film-based techniques and principles. Depending on the target materials, the active sensor layer may vary its properties. The layer materials are in general inorganic, organic, or hybrid materials, and this helps to enhance the sensitivity and specificity of sensors. This makes the thin film sensor capable of having remarkable progress in molecular design of sensors and their active materials. Metal-oxides, CNTs (carbon nanotubes), metal-oxide semiconductors (MOS), some polymeric materials are highly explored to develop thin film-based sensors. Organic semiconductors are in emerging future trends in large-area deposition of thin films, room temperature fabrication, and high diversity in molecular design [44, 64, 65].

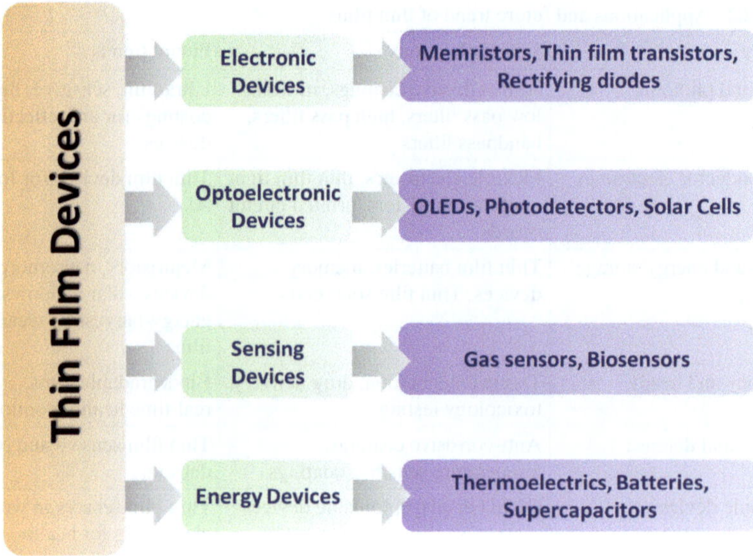

Fig. 12.4 Classification of thin film devices for potential applications

12.5.2 Optical Application

Thin film interference is the most modern thin film application in this era. This is an optical illusion which is created by the interference and reflection of light on the both surfaces of a film. When a light beam reflects partially from thin film surface, interference phenomenon occurs. The rest of light beam passes through reflecting off the bottom of the film and then interferes with the first beam on the top. Optical devices such as lenses and mirrors are familiar applications of thin film interference. This phenomenon is also seen in soap bubbles, credit cards, driver license, etc. Optical coatings and optoelectronics are the most familiar applications which can create optoelectronic semiconductor chips using MBE. In this method, a single atom is deposited at a time onto the substrate to prepare epitaxial films. To avoid the impurities in optical coatings fabrication, a highly pure and high vacuum environment is used to prepare the film for application in lenses, glasses, telescopes, mirrors.

To enhance or reduce the reflected light at a surface by using two refractive index materials, multilayer optical thin films were fabricated. These are also called high reflective (HR) mirror coating and antireflective (AR) coating. Two main components of the optical thin films are metal coating and dielectric coating. Wavelength selective low-pass, high-pass, and band-pass filters are important applications of dielectric coating. Fiber-optic communication is one of the famous applications of optical thin film technology. Ultralow reflection super-AR coating, wavelength selective edge filter super-PON filter are familiar examples of optical coatings. When light is incident on the film interference occurs. The interference pattern occurs between the light reflected inside the film interface and the light reflected outside the thin film. The most prominent example of this interference phenomenon is rainbow occurrence from soap bubbles and oil.

12.5.3 Energy Application

Thin film solar cells and energy harvesting devices are prominent in solar energy field which provides low-cost generation of energy. The main applications of this technology are photovoltaic systems and thermal energy power. The rechargeable batteries are key components of electronic devices like laptops and mobile phones. This technology has grown exponentially from past decades by integrating the thin films in more advanced way. Thin film batteries are thinner, lighter, and denser with high control in voltage and current. This helps to fabricate narrower ultra-thin laptops. These thin film batteries are in high demand in health care industry for power implantable medical devices. On another hand, silicon thin films are key components of solar cell technology. Now CuInGaSe (copper indium gallium selenide) perovskite has replaced silicon which has better efficiency and stability. They are prevalent in solar cell energy sector as the absorber layer has high absorption coefficient. This reduces the cost and enhances the conversion efficiency [22–28, 66, 67].

12.5.4 Biomedical Application

Thin film optical coatings are hugely used in medical devices of healthcare applications. For example, endoscopes, microscopes which improve the quality of images and increase the accuracy of diagnosis. This also reduces the eyeglass lenses reflectivity in a large amount which helps it toward better improvement of comfort and visual acuity for wearers. Thin films have grown as an idea surface for implanting permanent materials to increase the biodegradation resistance properties for better biocompatibility and mechanical strength. An ideal implantation ought to have the ability of resistance to degradation, prevent bacterial adhesion, stimulate osseointegration. The clinical success of implants relies on the cells and material surfaces interaction as a suitable and essential factor. Different types of coatings have been created on biomaterials as recent developments to overcome these issues.

Biocompatibility, bioactivity, and better mechanical strength have been achieved by these coatings. In this chapter, the recent trends in the surface modification in thin films are demonstrated. One of the most widely used techniques is spin coating for thin film fabrication over large areas which is smooth, homogeneous, and uniform surface. The freestanding film structures which are self-supported films have been developed for wide range of biomedical applications. Drug delivery, biosensing, and wound dressing are the main applications of this process. The structure, chemical composition, functionalities are the biological signal of coatings. A thin film layer of biomolecules is adsorbed on the surface of biomedical devices by CVD and Langmuir–Blodgett (LB) film. Performance depends on the interaction with the host medium, initiated and mediated by host proteins adsorbing onto the thin film surface upon implantation. When implanted surface gets hydrated in microsecond timescale, small molecules migrate toward the thin film surface in milliseconds, and protein in minutes. Subsequent cell attachment happens over the adsorbed protein layers which determines the long-term tissue response. Tissue scaffolds, drug discovery, regenerative medicine, and biosensors are key area of measurements [29–40].

In terms of biocompatibility, functionality, and customization, thin film offers unique advantages in drug delivery. There are various types of thin films used in the biomedical field, their fabrication techniques, and applications. This focuses on the materials like sputtering and evaporation target materials that play a significant role in these technologies.

a. *Tissue Regeneration for Wound Healing*: Ceramic and polymeric thin films are hugely used in wound care and tissue regeneration. Collagen or synthetic substitute polymers are utilized for the creation of scaffolds of tissue growth. Meanwhile it also helps in bone healing by hydroxyapatite ceramic films. These types of films are in general created by evaporation or sputtering which provides a controlled environment for cell growth and healing.
b. *Implantation in Medical Devices*: Thin films are used in medical implants for joint replacements, pacemakers, and dental implants. These coatings enhance the resistance of implants to avoid corrosion, decrease wear, and help better tissue integration. Titanium and hydroxyapatite thin films are most common materials used in this application. These coatings are often deposited by using sputtering targets. This method allows precise control over the film uniformity and thickness with constant performance and biocompatibility.
c. *Drug Delivery*: Biocompatible polymeric thin films are utilized in drug delivery systems. This provides controlled release of medication over the time. This helps in treating chronic diseases with the need of constant drug levels. Sputtering and evaporation methods are used to tailor and release profiles to ensure efficacy. Biodegradable films degrade in the body over time. This eliminates the requirement of surgical removal after the drug has been delivered. The polymeric thin films are ideal for a wide range of medical treatments, from cancer to chronic pain therapy.
d. *Biosensors and Biomedical Devices*: Au and Ag thin films are hugely used in diagnostic devices for biosensors in medical diagnostics and monitoring due to their excellent electrical behavior, biocompatibility, and high conductivity. Evaporation techniques including sputtering are used to deposit these films which allows the growth of thin and uniform layers that are perfect for highly efficient sensor [50, 51, 56–58, 67]. Figure 12.5 describes an illustration of the structure of a typical thin film biosensor.

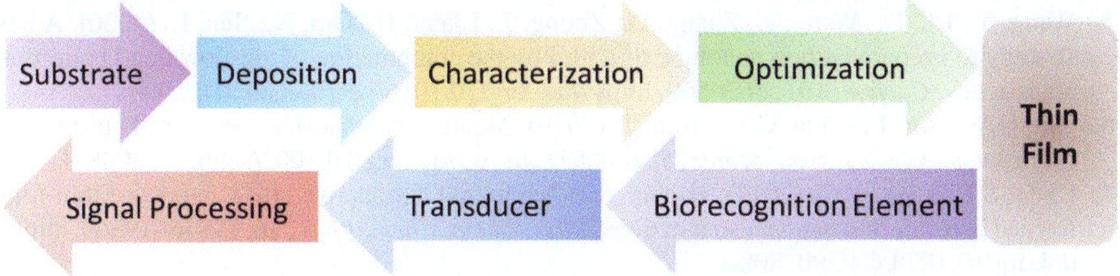

Fig. 12.5 Biomedical applications and biosensors: the diagram illustrates the structure of a typical thin film biosensor

12.6 Summary and Outlook of the Chapter

This chapter discusses the achievements and advances of thin film technology in research and industry over the past few decades. Different deposition and fabrication techniques are demonstrated, such as thermal evaporation technique, sputtering, chemical vapor deposition, electrochemical deposition, chemical precipitation. Thin film sensors for biomedical, pressure, chemical, temperature, antimicrobial, optical applications are explained. The thin film technology opens the pave to compete with the existing mainstream integrated photonics in applications for low-volume niche in sensing. The characterization techniques were explored to study the properties of and prospective developments of thin film industry. To improve the performance and quality of thin films, recent advancements of the deposition techniques have been also addressed. The advanced and hybrid deposition techniques and their developments aims to enhance the efficiency of thin film preparation. In conclusion, deposition techniques are highly dependent on the intended device application and required material. The future trend is to optimize the present techniques and discover new techniques for better materials, cost-effectiveness, and scalability for the scientific and industrial innovation.

References

1. Zhang, Z.; Lagally, M.G. (1997). Atomistic processes in the early stages of thin-film growth. *Science*, 276, 377–383. https://doi.org/10.1126/science.276.5311.377
2. Chiu, T.W.; Tonooka, K.; Kikuchi, N. (2010). Growth of b-axis oriented VO_2 thin films on glass substrates using ZnO buffer layer. *Appl. Surf. Sci.*, 256, 6834–6837. https://doi.org/10.1016/j.apsusc.2010.04.097
3. Cranston, R.R.; Lessard, B.H. (2021). Metal phthalocyanines: Thin-film formation, microstructure, and physical properties. *RSC Adv.*, 11, 21716–21737. https://doi.org/10.1039/D1RA03853B
4. Elanjeitsenni, V.P.; Vadivu, K.S.; Prasanth, B.M. (2022). A review on thin films, conducting polymers as sensor devices. *Mater. Res. Express*, 9, 022001. https://doi.org/10.1088/2053-1591/ac4aa1

5. Wang, Y.; Lu, D.; Wang, F.; Zhang, D.; Zhong, J.; Liang, B.; Gui, X.; Sun, L. (2020). A new strategy to prepare carbon nanotube thin film by the combination of top-down and bottom-up approaches. *Carbon*, 161, 563–569. https://doi.org/10.1016/j.carbon.2020.01.090
6. Tan, H.W.; An, J.; Chua, C.K.; Tran, T. (2019). Metallic nanoparticle inks for 3D printing of electronics. *Adv. Electron. Mater.*, 5, 1800831. https://doi.org/10.1002/aelm.201800831
7. Chen, S.P.; Liao, Y.C. (2014). Highly stretchable and conductive silver nanowire thin films formed by soldering nano mesh junctions. *Phys. Chem. Chem. Phys.*, 16, 19856–19860. https://doi.org/10.1039/C4CP02808B
8. George, S.M. (2009). Atomic layer deposition: An overview. *Chem. Rev.*, 110, 111–131. https://doi.org/10.1021/cr900056b
9. Chen, L.F.; Wang, Y.P.; Chiu, T.W.; Shih, W.C.; Wu, M.S. (2013). Fabrication of transparent $CuCrO_2$: Mg/ZnO p–n junctions prepared by magnetron sputtering on an indium tin oxide glass substrate. *Jpn. J. Appl. Phys.*, 52, 05EC02. https://doi.org/10.7567/JJAP.52.05EC02
10. Chernikova, V.; Shekhah, O.; Eddaoudi, M. (2016). Advanced fabrication method for the preparation of MOF thin films: Liquid phase epitaxy approach meets spin coating method. *ACS Appl. Mater. Interfaces*, 8, 20459–20464. https://doi.org/10.1021/acsami.6b04701
11. Ma, S.; Sui, J.; Cao, L.; Li, Y.; Dong, H.; Zhang, Q.; Dong, L. (2015). Synthesis of Cu2ZnSnS4 thin film through chemical successive ionic layer adsorption and reactions. *Appl. Surf. Sci.*, 349, 430–436. https://doi.org/10.1016/j.apsusc.2015.05.032
12. Vankova, S.; Zanarini, S.; Amici, J.; Camara, F.; Arletti, R.; Bodoardo, S.; Penazzia, N. (2015). WO_3 nanorolls self-assembled as thin films by hydrothermal synthesis. *Nanoscale*, 7, 7174–7177. https://doi.org/10.1039/C4NR07290A
13. Gole, A.; Jana, N.R.; Selvan, S.T.; Ying, J.Y. (2008). Langmuir-blodgett thin films of quantum dots: Synthesis, surface modification, and fluorescence resonance energy transfer (FRET) studies. *Langmuir*, 24, 8181–8186. https://doi.org/10.1021/la8000224
14. Carey, T.; Jones, C.; Moal, F.L.; Deganello, D.; Torrisi, F. (2018). Spray-Coating Thin Films on Three-Dimensional Surfaces for a Semitransparent Capacitive Touch Device. *ACS Appl. Mater. Interfaces*, 10, 19948–19956. https://doi.org/10.1021/acsami.8b02784
15. He, W.; Zhao, Y.; Xiong, Y. (2020). Bilayer Polyaniline-WO3 Thin Film Sensors Sensitive to NO_2. *ACS Omega*, 5, 9744–9751. https://doi.org/10.1021/acsomega.9b04122
16. Bruno, L.; Pijolat, C.; Lalauze, R. (1994). Tin dioxide thin-film gas sensor prepared by chemical vapour deposition Influence of grain size and thickness on the electrical properties. *Sens. Actuators B Chem.*, 18, 195–199. https://doi.org/10.1016/0925-4005(94)87083-7
17. Tampo, H.; Shibata, H.; Maejima, K.; Chiu, T.-W.; Itoh, H.; Yamada, A.; Matsubara, K.; Fons, P.; Chiba, Y.; Wakamatsu, T.; et al. (2009). Band profiles of ZnMgO/ZnO heterostructures confirmed by Kelvin probe force microscopy. *Appl. Phys. Lett.*, 94, 242107. https://doi.org/10.1063/1.3157149
18. Keefe, M.H.; Slone, R.V.; Hupp, J.T.; Czaplewski, K.F.; Snurr, R.Q.; Stern, C.L. (2000). Mesoporous thin films of 'Molecular Squares' as sensors for volatile organic compounds. *Langmuir*, 16, 3964–3970. https://doi.org/10.1021/la991252k
19. Rondinelli, J.M.; Spaldin, N.A. (2011). Structure and properties of functional oxide thin films: Insights from electronic-structure calculations. *Adv. Mater.*, 23, 3363–3381. https://doi.org/10.1002/adma.201101152
20. Mauchamp, V.; Yu, W.; Gence, L.; Piraux, L.; Cabioc, T.; Gauthier, V.; Eklund, P.; Dubois, S. (2013). Anisotropy of the resistivity and charge-carrier sign in nano laminated Ti2AlC: Experiment and ab initio calculations. *Phys. Rev. B*, 87, 235105. https://doi.org/10.1103/PhysRevB.87.235105
21. Solomon, G.; Kohan, M.G.; Landström, A.; Vomiero, A.; Concina, I. (2020). Semiconducting metal oxides empowered by graphene and its derivatives: Progresses and critical perspective on selected functional applications. *J. Appl. Phys.*, 128, 180905. https://doi.org/10.1063/5.0021826
22. He, Y.; Wang, X.; Gao, Y.; Hou, Y.; Wan, Q. (2018). Oxide-based thin film transistors for flexible electronics. *J. Semicond.*, 39, 011005. https://doi.org/10.1088/1674-4926/39/1/011005

23. Yoon, B.; Lee, B.H.; George, S.M. (2012). Highly Conductive and Transparent Hybrid Organic–Inorganic Zincone Thin Films Using Atomic and Molecular Layer Deposition. *J. Phys. Chem. C*, 116, 24784–24791. https://doi.org/10.1021/jp3057477
24. Tynell, T.; Terasaki, I.; Yamauchi, H.; Karppinen, M. (2013). Thermoelectric characteristics of (Zn, Al)O/hydroquinone superlattices. *J. Mater. Chem. A*, 1, 13619–13624. https://doi.org/10.1039/C3TA12909H
25. Michiardi, A.; Engel, E.; Aparicio, C.; Planell, J. A.; Gil, F. J. (2008). Oxidized NiTi surfaces enhance differentiation of osteoblast-like cells, *J. Biomed. Mater. Res. Part A*, 85A, 108–114. https://doi.org/10.1002/jbm.a.31486.
26. Hallab, N.J.; Vermes, C.; Messina, C.; Roebuck, K.A.; Glant, T.T.; Jacobs, J.J. (2002). Concentration and composition-dependent effects of metal ions on human MG-63 osteoblasts, *J. Biomed. Mater. Res.*, 60, 420–433. https://doi.org/10.1002/jbm.10106.
27. Möller, B.; Terheyden, H.; Açil, Y.; Purcz, N.M.; Hertrampf, K.; Tabakov, A.; Behrens, E.; Wiltfang, J. (2012). A comparison of biocompatibility and osseointegration of ceramic and titanium implants: an in vivo and in vitro study, *Int. J. Oral Maxillofac. Surg.*, 41, 638–645. https://doi.org/10.1016/j.ijom.2012.02.004.
28. Aun, D.P.; Houmard, M.; Mermoux, M.; Latu-Romain, L.; Joud, J.-C.; Berthomé, G.; Buono, V.T.L. (2016). Development of a flexible nanocomposite TiO_2 film as a protective coating for bioapplications of superelastic NiTi alloys, *Appl. Surf. Sci.*, 375, 42–49. https://doi.org/10.1016/j.apsusc.2016.03.064.
29. Shabalovskaya, S. A. (1996). On the nature of the biocompatibility and on medical applications of NiTi shape memory and superelastic alloys, *Biomed. Mater. Eng.*, 6, 267–289. https://doi.org/10.3233/BME-1996-6405.
30. Kurita, T.; Matsumoto, H.; Sakamoto, K.; Tanji, K.; Abe, H. (2005). Effect of aluminum addition on the transformation of NiTi alloy, *J. Alloy. Compd.*, 396, 193–196. https://doi.org/10.1016/j.jallcom.2004.12.032.
31. Crawford, G.; Chawla, N.; Das, K.; Bose, S.; Bandyopadhyay, A. (2007). Microstructure and deformation behavior of biocompatible TiO_2 nanotubes on titanium substrate, *Acta Biomater.*, 3, 359–367. https://doi.org/10.1016/j.actbio.2006.08.004.
32. Es-Souni, M.; Es-Souni, M.; Fischer-Brandies, H. (2005). Assessing the biocompatibility of NiTi shape memory alloys used for medical applications, *Anal. Bioanal. Chem.*, 381, 557–567. https://doi.org/10.1007/s00216-004-2888-3.
33. Lin, H.-C.; Chang, Y.-L.; Han, Y.-Y.; Yang, K.-C.; Chen, M.-C. (2019). Atomic layer deposited Al2O3 films on NiTi shape memory alloys for biomedical applications, *Procedia Manuf.*, 37, 431–437. https://doi.org/10.1016/j.promfg.2019.12.070.
34. Paula, A.S.; Canejo, J.P.H.G.; Martins, R.M.S.; Braz Fernandes, F.M. (2004). Effect of thermal cycling on the transformation temperature ranges of a Ni–Ti shape memory alloy, *Mater. Sci. Eng. A*, 378, 92–96. https://doi.org/10.1016/j.msea.2003.11.057.
35. Pan, C.-J.; Liu, H.-Q.; Wang, Y.-N.; Ding, H.-Y. (2014). Phase transformation and mechanical behaviors of vena cava filter modified by deposition of copper–titanium coating, *Mater. Res. Bull.*, 60, 217–221. https://doi.org/10.1016/j.materresbull.2014.08.042.
36. Tavares, C.J.; Marques, S.M.; Lanceros-Méndez, S.; Sencadas, V.; Teixeira, V.; Carneiro, J.O.; Martins, A.J.; Fernandes, A.J. (2008). Strain analysis of photocatalytic TiO2 thin films on polymer substrates, *Thin Solid Films*, 516, 1434–1438. https://doi.org/10.1016/j.tsf.2007.03.134.
37. Baradaran, S.; Basirun, W.J.; Zalnezhad, E.; Hamdi, M.; Sarhan, A.A.D.; Alias, Y. (2013). Fabrication and deformation behaviour of multilayer Al2O3/Ti/TiO2 nanotube arrays, J. Mech. Behav. *Biomed. Mater.*, 20, 272–282. https://doi.org/10.1016/j.jmbbm.2013.01.020.
38. Snaith, H. J.; Moule, A. J.; Klein, C.; Meerholz, K.; Friend, R. H.; Grätzel, M. (2007). Efficiency enhancements in solid-state hybrid solar cells via reduced charge recombination and increased light capture. *Nano Lett*, 7, 3372. https://doi.org/10.1021/nl071656u
39. Wang, M.; Xu, M.; Shi, D.; Li, R.; Gao, F.; Zhang, G., et al. (2008). High-performance liquid and solid dye-sensitized solar cells based on a novel metal-free organic sensitizer. *Adv Mater*, 20, 4460. https://doi.org/10.1002/adma.200801178

40. Bach, U.; Lupo, D.; Comte, P.; Moser, J.; Weissörtel, F.; Salbeck, J. et al. (1998). Solid-state dye-sensitized mesoporous TiO_2 solar cells with high photon-to-electron conversion efficiencies. *Nature*, 395, 583. https://doi.org/10.1038/26936
41. Wang, M.; Liu, J.; Cevey-Ha, N. L.; Moon, S. J.; Liska, P.; Humphry-Baker, R. et al. (2010). High efficiency solid-state sensitized heterojunction photovoltaic device. *Nano Today*, 5, 169. https://doi.org/10.1016/j.nantod.2010.04.001
42. Wang, M.; Moon, S. J.; Zhou, D.; Le Formal, F. Cevey-Ha N. L.; Humphry-Baker, R. et al. (2010). Enhanced-light-harvesting amphiphilic ruthenium dye for efficient solid-state dye-sensitized solar cells. *Adv Funct Mater*, 20, 1821. https://doi.org/10.1002/adfm.200902396
43. Wang, M.; Moon, S. J.; Xu, M.; Chittibabu, K.; Wang, P.; Cevey-Ha, N. L. et al. (2010). Efficient and stable solid-state dye-sensitized solar cells based on a high-molar-extinction-coefficient sensitizer. *Small*, 6, 319. https://doi.org/10.1002/smll.200901317
44. Lu, J.; Chang, Y. C.; Cheng, H. Y.; Wu, H. P.; Cheng, Y.; Wang, M. et al. (2015). Molecular engineering of organic dyes with a hole-extending donor tail for efficient all-solid-state dye-sensitized solar cells. *ChemSusChem*, 8, 2529. https://doi.org/10.1002/cssc.201500309
45. Xue, Z; Wang, L; Liu, W.; Liu, B. (2014). Solid-state D102 dye sensitized/poly(3-hexylthiophene) hybrid solar cells on flexible Ti substrate. *Renew Energy*, 72, 22. https://doi.org/10.1016/j.renene.2014.06.043
46. Ali Umar, A.; Nafisah, S.; Md Saad, S. K.; Tee Tan, S.; Balouch, A.; Mat Salleh, M. et al. (2014). Poriferous microtablet of anatase TiO_2 growth on an ITO surface for high efficiency dye-sensitized solar cells. *Sol Energy Mater Sol Cells*, 122, 174. https://doi.org/10.1016/j.solmat.2013.12.002
47. Hagfeldt, A.; Boschloo, G.; Sun, L.; Kloo, L.; Pettersson, H. (2010). Dye-sensitized solar cells. *Chem Rev*, 110, 6595. https://doi.org/10.1021/cr900356p
48. Sangiorgi, A.; Bendoni, R.; Sangiorgi, N.; Sanson, A.; Ballarin, B. (2014). Optimized TiO_2 blocking layer for dye-sensitized solar cells. *Ceram Int*, 40, 10727. https://doi.org/10.1016/j.ceramint.2014.03.060
49. Liu, Y.; Sun, X.; Tai, Q.; Hu, H.; Chen, B.; Huang; N. et al. (2011). Influences on photovoltage performance by interfacial modification of FTO/mesoporous TiO_2 using ZnO and TiO_2 as the compact film. *J Alloy Compd*, 509, 9264. https://doi.org/10.1016/j.jallcom.2011.07.018
50. Choudhury, B.; Choudhury, A. (2012). Dopant induced changes in structural and optical properties of Cr3þ doped TiO_2 nanoparticles. *Mater Chem Phys*, 132, 1112. https://doi.org/10.1016/j.matchemphys.2011.12.083
51. Mathews, N. R.; Colín García, C.; Torres, I. Z. (2013). Effect of annealing on structural, optical and electrical properties of pulse electrodeposited tin sulfide films. *Mater Sci Semiconduct Process*, 16, 29–37. https://doi.org/10.1016/j.mssp.2012.07.003
52. Zhang, Q.; Cao, G. (2011). Nanostructured photoelectrodes for dye-sensitized solar cells. *Nano Today*, 6, 91–109. https://doi.org/10.1016/j.nantod.2010.12.007
53. Omar, A.; Abdullah, H. (2014). Electron transport analysis in zinc oxide-based dye-sensitized solar cells: a review. *Renew Sustain Energy Rev*, 31, 149. https://doi.org/10.1016/j.rser.2013.11.031
54. Chu, J. B.; Huang, S. M.; Zhang, D. W.; Bian, Z. Q.; Li, X. D.; Sun, Z. et al. (2009). Nanostructured ZnO thin films by chemical bath deposition in basic aqueous ammonia solutions for photovoltaic applications. *Appl Phys A*, 95, 849. https://doi.org/10.1007/s00339-009-5084-7
55. Liu, Z.; Li, Y.; Liu, C.; Ya, J.; Zhao, W.; Lei, E. et al. (2011). Performance of ZnO dye-sensitized solar cells with various nanostructures as anodes. *Solid State Sci*, 13, 1354. https://doi.org/10.1016/j.solidstatesciences.2011.04.005
56. Ghanbari Niaki, A. H., Bakhshayesh, A. M., Mohammadi, M. R. (2014). Double-layer dye-sensitized solar cells based on Zn-doped TiO_2 transparent and light scattering layers: improving electron injection and light scattering effect. *Sol Energy*, 103, 210. https://doi.org/10.1016/j.solener.2014.01.041
57. Kuang, D.; Brillet, J.; Chen, P.; Takata, M.; Uchida, S.; Miura, H. et al. (2008). Application of highly ordered TiO_2 nanotube arrays in flexible dye-sensitized solar cells. *ACS Nano*, 2, 1113. https://doi.org/10.1021/nn800174y

58. Yang, J.; Bark, C.; Kim, K.; Choi, H. (2014). Characteristics of the dye-sensitized solar cells using TiO_2 nanotubes treated with TiCl4. *Materials*, 7, 3522. https://doi.org/10.3390/ma7053522
59. Wahab, R.; Ansari, S. G.; Kim, Y. S.; Seo, H. K.; Kim, G. S.; Khang, G. et al. (2007). Low temperature solution synthesis and characterization of ZnO nano-flowers. *Mater Res Bull*, 42, 1640. https://doi.org/10.1016/j.materresbull.2006.11.035
60. Kakiuchi, K.; Hosono, E.; Kimura, T.; Imai, H.; Fujihara, S. (2006). Fabrication of mesoporous ZnO nanosheets from precursor templates grown in aqueous solutions. *J Sol-Gel Sci Technol*, 39, 63–72. https://doi.org/10.1007/s10971-006-6321-6
61. Cheng, H. M.; Chiu, W. H.; Lee, C. H.; Tsai, S. Y.; Hsieh, W. F. (2008). Formation of branched ZnO nanowires from solvothermal method and dye-sensitized solar cells applications. *J Phys Chem C*, 112, 16359. https://doi.org/10.1021/jp805239k
62. Woo Choi, H.; Lee, K. S.; David Theodore, N.; Alford, T. L. (2013). Improved performance of ZnO nanostructured bulk heterojunction organic solar cells with nanowire density modified by yttrium chloride introduction into solution. *Sol Energy Mater Sol Cells*, 117, 273. https://doi.org/10.1016/j.solmat.2013.06.036
63. Hsu, M. H.; Chang, C. J. (2014). Ag-doped ZnO nanorods coated metal wire meshes as hierarchical photocatalysts with high visible-light driven photoactivity and photostability. *J Hazard Mater*, 278, 444. https://doi.org/10.1016/j.jhazmat.2014.06.038
64. Chae, Y.; Park, J. T.; Koh, J. K.; Kim, J. H.; Kim, E. (2013). All-solid, flexible solar textiles based on dye-sensitized solar cells with ZnO nanorod arrays on stainless steel wires. *Mater Sci Eng: B*, 178, 1117. https://doi.org/10.1016/j.mseb.2013.06.018
65. Chou, T. P.; Zhang, Q.; Cao, G. (2007). Effects of dye loading conditions on the energy conversion efficiency of ZnO and TiO_2 dye-sensitized solar cells. *J Phys Chem C*, 111, 18804. https://doi.org/10.1021/jp076724f
66. Gondoni, P.; Mazzolini, P.; Russo, V.; Petrozza, A.; Srivastava, A. K.; Li Bassi, A. et al. (2014). Enhancing light harvesting by hierarchical functionally graded transparent conducting Al-doped ZnO nano- and mesoarchitectures. *Sol Energy Mater Sol Cells*, 128, 248. https://doi.org/10.1016/j.solmat.2014.05.035
67. Chou, C. Y.; Li, C. T.; Lee, C. P.; Lin, L.Y.; Yeh, M. H.; Vittal, R. et al. (2013). ZnO nanowire/nanoparticles composite films for the photoanodes of quantum dot sensitized solar cells. *Electrochim Acta*, 88, 35–43. https://doi.org/10.1016/j.electacta.2012.09.121
68. Feng, J.; Fan, J.; Zhang, Z.; Gao, Y.; Xue, S.; Cai, G.; Zhao, J.S. (2024). Spin-Coating Deposited SnS_2 Thin Film-Based Memristor for Emulating Synapses. *Adv. Funct. Mater.*, 34, 2401228. https://doi.org/10.1002/adfm.202401228

Dr. Piyali Maity is a Postdoctoral Researcher at the City University of New York, New York, USA. She completed her Ph.D. in Physics at the renowned Institute of Science at Banaras Hindu University in India. Dr. Maity also obtained her Master of Science in Physics from Banaras Hindu University and her Bachelor of Science in Physics from Bethune College, University of Calcutta. She also taught at City College of New York (CCNY) of the City University of New York as an adjunct assistant professor. Her research is interdisciplinary, blending Physics, Chemistry, and Materials Science, with a specific emphasis on quantum materials, non-linear optics, and functional devices. Her current research focuses on exploring light-matter interactions in nanoscale material systems. She concentrates on domain dynamics of thin films and is working on innovative applications in optoelectronics, storage devices and nonlinear optics that rely on the mechanism of photons and charge carriers.

Index

S. Sharma (ed.), *Multifunctional Heterostructure Thin Films Handbook*,
https://doi.org/10.1007/978-3-032-18692-8

MIX
Papier aus verantwortungsvollen Quellen
Paper from responsible sources
FSC® C105338

If you have any concerns about our products,
you can contact us on
ProductSafety@springernature.com

In case Publisher is established outside the EU,
the EU authorized representative is:
Springer Nature Customer Service Center GmbH
Europaplatz 3, 69115 Heidelberg, Germany

Printed by Libri Plureos GmbH
in Hamburg, Germany